Functions of Plant Secondary Metabolites and their Exploitation in Biotechnology

Annual Plant Reviews

A series for researchers and postgraduates in the plant sciences. Each volume in this annual series will focus on a theme of topical importance and emphasis will be placed on rapid publication.

Titles in the Series:

1. Arabidopsis
Edited by M. Anderson and J. Roberts.

2. Biochemistry of Plant Secondary Metabolism
Edited by M. Wink.

3. Functions of Plant Secondary Metabolites and their Exploitation in Biotechnology
Edited by M. Wink.

Functions of Plant Secondary Metabolites and their Exploitation in Biotechnology

Edited by

MICHAEL WINK
Professor of Pharmaceutical Biology
University of Heidelberg
Germany

Sheffield
Academic Press

CRC Press

First published 1999

Published by
Sheffield Academic Press Ltd
Mansion House, 19 Kingfield Road
Sheffield S11 9AS, England

ISBN 1-84127-008-3
ISSN 1460-1494

Published in the U.S.A. and Canada (only) by
CRC Press LLC
2000 Corporate Blvd., N.W.
Boca Raton, FL 33431, U.S.A.
Orders from the U.S.A. and Canada (only) to CRC Press LLC

U.S.A. and Canada only:
ISBN 0-8493-4086-1
ISSN 1097-7570

Printed on acid-free paper in Great Britain by
Bookcraft Ltd, Midsomer Norton, Bath

British Library Cataloguing-in-Publication Data:
A catalogue record for this book is available from the British Library

Library of Congress Cataloging-in-Publication Data:
Functions of plant secondary metabolites and their exploitation in
biotechnology/ edited by Michael Wink.
p. cm.
Includes bibliographical references and index.
ISBN 0-8493-4086-1 (alk. paper)
1. Plant metabolites. 2. Metabolism, Secondary. 3. Plant
biotechnology I. Wink, Michael.
QK887.F86 1999
572'.42--dc21 99-11778
CIP

Preface

Secondary metabolites, of which over 60 000 low molecular weight compounds, of enormous variety, have already been isolated and identified, play an important role in the organisms producing them. They may function as signal molecules in plant-plant, plant-herbivore, plant-microbe, animal-animal, and animal-predator relationships. More often, secondary metabolites serve as chemical defence compounds against herbivores and predators, microbes, viruses or competing plants. Secondary metabolites are therefore of ultimate importance for the fitness of the organisms producing or sequestering them.

To understand the structural variability of secondary metabolites, their ecological function and their potential exploitation in biotechnology (utilisation in pharmacology as molecular probes, in medicine as therapeutic agents, or in agriculture as biorational pesticides), detailed information is required on their biochemistry (production and accumulation in space and time), pharmacology (toxicity and modes of action) and ecology (functions in plants and animals).

In this volume of *Annual Plant Reviews*, we have tried to provide an up-to-date survey of the functions of secondary metabolites and their utilisation in biotechnology. A companion volume—M. Wink (ed.) *Biochemistry of Plant Secondary Metabolism*—published simultaneously, provides overviews of the biochemistry, biosynthesis and storage of alkaloids, cyanogenic glycosides, glucosinolates, nonprotein amino acids, phenylpropanoids, flavonoids, mono/sesqui/diterpenes, saponins and cardiac glycosides in detail. To achieve a comprehensive and up-to-date account, we have invited scientists who are specialists in their particular areas to review current understanding. This volume draws together a broad range of topics and it cannot be exhaustive on such a large and diverse group of constituents. Emphasis was therefore placed on new results and concepts which have emerged over the last decade.

This volume starts with an overview of the modes of action of defensive secondary metabolites, followed by detailed surveys of chemical defence in marine ecosystems, the biochemistry of induced defence, plant-microbe interactions, and medical applications. A chapter is included on biotechnological aspects of producing valuable secondary metabolites in plant cell and organ cultures.

The book is designed for use by advanced students, researchers and professionals in plant biochemistry, physiology, molecular biology, genetics, pharmacology, medicine, pharmacy and agriculture working in the academic and industrial sectors, including the pesticide and pharmaceutical industries.

Contributions were commissioned from friends and colleagues in many parts of the world. In several instances we had to ask two or more colleagues working in a related field to contribute to a common chapter, instead of writing individual chapters. The Publishers and the Editor are aware that this was not an easy task but, as the reader will be able to judge, our authors were successful in rising to the challenge. As Editor, I would like to thank all those who have taken part in the writing and preparation of this book. Special thanks go to the Publisher, Dr Graeme MacKintosh, and his team for their interest, support and encouragement.

I would like to thank my wife, Dr Coralie Wink, for her help in preparation of the index.

Michael Wink
Heidelberg

Contributors

Professor Dr A. Wilhelm Alfermann	Institut für Entwicklungs-& Molekularbiologie der Pflanzen, Heinrich-Heine-Universität Düsseldorf, Universitätsstrasse 3, D-40225 Düsseldorf, Germany
Professor Dr Ian Baldwin	Max-Planck-Institut für Chemische Ökologie, Tatzendpromenade 1A, 07745 Jena, Germany
Professor Dr Rudolf Bauer	Institut für Pharmazeutische Biologie, Heinrich-Heine-Universität Düsseldorf, Universitätsstrasse 1, D-40225 Düsseldorf, Germany
Dr Jörg Heilmann	Departement Pharmazie, ETH Zürich, Winterthurerstrasse 190, CH-8057 Zürich, Switzerland
Professor Dr Peter Proksch	Julius-von-Sachs-Institut für Biowissenschaften, Universität Würzburg, Julius-von-Sachs-Platz 2, D-97082 Würzburg, Germany
Professor Dr Jürgen Reichling	Institut für Pharmazeutische Biologie, Ruprecht-Karls-Universität Heidelberg, Im Neuenheimer Feld 364, 69120 Heidelberg, Germany
Dr Michael J. C. Rhodes	c/o Food Safety Science Division, Institute of Food Research, Norwich Research Park, Colney, Norwich NR4 7UA, UK
Professor Dr Oskar Schimmer	Universität Erlangen-Nürnberg, Lehrstuhl für Pharmazeutische Biologie, Staudtstrasse 5, D-91058 Erlangen, Germany
Dr Nicholas J. Walton	Food Safety Science Division, Institute for Food Research, Norwich Research Park, Colney, Norwich NR4 7UA, UK

Professor Dr Michael Wink Institut für Pharmazeutische Biologie, Universität Heidelberg, Im Neuenheimer Feld 364, D-69120 Heidelberg, Germany

Contents

1 Introduction

Michael Wink

1.1 Function of secondary metabolites

A typical trait of plants is the production of a high diversity of secondary metabolites (the number of identified substances exceeds 100 000 at present), including many nitrogen-free (such as terpenes, polyketides, saponins and polyacetylenes) and nitrogen-containing compounds (such as alkaloids, amines, cyanogenic glycosides, non-protein amino acids and glucosinolates). Several major secondary metabolites are commonly accompanied by dozens of minor components. Complex mixtures, which differ from organ to organ, sometimes between individual plants and regularly between species, are the result. These compounds are synthesised in plants in a tissue-, organ- and developmental-specific way by specific biosynthetic enzymes. The corresponding genes are regulated accordingly and gene regulation shows all the complexity known for genes encoding enzymes of primary metabolism. It is a particular feature of secondary metabolites that they are accumulated and stored in high concentrations in the plants producing them; levels of 1–3% dry weight are regularly seen. In general, hydrophilic compounds are stored in the vacuole whereas lipophilic substances are deposited in resin ducts, laticifers, trichomes, oil cells, or in the cuticle. As sites of synthesis are not necessary the sites of storage, long-distance transport by xylem, phloem or via the apoplast have been discovered in some instances.

Although secondary metabolites (SMs) have been used by mankind for thousands of years (Mann, 1992; Roberts and Wink, 1998) as dyes (e.g. indigo, shikonin), flavours (e.g. vanillin, capsaicin, mustard oils), fragrances (e.g. rose oil, lavender oil and other essential oils), stimulants (e.g. caffeine, nicotine, ephedrine), hallucinogens (e.g. morphine, cocaine, scopolamine, tetrahydrocannabinol), insecticides (e.g. nicotine, piperine, pyrethrin), vertebrate and human poisons (e.g. coniine, strychnine, aconitine), and even as therapeutic agents (e.g. atropine, quinine, cardenolides, codeine, etc.), their putative functions have been a matter of controversy.

Whereas most animals can run or fly away in case of danger (Edmunds, 1974), or possess an immune system to protect them against invading microbes or parasites, these means are apparently not available for plants when attacked by herbivores, microbes (bacteria, fungi) and even other plants competing for light, space and nutrients. In contrast to most animals, plants can replace the parts which have been diseased, wounded or browsed. This capacity for open growth and regeneration, which is most prominent in perennials, allows a certain tolerance towards herbivores and microbes. A number of plants employ mechanical and morphological protection, such as thorns, spikes, glandular and stinging hairs (often filled with noxious chemicals), or develop an almost impenetrable bark (especially woody perennials); these features can be interpreted as antipredatory means (in analogy to weapons and shells in animals). Sessile or slow-moving animals, such as sponges, nudibranch molluscs, corals (see Chapter 3 by P. Proksch in the present volume) and amphibia (e.g. salamanders, poisonous frogs, toads) are infamous for their ability to produce a wide range of chemicals that are usually toxic (for reviews see Braekman *et al.*, 1998; Proksch and Ebel, 1998). Some insects either produce SMs themselves or sequester them from their host plants (for overviews see Blum, 1981; Duffey, 1980; Bernays and Chapman, 1994). Zoologists have never doubted that these compounds serve for chemical defence against predators. Surprisingly, the defence function of SMs in plants has been and sometimes still is controversial.

It has often been argued that SMs are waste products or have no function at all. This hypothesis fails to explain several observations: 1) Waste products are characteristic and necessary for heterotrophic animals that cannot degrade their food completely for energy production. These organisms excrete waste products that are often rich in nitrogen (i.e. urea, uric acid). However, plants are essential autotrophs and, therefore, do not need elaborate excretory mechanisms. Furthermore, nitrogen is a limiting nutrient for plants. Consequently, the production of nitrogen-containing excretions, such as alkaloids, would be difficult to explain. In addition, alkaloids are often found in young or metabolically-active tissues but not in dying or senescing cells, as would be expected according to the waste product hypothesis. 2) Secondary metabolites are often not inert end-products of metabolism (an expected trait of waste products) but many of them can be metabolized by plant cells. For example, nitrogenous secondary metabolites, such as alkaloids, non-protein amino acids (NPAAs) and cyanogenic glycosides, are often stored in considerable quantities in leguminous seeds. During germination, a degradation of these compounds can be seen, indicating that their nitrogen is reused by the seedling. 3) Secondary metabolism is often highly complex and regulated in a tissue- and developmentally-specific manner, which would be surprising for a waste product without function.

Alternatively, it was argued as long as 100 years ago by E. Stahl in Jena (Germany) that secondary metabolites serve as defence compounds against herbivores. This hypothesis has been elaborated during recent decades (Fraenkel, 1959; Ehrlich and Raven, 1964; Levin, 1976; Swain,

1977) and a large body of experimental evidence supports the concept that follows (for reviews see Harborne, 1993; Wink, 1988, 1992, 1993b; Bernays and Chapman, 1994). Several secondary metabolites have evolved for protection against viruses, bacteria, fungi, competing plants and, importantly, against herbivores (e.g. slugs and snails, arthropods and vertebrates). In addition, secondary metabolites can serve as signal compounds to attract animals for pollination (fragrant monoterpenes, coloured anthocyanins or carotinoids) and for seed dispersal (reviewed by Cipollini and Levey, 1997) (Fig. 1.1). In several instances, both activities are exhibited by the same compounds: anthocyanins or monoterpenes can be insect attractants in flowers but are insecticidal and antimicrobial at the same time in leaves. In addition, some secondary metabolites concomitantly exhibit physiological functions, for example they can serve as mobile and toxic nitrogen transport and storage compounds or UV-protectants. These multiple functions are typical and do not contradict their main role as chemical defence and signal compounds. If a trait can serve multiple functions, it is more likely to be maintained by natural selection. In the present volume, Chapter 2 by M. Wink and O. Schimmer and Chapter 5 by J. Reichling review some of these aspects.

1.2 Presence of defence and signal compounds at the right time and place

In most plants, synthesis and accumulation of secondary metabolites is regulated in space and time. As a rule, vulnerable tissues are defended more than old, senescing tissues. For example, it is usually observed that seeds, seedlings, buds and young tissues either sequester large amounts of a compound or actively synthesize them-'optimal defence theory'. Organs that are important for survival and multiplication, such as flowers, fruits and seeds, are nearly always a rich source of defence chemicals.

The specific localisations of SMs make sense if their role as defence and/or signal compounds is accepted. Trichomes and glandular hairs are always on the surface of the plant; a herbivore cannot avoid direct contact with them if it tries to feed on the plant. If membrane-active terpenes reach their lips, tongue or mandibles, many herbivores can be deterred before they actually start feeding on the plant. Another example is the sequestration of high concentrations of SMs in vacuoles, which are often positioned in a favourable site for defence, as many of them are stored in epidermal and subepidermal cells (Kojima *et al.*, 1979; Saunders and Conn, 1978; Gruhnert *et al.*, 1994; Matile, 1984; Werner and Matile, 1985; Wink *et al.*, 1984; Wink, 1992, 1997). If a small herbivore or

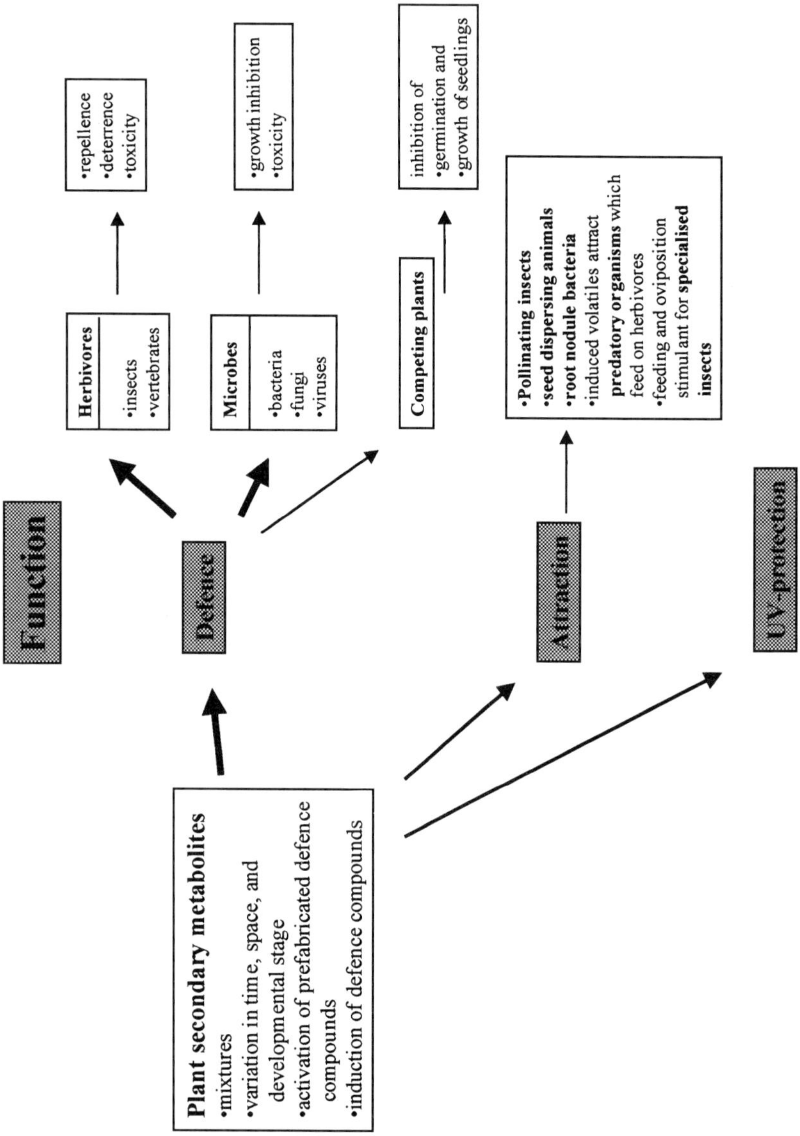

Figure 1.1 Functions of plant secondary metabolites.

microbe attacks such a plant, it will encounter a high SM concentration immediately at the periphery when wounding or entering the tissue, which might deter further feeding. Compounds that are sequestered in resin ducts or laticifers are often under high pressure and readily squirt out when these elements become wounded. For a small herbivorous insect, this will be a dangerous situation, since these effluents will make their mandibles sticky. A few 'clever' beetles and caterpillars cut the veins of leaves upstream to the area on which they want to feed. The fluids emerge from the cuts but can no longer reach the parts downstream, which are eaten later (Dussourd and Eisner, 1987; Becerra, 1994).

Several defence compounds are transported via the phloem from the site of synthesis to other plant organs. Since the phloem is a target for many sucking insects, such as aphids, these insects encounter a high load of alkaloids in the plants producing them. For lupins, in which alkaloid-rich and almost alkaloid-free varieties (sweet lupins) are known, it could be shown that aphid generalists (e.g. *Myzus persicae*) sucked only on 'sweet' lupins but never on alkaloid-rich varieties, with high alkaloid contents (Table 1.1) in the phloem (Wink, 1992). Moreover, many other animals, from leaf miners (Agromyzidae) to rabbits (*Oryctolagus cuniculus*) showed a similar discrimination, in that alkaloid-rich plants were left alone, while 'alkaloid-free' cultivars were highly susceptible. The only exception is a specialized aphid, *Macrosiphum albifrons*, which lives on lupins and sequesters the dietary alkaloids, using them for defence against predators (Wink and Römer, 1986).

In general, a series of related compounds is found in each plant; often a few major metabolites and several minor components, which differ in

Table 1.1 Relationship between alkaloid content and percentage herbivory by aphids (generalists and specialist) and other herbivores (after Wink, 1987a, 1988, 1992; Wink and Römer, 1986; Wink and Witte, 1991)

Species	Alkaloid content mg/g FW	Herbivory (%) by			*Macrosiphum albifrons*
		Myzus spp.	Leaf miners	Rabbits	
Lupinus albus					
var. *lucky*	< 0.01	20	100	100	< 5
var. *lublanc*	< 0.01	15	100	100	< 5
var. *multolupa*	0.03	15	100	80	< 10
from Syria	2.0	0	< 1	< 10	100
from Crete	2.2	0	< 1	n.d.	100
L. luteus	0.01	100	n.d.	n.d.	n.d.
	0.25	50	n.d.	n.d.	n.d.
	0.7	< 1	n.d.	< 5	0
L. polyphyllus	1.0	0	< 1	< 5	80
L. angustifolius	1.5	0	< 1	< 10	100

Abbreviation: n.d., not determined; FW, fresh weight.

the position of their substitutents. The profile usually varies between plant organs, within developmental periods and sometimes even diurnally, e.g. in lupin alkaloids (Wink and Witte, 1984). Furthermore, marked differences can usually be seen between individual plants of a single population and even more so between members of different populations. This variation, which is part of the apparent evolutionary 'arms race' between plants and herbivores, makes adaptation by herbivores more difficult, since even small changes in chemistry can be the basis for a new pharmacological activity (for more details see Chapter 2 by M. Wink and O. Schimmer in the present volume).

Defence against herbivores and pathogens is not necessarily constitutive. Research in recent decades has shown that wounding and infection triggers several events in plants. For example, wounding can lead to a decompartmentalization, thus releasing prefabricated defence chemicals (protoxins such as glucosinolates, cyanogenic glycosides, bidesmosidic saponins, alliin, ranunculin, coumaroylglycosides) and mixing them with hydrolyzing enzymes, such as β-glycosidase, myrosinase, nitrilase or alliinase (Matile, 1980) (Fig. 1.2). Active allelochemicals are the result. In other instances, it has been shown that the level of existing defence chemicals is increased substantially within hours or days after wounding or infection; for example, nicotine in *Nicotiana tabacum* (Baldwin, 1994) or lupin alkaloids in *Lupinus polyphyllus* (Wink, 1983). After infection, in particular, new compounds with antifungal, antibacterial or herbivore-deterring activities are made and sequestered; phytopathologists have

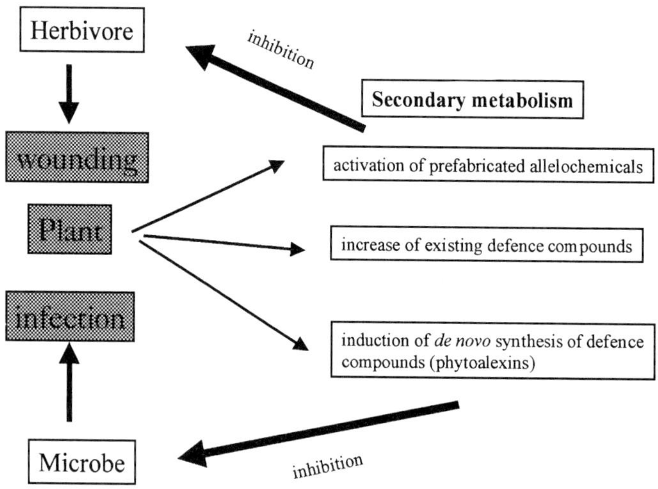

Figure 1.2 Examples of induced defence in plants.

termed these compounds 'phytoalexins' (Fig. 1.2). These compounds include, among others, several isoflavones, pterocarpans, furocoumarins, chalcones and stilbenes. Many of these metabolites have antifungal properties, so that they are sometimes considered to be part of the specific antimicrobial defence system of plants. However, since most of these compounds also affect herbivores, the plant defence induced appears to be a more general phenomenon (see also Chapter 4 and Chapter 5 in the present volume).

Recent research has shown that elicitors, receptors, ion channels, salicylic acid and the pathway leading to jasmonic acid and methyljasmonate are important elements in converting the external signal into a cellular response (Creelman and Mullet, 1997). In the present Volume, Chapter 4 by I. Baldwin summarizes the corresponding biochemical events and complexity of interactions.

The SM defence system works in general but a number of herbivores and microorganisms have evolved that have overcome the defence barrier (analogous to the situation in which some viruses, bacteria or parasites overwhelm the human immune system). In these organisms, a series of adaptations can be observed, allowing them to tolerate or even use the dietary defence chemicals (a schematic overview is presented in Fig. 1.3) (for reviews see Ahmad, 1983; Brattsten and Ahmad, 1986; Bernays and

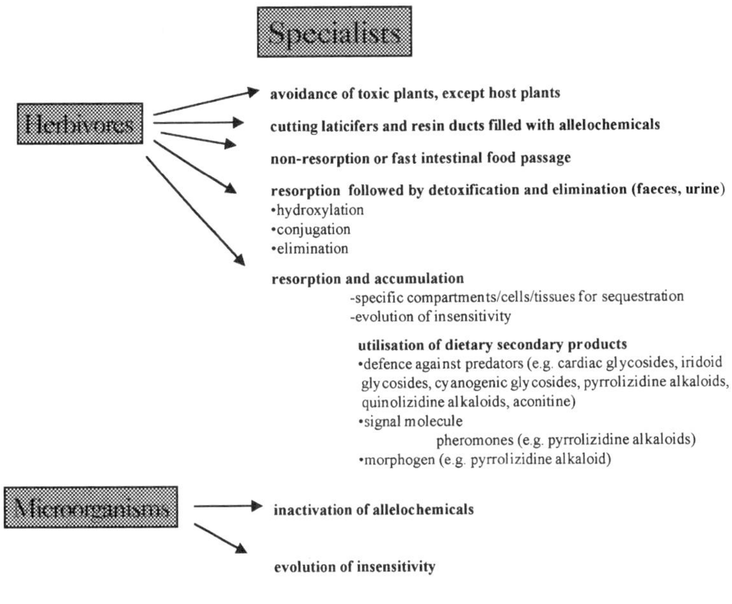

Figure 1.3 Adaptations of specialist herbivores and pathogens.

Chapman, 1994; Rosenthal and Berenbaum, 1991, 1992; Wink, 1993b; Brown and Trigo, 1995; Hartmann and Witte, 1995).

Several volatiles are produced by plants when wounded, including aldehydes, esters, amines or ethylene. It has been proposed that some of these volatiles can alert the defence system of neighbouring plants. In addition, they can attract predatory arthropods. A well-studied example is that of spider mites (*Tetranyches urticae*) on *Phaseolus lunatus* leaves. Volatiles from infested plants attract predatory mites (*Phytoseiulus persimilis*), which prey on the mites that induced the reaction in the first place (Dicke *et al.*, 1990; De Moraes *et al.*, 1998). It is likely that more tritrophic systems work in this way; many of them still await discovery.

Chemical defence is not only obvious in terrestrial ecosystems but is of major importance in the survival of marine organisms. In the present volume, Chapter 3 by P. Proksch provides an overview of this exciting and rapidly growing research field.

If defence compounds inhibit the growth of microbes or herbivores or are otherwise toxic to them, they must interfere with the physiology and biochemistry of these organisms. A large body of pharmacological and toxicological literature clearly documents that these activities do exist (Teuscher and Lindequist, 1994; Wink, 1993, 1999). Typical organ systems that are often affected by SMs in animals are schematically illustrated in Figure 1.4. In many instances, the mechanisms which

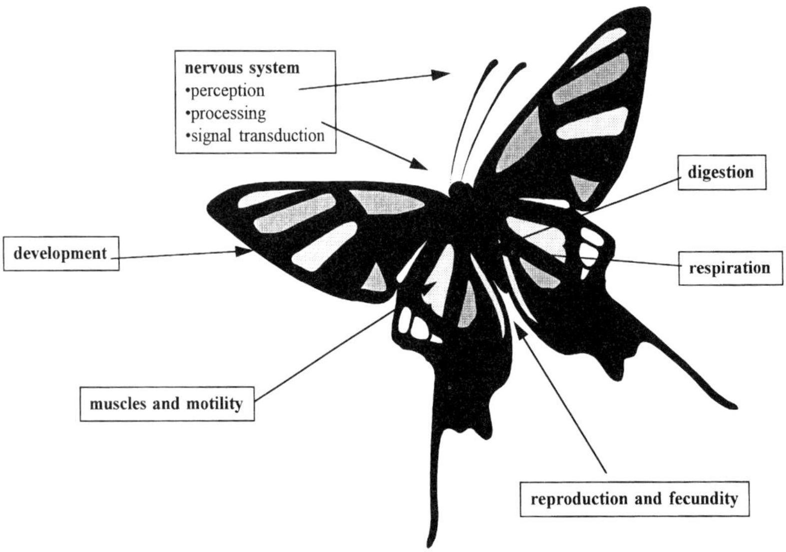

Figure 1.4 Targets for allelochemicals in animals.

underlie these effects have been elucidated; often specific interactions with one or several of the molecular targets shown in Figure 1.5 can be observed. It has been argued that defence compounds have been shaped during evolution to specifically interact with particular targets in a process termed 'evolutionary molecular modelling' (Wink, 1997). In the present volume, Chapter 2 by M. Wink and O. Schimmer and Chapter 5 by J. Reichling explore this topic in more detail.

1.3 Biotechnology of secondary metabolites

Since secondary metabolites have evolved as compounds that are important for the fitness of the organisms producing them, many of them interfere with the pharmacological targets which make them interesting for several biotechnological applications (an overview is presented in Fig. 1.6). A main area is phytomedicine, and several thousand plants are in use worldwide to treat human ailments and diseases (Fig. 1.6). In addition to isolated substances with established pharmacological profiles (including potent antineoplasmic drugs, such as the alkaloids, vinblastine, vincristine or taxol), complex extracts are often used. Controlled clinical studies have shown the efficacy of several, e.g. extracts from *Ginkgo biloba*, *Hypericum perforatum*, *Piper methysticum*, *Chamomilla recutita*, *Melissa officinalis*, *Mentha* x *piperita*, *Valeriana officinalis* (Wagner and Wiesenauer, 1995). The use of stimulants (such as caffeine, nicotine, ephedrine), fragrances (several essential oils), flavours (essential oils, capsaicin, piperine, etc.), natural dyes, poisons (strychnine) and hallucinogens (morphine, heroin, cocaine, tetrahydrocannabinol) is based on secondary metabolites (Fig. 1.6). In the present volume, Chapter 5 and Chapter 6 explore this wide field in more detail.

Since many SMs are insecticidal, fungicidal and phytotoxic, they may be used in agriculture as natural plant protectants. Before the advent of synthetic pesticides about 50 years ago, plant-derived insecticides (including nicotine, rotenone, quassin, ryanodine, pyrethrins and azadirachtins) were a common theme (Jacobson and Crosby, 1971; Wink, 1993c). Applications unequivocally showed that these natural insecticides worked. One ecological advantage, i.e. that SMs are readily degraded in plants and in soil, is also their disadvantage and synthetic pesticides are more resistant and persistent. Moreover, modern pesticides are usually more potent than biopesticides. On the other hand, plants are easy to grow and biopesticides could be a sustainable source of plant protectants for farmers in countries that do not have access to Western synthetic pesticides. Unfortunately, legislation does not favour mixtures of compounds to be used as pesticides, therefore the development of

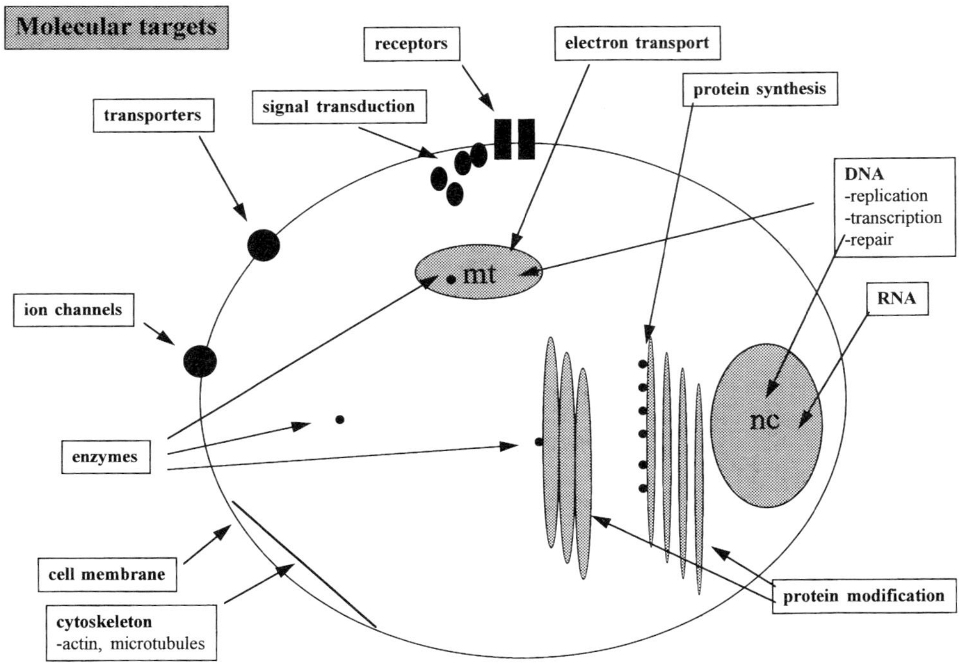

Figure 1.5 Molecular targets of defence chemicals in animal cells. Abbreviations: mt, mitochondria; nc, nucleus; RNA, ribonucleic acid; DNA, deoxyribonucleic acid.

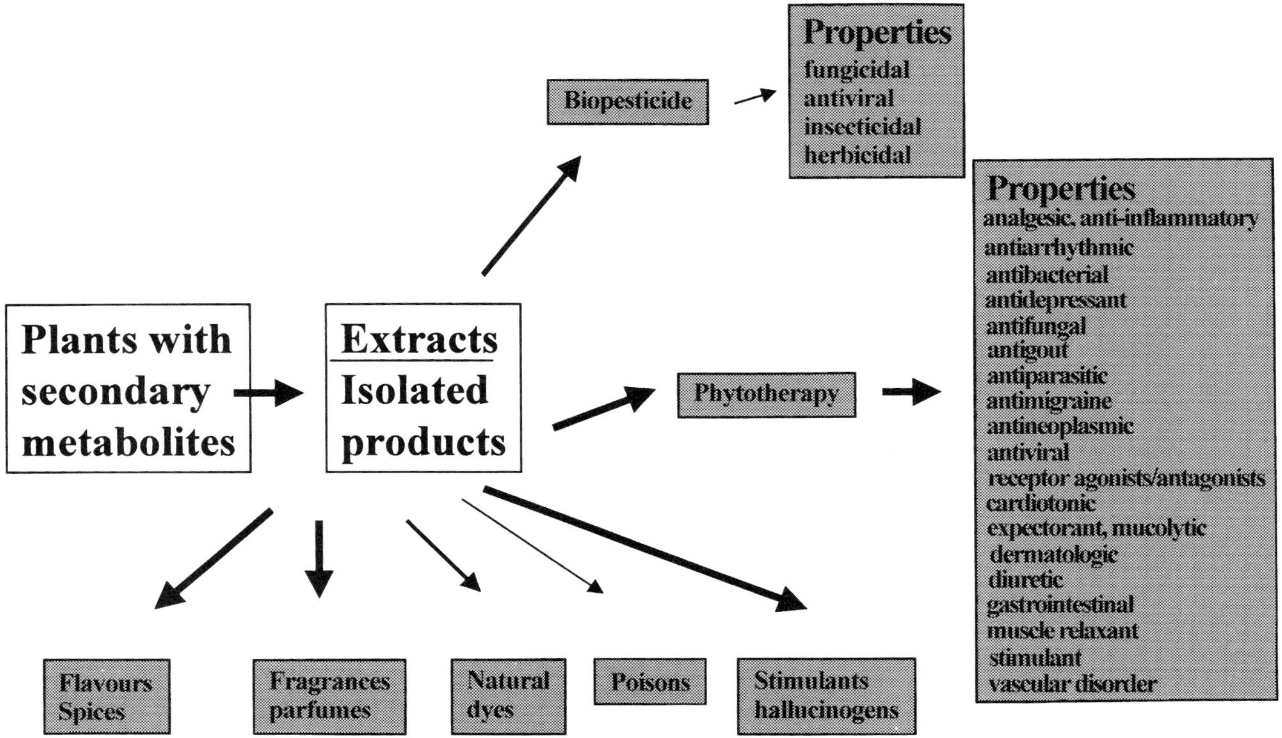

Figure 1.6 Utilisation of secondary metabolites in biotechnology.

biorational pesticides has to face many obstacles. Nevertheless, natural compounds do provide an underexplored alternative (for review see Wink, 1993c).

As a consequence of these various applications, a world market for plant extracts and isolated SMs exists, which exceeds 10 billion US dollars annually (Balandrin *et al.*, 1985). Therefore, it is a challenge for biotechnologists to find ways to produce these compounds in sufficient quantity and quality. The main and traditional way is to grow the respective plants in the field or in greenhouses and to extract the products from them (Fig. 1.7). For several species, new varieties have been selected with improved yields and quality. In this context, cell and organ culture are important techniques for *in vitro* propagation. In a few instances, genetic engineering of secondary metabolism has already had a direct influence; e.g. when *Atropa belladonna* plants were transformed with the gene that encodes the enzymes converting L-hyoscyamine into L-scopolamine, new plants were generated which produced scopolamine as the major product (Hashimoto and Yamada, 1992). More often, flavonoid metabolism has been altered genetically, producing plants with different flower colours (Mol *et al.*, 1998). It is a challenge for future research to isolate the genes of biosynthetic pathways and to express them either in transgenic plants or in microbes. If successful, recombinant bacteria or yeasts might be grown some day, which will produce valuable plant secondary metabolites (Wink, 1989; Kutchan, 1995) (Fig. 1.7); combinatorial biosynthesis might then be an open field. Using genes encoding enzymes for the biosynthesis of antibiotics, this strategy has already been successful (Katz and Hutchinson, 1992; Katz and Donadio, 1993; McDaniels *et al.*, 1993). If the corresponding secondary metabolites (both from plant or microbial origin) confer resistance to insects or pathogens, genetic transformation of susceptible crop plants could be another valuable avenue for exploitation.

For more than two decades, scientists around the world have tried to produce valuable secondary metabolites in cell or organ cultures (for overviews see Constabel and Vasil, 1987; Neumann *et al.*, 1985; Kurz, 1989; Charlwood and Rhodes, 1990). Whereas undifferentiated cell cultures have often failed to produce such a compound in reasonable yields, differentiated organ cultures (e.g. transformed root cultures) are often as active as the intact plant (Rhodes *et al.*, 1990). Cell- and tissue-specific gene expression appear to control these processes. In the present volume, Chapter 7 addresses the production of SMs *in vitro*. It is possible that genetic engineering may help to improve plant cell cultures as biotechnological production systems in the future.

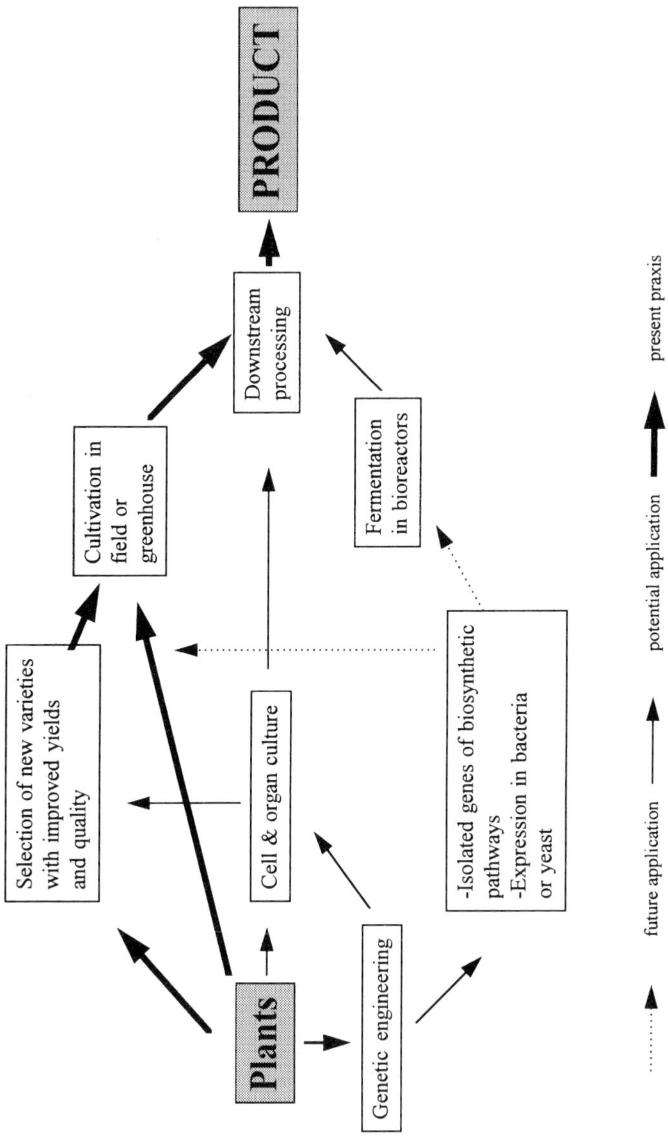

Figure 1.7 Production of secondary metabolites.

1.4 Conclusions

Plant secondary metabolism is still an interesting and challenging field of scientific endeavour ranging from botany, plant physiology and biochemistry, chemistry, pharmacology and medicine to evolution, ecology and biotechnology. It is the aim of the present volume to highlight recent results and to stimulate young researchers to enter a field that looks back on a long history but nevertheless provides an interesting presentday research topic and will hopefully experience an exciting future.

References

Ahmad, S. (1983) Mixed-function oxidase activity in a generalist herbivore in relation to its biology, food plants and feeding history. *Ecology*, **64** 235-43.

Balandrin, M.F., Klocke, J.A., Wurtele, E.S. and Bollinger, W.H. (1985) Natural plant chemicals: sources of industrial and medicinal materials. *Science*, **228** 1154-60.

Baldwin, I. (1994) Chemical changes rapidly induced by folivory, in *Insect Plant Interactions* (ed. E.A. Bernays), CRC Press, Boca Raton, pp. 1-23.

Becerra, J.X. (1994) Squirt-gun defense in *Bursera* and the chrysomelid counterploy. *Evolution*, **75** 1991-96.

Bernays, E.A. and Chapman, R.F. (1994) *Host-Plant Selection by Phytophagous Insects*. Chapman & Hall, New York, p. 312.

Blum, M.S. (1981) *Chemical Defenses of Arthropods*. Academic Press, New York, p. 562.

Braekman, J.C., Daloze, D. and Pasteels, J.M. (1998) Alkaloids in animals, in *Alkaloids: Biochemistry, Ecology and Medicinal Applications* (eds. M.F. Roberts and M. Wink), Plenum, New York, pp. 349-78.

Brattsten, L.B. and Ahmad, S. (1986) *Molecular Aspects of Insect-Plant Associations*. Plenum, New York.

Brown, K. and Trigo, J.R. (1995) The ecological activity of alkaloids, in *The Alkaloids* (ed. G.A. Cordell), Vol. 47, pp. 227-354.

Charlwood, B.V. and Rhodes, M.J. (1990) *Secondary Products From Plant Tissue Culture*. Clarendon Press, Oxford.

Cipollini, M.L. and Levey, D.J. (1997) Secondary metabolites of fleshy vertebrate-dispersed fruits: adaptive hypotheses and implications for seed dispersal. *Amer. Naturalist*, **150** 346-73.

Constabel, F. and Vasil, I. (1987) *Cell Culture and Somatic Cell Genetics of Plants*. Vol. 4. Cell Culture in Phytochemistry. Academic Press, San Diego.

Creelman, R.A. and Mullet, J.E. (1997) Biosynthesis and action of jasmonates in plants. *Annu. Rev. Plant Physiol. Plant Mol. Biol.*, **48** 355-81.

De Moraes, C.M., Lewis, W.J., Paré, P.W., Alborn, H.T. and Tumlinson, J.H. (1998) Herbivore-infested plants selectively attract parasitoids. *Nature*, **393** 570–73.

Dicke, M., Sabelius, M.W., Takabayashi, J., Bruin, J. and Posthumus, M.A. (1990) Plant strategies of manipulating predator-prey interactions through allelochemicals: prospects for application in pest control. *J. Chem. Ecol.*, **16** 3091-118.

Duffey, J. (1980) Sequestration of plant natural products by insects. *Annu. Rev. Entomol.*, **25** 447-77.

Dussourd, D.E. and Eisner, T. (1987) Vein-cutting behavior: insect counterploy to latex defence of plants. *Science*, **237** 898-901.

Edmunds, M. (1974) *Defense in Animals*. Longman, Harlow.
Ehrlich, P.R. and Raven, P.H. (1964) Butterflies and plants: a study of coevolution. *Evolution*, **18** 586-608.
Fraenkel, G. (1959) The raison d'être of secondary substances. *Science*, **129** 1466-70.
Gruhnert, C., Biehl, B. and Selmar, D. (1994) Compartmentation of cyanogenic glucoside and their degrading enzymes. *Planta*, **195** 36-42.
Harborne, J.B. (1993) *Introduction to Ecological Biochemistry*, 4th Edn, Academic Press, London.
Hartmann, T. and Witte, L. (1995) Chemistry, biology and chemoecology of the pyrrolizidine alkaloids, in *Alkaloids: Chemical and Biological Perspectives* (ed. S.W. Pelletier), Vol. 9, Pergamon, Oxford, pp. 155-233.
Hashimoto, T. and Yamada, Y. (1992) Biosynthesis of scopolamine and an application for genetic engineering of medicinal plants, in *Plant Tissue Culture and Gene Manipulation for Breeding and Formations of Phytochemicals*. (eds. K. Oono *et al.*), NIAR, Tsukuba, pp. 255-59.
Jacobson, M. and Crosby, D.G. (1971) *Naturally Occurring Insecticides*. Marcel Dekker, New York, p. 585.
Katz, L. and Donadio, S. (1993) Polyketide synthesis: prospects for hybrid antibiotics. *Annu. Rev. Microbiol.*, **47** 875-912.
Katz, L. and Hutchinson, R. (1992) Genetic engineering of antibiotic producing organisms. *Annu. Rep. Medic. Chem.*, **27** 129-38.
Kojima, M., Poulton, J.E., Thayer, S. and Conn, E.E. (1979) Tissue distribution of dhurrin and of enzymes involved in its metabolism in leaves of *Sorghum bicolor*. *Plant Physiol.*, **63** 1022-28.
Kurz, W. (1989) *Primary and Secondary Metabolism in Cell Cultures*. II. Springer, Heidelberg.
Kutchan, T.M. (1995) Alkaloid biosynthesis: the basis for metabolic engineering of medicinal plants. *The Plant Cell*, **7** 1959-70.
Levin, D.A. (1976) The chemical defences of plants to pathogens and herbivores. *Annu. Rev. Ecol. Syst.*, **7** 121-59.
Mann, J. (1992) *Murder, Magic and Medicine*. Oxford University Press, London.
Matile, P. (1980) The 'mustard oil bomb': compartmentation of myrosinase systems. *Biochem. Physiol. Pflanzen*, **175** 722-31.
Matile, P. (1984) Das toxische Kompartiment der Pflanzenzelle. *Naturwissenschaften*, **71** 18-24.
McDaniels, R., Ebert-Koshla, S., Hopwood, D.A. and Koshla, C. (1993) Engineered biosynthesis of novel polyketides. *Science*, **262** 1546-50.
Mol, J., Grotewold, E. and Koes, R. (1998) How genes paint flowers and seeds. *Trends Plant Sci.*, **3**, 212-17.
Neumann, K.H., Barz, W. and Reinhard, E. (1985) *Primary and Secondary Metabolism of Plant Cell Cultures*. Springer, Heidelberg.
Proksch, P. and Ebel, R. (1998) Ecological significance of alkaloids from marine invertebrates, in *Alkaloids:- Biochemistry, Ecological Functions and Medical Applications*. (eds. Roberts, M.F. and Wink, M.), Plenum, New York, pp. 379-94.
Rhodes, M.J.C., Robins, R.J., Hamill, J.D., Parr, A.J., Hilton, M.G. and Walton, N.J. (1990) Properties of transformed root cultures, in *Secondary Products From Plant Tissue Culture* (eds. B.V. Charlwood and M.J.C. Rhodes) Clarendon Press, Oxford, pp. 201-25.
Roberts, M.F. and Wink, M. (1998) *Alkaloids:-Biochemistry, Ecological Functions and Medical Applications*. Plenum, New York.
Rosenthal, G.A. and Berenbaum, M.R. (1991/1992) *Herbivores: Their Interactions With Secondary Plant Metabolites*, Vol. 1, The Chemical Participants; Vol. 2, Ecological and Evolutionary Processes. Academic Press, San Diego.

Saunders, G.A. and Conn, E.E. (1978) Presence of the cyanogenic glycoside dhurrin in isolated vacuoles from Sorghum. *Plant Physiol.*, **61** 154-57.

Swain, T. (1977) Secondary compounds as protective agents. *Annu. Rev. Plant Physiol.*, **28** 479-501.

Teuscher, E. and Lindequist, U. (1994) *Biogene Gifte*, Fischer, Stuttgart.

Wagner, H. and Wiesenauer, M. (1995) *Phytotherapie*, Fischer, Stuttgart.

Weißenboeck, G., Hedrich, R. and Sachs, G. (1986) Secondary phenolic products in isolated guard cell, epidermal cell and mesophyll cell protoplasts from pea (*Pisum sativum*(L.)) leaves: distribution and determination. *Protoplasma*, **134** 141-48.

Werner, C. and Matile, P. (1985) Accumulation of coumaroylglucosides in vacuoles of barley mesophyll protoplasts. *J. Plant Physiol.*, **118** 237-49.

Wink, M. (1983) Wounding-induced increase of quinolizidine alkaloid accumulation in lupin leaves. *Z. Naturforsch.*, **38c** 905-909.

Wink, M. (1987a) Physiology of the accumulation of secondary metabolites with special reference to alkaloids, in *Cell Culture and Somatic Cell Genetics of Plants*, Vol 4, Cell Culture in Phytochemistry. (eds. F. Constabel and I. Vasil), Academic Press, San Diego, pp. 17-41.

Wink, M. (1988) Plant breeding: importance of plant secondary metabolites for protection against pathogens and herbivores. *Theor. Appl. Gen.*, **75** 225-33.

Wink, M. (1989) Genes of secondary metabolism: differential expression in plants and *in vitro* cultures and functional expression in genetically-transformed microorganisms, in *Primary and Secondary Metabolism in Cell Cultures*, (ed. W. Kurz), Springer, Heidelberg, pp. 239-51.

Wink, M. (1992) The role of quinolizidine alkaloids in plant insect interactions, in *Insect-Plant Interactions* (ed. E.A. Bernays), CRC Press: Boca Raton, Vol. IV, pp. 133-69.

Wink, M. (1993b) Allelochemical properties and the raison d'être of alkaloids, in *The Alkaloids* (ed. G. Cordell), Academic Press, Orlando, Vol. 43, 1-118.

Wink, M. (1993c) Production and application of phytochemicals from an agricultural perspective, in *Phytochemistry and Agriculture* (eds. T.A. van Beek and H. Breteler), Clarendon Press, London, pp. 171-213.

Wink, M. (1997) Compartmentation of secondary metabolites and xenobiotics in plant vacuoles. *Advances Bot. Research*, **25** 141-69.

Wink, M. (1999) Interference of alkaloids with neuroreceptors and ion channels, in *Bioactive Natural Products*, (Atta-Ur-Rahman ed.), Elsevier, pp. 1-127, (in press).

Wink, M., Heinen, H.J., Vogt, H. and Schiebel, H.M. (1984) Cellular localization of quinolizidine alkaloids by laser desorption mass spectrometry (LAMMA 1000). *Plant Cell Rep.*, **3** 230-33.

Wink, M. and Römer, P. (1986) Acquired toxicity: the advantages of specializing on alkaloid-rich lupins to *Macrosiphum albifrons* (Aphidae). *Naturwissenschaften*, **73** 210-12.

Wink, M. and Witte, L. (1984) Turnover and transport of quinolizidine alkaloids: diurnal variation of lupanine in the phloem sap, leaves and fruits of *Lupinus albus* L. *Planta*, **161** 519-24.

Wink, M. and Witte, L. (1991) Storage of quinolizidine alkaloids in *Macrosiphum albifrons and Aphis genistae* (Homoptera: Aphididae). *Entomol. Gener.*, **15** 237-54.

2 Modes of action of defensive secondary metabolites

Michael Wink and Oskar Schimmer

2.1 Introduction

Since only autotrophic plants can convert light energy via photosynthesis into organic compounds, heterotrophic animals and most microorganisms depend on plant material as an energy source or for vital precursors and vitamins. In order to survive, plants have had to develop defence strategies against herbivores, microbes (bacteria, fungi), viruses and even against other plants competing for light, space and nutrients. Efficient defence strategies must exist because the world is still green, despite a multitude of herbivores and microorganisms (Hartley and Jones, 1997). Many plants are avoided by herbivores (an obvious exception being crop plants, in which chemical defence compounds have been selected away).

Plants are always a rich source of compounds that do not appear essential for primary metabolism, including thousands of secondary metabolites and several macromolecules, such as peptides, proteins, enzymes, lignin, callose, cellulose or cuticular waxes. In addition to their function in physiology or in structural maintenance, many serve for defence against microbes or herbivorous animals. In addition, some secondary metabolites (e.g. flavonoids, anthocyanins, tetraterpenes and monoterpenes) function as signal compounds to attract pollinating and seed-dispersing animals. Some of these compounds exhibit both defence and signal functions (for reviews see Bernays and Chapman, 1994; Harborne, 1993; Wink, 1992, 1993a,b,c, 1997).

A large body of experimental, toxicological data and circumstantial evidence clearly shows that many alkaloids, cyanogenic glycosides, glucosinolates, terpenes, saponins, tannins, anthraquinones, polyacetylenes and other allelochemicals are toxic or deterrent to animals (insects, vertebrates), and several display antibiotic or even allelopathic activities (for overviews see Bernays and Chapman, 1994; Rosenthal and Janzen, 1979; Harborne, 1993; Wink, 1988, 1993a,b,c; Rosenthal and Berenbaum, 1991, 1992; Waller, 1987; Swain, 1977; Levin, 1976; Roberts and Wink, 1998).

Allelochemicals can only function as chemical defence compounds if they are able to influence herbivores or microbes in a negative way. A closer analysis shows clearly that most allelochemicals interfere with one or several molecular targets in animals and microbes (Wink, 1992, 1993a, 1998; Wink *et al.*, 1998a,b). Structures of allelochemicals appear to have been shaped during evolution in such a way that they can mimic the structures of endogenous substrates, hormones, neurotransmitters or other ligands; this process can be termed 'evolutionary molecular modelling'. Other metabolites intercalate or alkylate deoxyribonucleic acid (DNA), inhibit DNA- and ribonucleic acid (RNA)-related enzymes, protein biosynthesis or other metabolic enzymes, or disturb membrane stability.

This chapter will provide a short overview of the modes of action of some important groups of secondary metabolites (Teuscher and Lindequist, 1994; Roberts and Wink, 1998; Wink, 1993a, 1999), followed by a more detailed analysis of the interactions of allelochemicals with neuronal signal transduction and DNA.

In general, alkaloids are infamous as animal toxins and certainly serve mainly as defence chemicals against predators (herbivores, carnivores) and to a lesser degree against bacteria, fungi and viruses (Wink, 1988; 1992, 1993a; Harborne, 1993; Swain, 1977; Bernays and Chapman, 1994; Hartmann, 1991; Levin, 1976; Roberts and Wink, 1998). Alkaloids and amines often affect neuroreceptors as agonists or antagonists, or they modulate other steps in neuronal signal transduction, such as ion channels or enzymes which take up or degrade neurotransmitters or second messengers. Since alkaloids often derive from the same amino acid precursor as the neurotransmitters, acetylcholine, serotonin, noradrenaline, dopamine, gamma aminobutyric acid (GABA), glutamic acid or histamine, their structures can frequently be superimposed on those of neurotransmitters. Other alkaloids intercalate DNA, alkylate DNA (see section 2.2), induce apoptosis or inhibit carbohydrate processing enzymes (Goss *et al.*, 1995). It is apparent that the toxicity of alkaloids is correlated with their interactions with particular molecular targets (Wink, 1993a, 1999b; Wink *et al.*, 1998a,b; Roberts and Strack, 1999).

Nonprotein amino acids (NPAAs) can be considered as structural analogues to one of the 20 protein amino acids. NPAAs frequently block the uptake and transport of amino acids or disturb their biosynthetic feedback regulations. Some NPAAs are even incorporated into proteins, since transfer ribonucleic acid (tRNA) transferases cannot usually discriminate between a protein amino acid and its analogue; resulting in defective or malfunctioning proteins (Rosenthal, 1982). Other NPAAs interfere with neuronal signal transduction or enzymatic processes (Teuscher and Lindequist, 1994; Selmar, 1999).

Cyanogenic glycosides are stored in the vacuole as prefabricated allelochemicals. If tissue decomposition occurs due to wounding by a herbivore or a pathogen, then a β-glucosidase comes into contact with the cyanogenic glycosides, which are split into a sugar and a nitrile moiety that is further hydrolyzed to hydrocyanic acid (HCN) and an aldehyde. HCN effectively blocks mitochondrial respiration and, thus, adenosine triphosphate (ATP) production, and functions as a strong poison in most animals (Conn, 1980; Selmar, 1999).

Glucosinolates also function as prefabricated vacuolar defence compounds. The resulting mustard oil, which is released after cleavage by myrosinase, is highly lipophilic and can disturb the fluidity of biomembranes (thereby exhibiting a substantial antimicrobial effect) and bind to some enzymes, receptors or other macromolecules, such as DNA (see section 2.3.2) (Selmar, 1999).

Terpenes (mono-, sesqui-, di- and triterpenes) are usually highly hydrophobic substances and are stored in resin ducts, oil cells or glandular trichomes. Most of them readily interact with biomembranes. They can increase the fluidity of the membranes, which can lead to uncontrolled efflux of ions and metabolites, modulation of membrane proteins and receptors or even to cell leakage, resulting in cell death. This membrane activity is rather nonspecific; therefore, terpenes show activities against a wide range of organisms, ranging from bacteria and fungi to insects and vertebrates. Even if the concentrations were not critical for a large vertebrate herbivore, terpene-rich food is usually avoided, since these terpenes would inhibit the growth of rumen microorganisms, which are important for the breakdown of cellulose. A number of terpenes have special additional activities because their structures figure as analogues to natural substrates, hormones (e.g. steroidal hormones, sex hormones, ecdysone, juvenile hormone) or neurotransmitters.

Saponins are the glycosides of triterpenes or steroids and include the group of cardiac glycosides and steroidal alkaloids. Some of them are stored as bidesmosidic compounds in the vacuole, which are cleaved to monodesmosidic compounds by β-glucosidase upon wounding-induced decompartmentation. Monodesmosidic saponins are amphiphilic compounds, which can complex cholesterol in biomembranes with their lipophilic terpenoid moiety and bind to surface glycoproteins and glycolipids with their sugar side chain. This leads to a severe tension of the biomembrane and leakage. This activity can easily be demonstrated with erythrocytes, which lose their haemoglobin when in contact with saponins. This membrane activity is rather unspecific and effects a wide set of organisms from microbes to animals. Some saponins have additional functional groups, such as cardiac glycosides (carrying a 5 or 6 membered cardenolide or bufadienolide ring), which enable them to inhibit one of the most important molecular targets of animal cells, the Na^+-, K^+-ATPase (Gershenzon and Kreis, 1999).

Polyketides include anthraquinones, which produce severe diarrhoea in vertebrates by interfering with intestinal Na^+,K^+-ATPase and adenylyl cyclase. These compounds can also interact with DNA (see section 2.3.2).

Flavonoids and phenylpropanoids (including coumarins, furocoumarins and tannins) are widespread in plants (Petersen *et al.*, 1999). They exhibit a wide range of biological activities. In several instances, they act as analogues of cellular signal compounds or substrates. Afflicted mechanisms range from prostaglandin and leukotriene formation, enzyme inhibition, oestrogenic properties (coumarins, isoflavones, stilbenes) to DNA alkylation (e.g. by furocoumarins) (see section 2.3.2). These molecules usually have several phenolic hydroxyl groups in common, which can form hydrogen bonds with proteins and peptides. The higher the number of hydroxyl groups, the stronger the astringent and denaturing effect. Tannins inhibit enzymatic activities very effectively; however, most digestive enzymes of herbivores have apparently adapted to tannins and are less sensitive than other enzymes.

2.2 Interference of secondary metabolites with neuronal signalling

In the present chapter, special emphasis has been laid on interactions between nitrogen-containing secondary metabolites and major neuroreceptors, such as cholinergic, adrenergic, serotonergic, and GABAergic neuroreceptors, and Na^{+}-, K^{+}-, Cl^{-}- and Ca^{2+}- channels, although other elements of the neuronal signal transduction (acetylcholine esterase, monoamine oxidase [MAO], adenylyl cyclase, phosphodiesterase, phospholipase, protein kinase and neurotransmitter transport) have also been addressed. Since any substantial interference at the neuroreceptors (e.g. competitive inhibition of ligand binding by antagonistic secondary metabolites or agonistic receptor activation by a defence substance with structural similarity to the natural ligand) will influence neuronal signal transduction (including muscular and central nervous system (CNS) activity), the intake of a larger dose should lead to substantial physiological disturbance.

It is outside the scope of the present review to provide a complete overview of all the relevant interactions that have been published. Rather, one aim is to highlight the body of information that has accumulated, especially during the last decade, due to a number of technical breakthroughs, such as: patch clamp techniques (Neher and Sakman, 1992) allowing direct measurements of ion channel activities; and receptor ligand assays (Yamamura and Snyder, 1974) with a wide number of cloned receptors and specifically-labelled ligands.

2.2.1 Basic elements of neuronal signalling

The nervous system consists of the central (brain and spinal cord) and the peripheral system (afferent sensory and efferent motor nerves); it regulates all aspects of bodily function and is staggering in its complexity. Another distinction is between the somatic and the autonomic nervous system, which regulates heart and blood circulation, respiration, motility of the gastrointestinal tract, smooth muscles of the gall and urinary bladder, ureter and uterus, and also glandular secretion. The somatic nervous system innervates the skeletal muscles. The autonomic nervous system is further divided into a sympathetic and parasympathetic part, which often regulate the same organ in opposite ways. The basic elements of the nervous systems are neurons, which communicate with each other via chemical synapses. Whereas somatic nerves are usually monosynaptic, two neurons, which communicate via a ganglionic synapse are found in sympathetic and parasympathetic nerves. When the postsynaptic cell is a muscle cell, the synapse is called a neuromuscular junction or motor end-plate.

The presynaptic terminal contains vesicles that are filled with neurotransmitters. Presynapse and postsynapse are separated by a narrow synaptic cleft, into which the neurotransmitters are released from the vesicles via exocytosis. Transmitters diffuse across the synaptic cleft and bind to a neuroreceptor on the postsynaptic cell. The ion permeability of the postsynaptic membrane is changed in the next step, causing a sudden change in the corresponding membrane potential. In neurons, this electric disturbance can induce an action potential. At the motor end-plate, the change of membrane potential leads to muscle contraction; in gland cells, it may induce hormone secretion. Many nerve and most nerve-muscle synapses are excitatory. Binding of neurotransmitters to inhibitory receptors on the postsynapse causes an opening of K^+- and Cl^-- channels that hyperpolarize the membrane and, thus, block the generation of an action potential. Neuroreceptors are found at the post- and presynaptic membrane. Activation of presynaptic receptors usually leads to an inhibition of neurotransmitter release, whereas their inhibition results in an enhanced release of neurotransmitters.

Thus, neurotransmitters and neuroreceptors are the basic elements for signal transduction in synapses of the central and peripheral nervous system and in neuromuscular junctions. The corresponding neurotransmitters (ligands) include acetylcholine (ACh), noradrenaline (NA), adrenaline, serotonin (i.e. 5-hydroxytryptamine, 5-HT), dopamine, histamine, glycine, glutamate/aspartate, GABA, ATP and several peptides. Two classes of membrane-residing neuroreceptors can be distinguished: a fast ligand-gated channel (class I) and a slower G protein linked receptor (class II), which are very similar across a wide range of animals. The class I neuroreceptor is part of an ion-channel complex (Fig. 2.1). When a neurotransmitter binds, a conformational change induces the opening of an ion channel. According to the geometry and polarity of the 'gate', a selective permeability is achieved for Na^+,

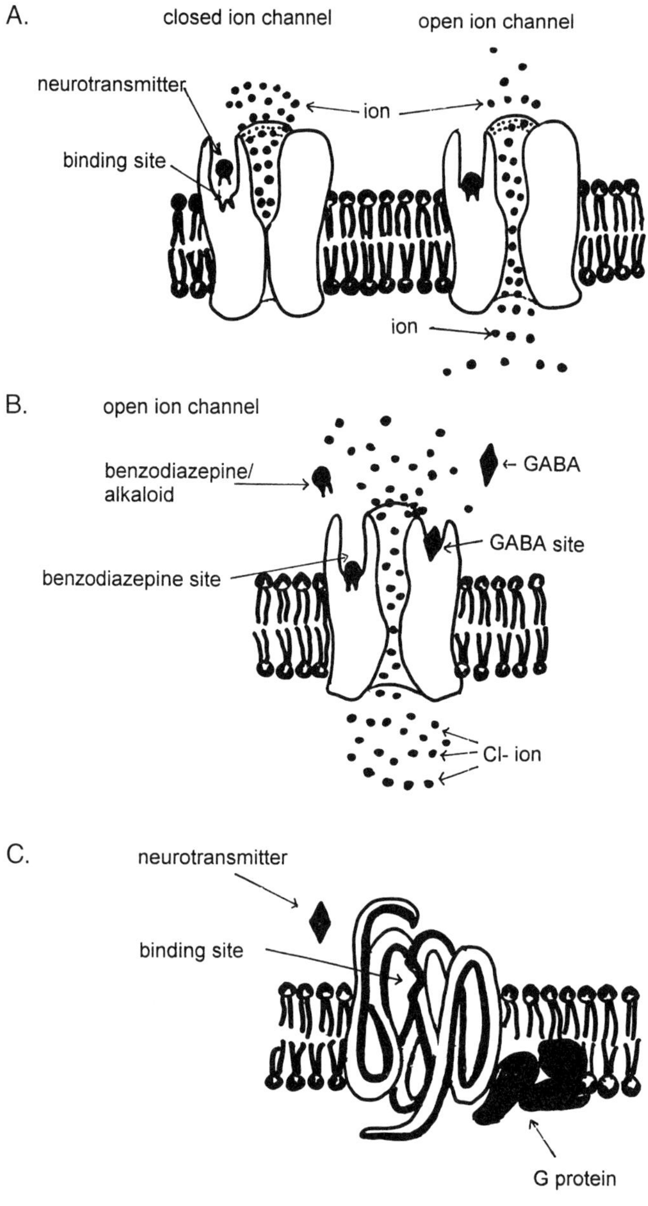

Figure 2.1 Schematic illustration of ligand-gated ion channels and G-protein linked neuroreceptors A) Could be a nicotinic acetylcholine receptor; B) a gamma aminobutyric acid (GABA) receptor; and C) a dopamine or noradrenaline receptor.

K^+, Ca^{2+} and Cl^- ions. The driving force is provided via the ion concentration inside the cell or the extracellular space. Class I ligand-operated ion channels include the excitatory nicotinic acetylcholine, glutamate/aspartate, ATP_{P2z}, and the 5-HT_3 receptor and the inhibitory glycine and $GABA_A$ receptor. In the well-studied nicotinic ACh-receptor, the five subunits, consisting of two α-subunits that bind acetylcholine and of one β-, γ- and δ-subunit each, form the ligand-gated ion channel (Changeux, 1993) (Fig. 2.1).

The G protein linked receptors (class II) (Fig. 2.1) are far more numerous and more complex than the ligand-operated ion channels. They include muscarinic acetylcholine, adenosine, adrenergic, serotonergic (except 5-HT_3; see above), $GABA_B$, glutamate (mGluR), histamine, and opiate receptors. They share a common architecture, having seven transmembrane domains and three internal and three external loops each (Fig. 2.1). When the corresponding neurotransmitter binds, the receptor changes its conformation, inducing a conformational change in an adjacent G protein molecule, consisting of three subunits α, β and γ. G proteins function as an on/off switch, which is off when the α-subunit binds guanosine diphosphate (GDP). Binding of a ligand to the receptor causes the G protein to release its bound GDP and to bind guanosine triphosphate (GTP), converting the α-subunit to the 'on' state. The α-subunit dissociates and then either interacts with an ion channel or activates/inhibits the enzymes of second messenger formation (Fig. 2.2), such as adenylyl cyclase (making cyclic adenosine monophosphate (cAMP), an allosteric regulator of protein kinases and other proteins), or phospholipase C (splitting phosphatidylinositol-4,5-diphosphate (PIP_2) into inositol-1,4,5-triphosphate (IP_3)) (which activates Ca^{2+} release channels in the endoplasmic reticulum (ER) setting free the second messenger Ca^{2+}) and diacylglycerol (DAG), which activates proteinkinase C (PKC). Whereas the hydrolysis of GTP (bound to the α-subunit) switches the G protein back to the inactive ('off') state, the second messenger can then regulate various ion channels, protein kinases and other proteins (Fig. 2.2). The second messengers, cAMP and cyclic guanosine monophosphate (cGMP), are inactivated by specific phosphodiesterases (PDE); IP_3 through hydrolysis to the inactive inositol 1,4-biphosphate or in some cells via phosporylation and hydrolysis to the inactive inositol 1,3,4-triphosphate.

Voltage-gated ion channels are also integral parts of the signal pathway. When an action potential approaches the axon terminal, voltage-gated Ca^{2+}- channels (N-type) open and Ca^{2+} enters the presynapse. Ca^{2+} ions bind to proteins that connect the synaptic vesicle with the plasma membrane (acronym SNAP), inducing membrane fusion and, consequently, exocytosis of the neurotransmitter, acetylcholine, into

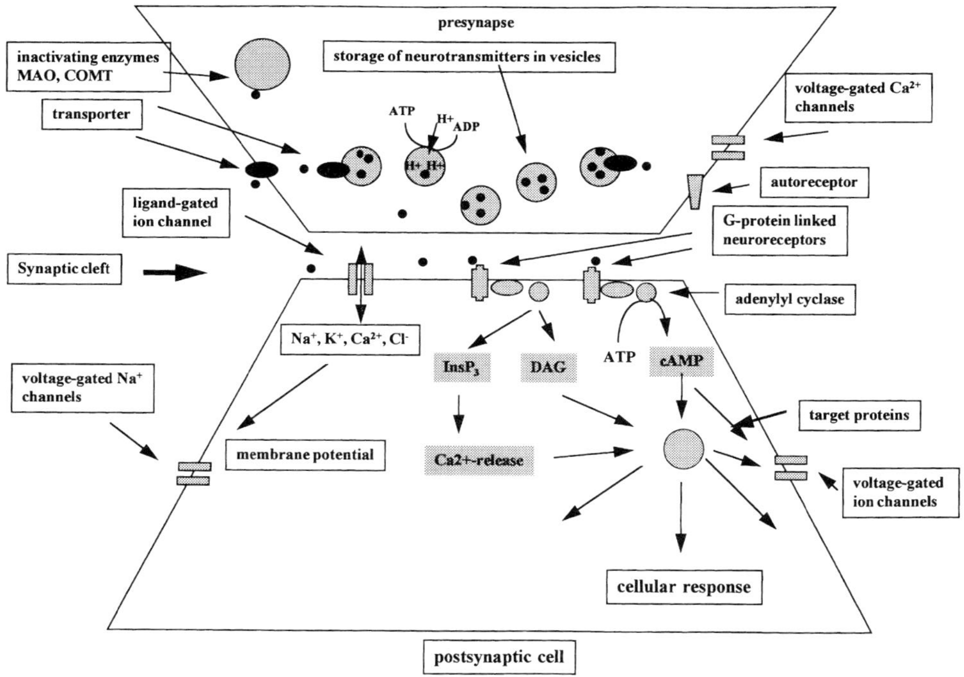

Figure 2.2 Schematic illustration of signalling in neuronal synapses (for explanation see text). Abbrevations: ATP, adenosine triphosphate; ADP, adenosine diphosphate; cAMP, cyclic adenosine monophosphate; MAO, monoamine oxidase; COMT, catecholamine-*O*-methyltransferase; IP_3, inositol-1,4,5-triphosphate; DAG, diacylglycerol.

the synaptic cleft. Acetylcholine activates the nicotinic acetylcholine receptor (nAChR), resulting in a rapid influx of Na^+ in the postsynapse, which open voltage-gated Na^+-channels. This event generates an action potential, which spreads along the membrane via voltage-gated Na^+-channels.

At the neuromuscular junction, the change in membrane potential induces a conformational change in the dihydropyrine-coupled ryanodine receptor, so that Ca^{2+} ions are released from the sarcoplasmic reticulum (SR) into the cytosol. The released Ca^{2+} ions activate the actin-myosin system of the muscle cell, leading to muscle contraction. The ER membranes in cardiac muscle cells and neurons also contain ryanodine-sensitive Ca^{2+} release channels that do not associate directly with a receptor in the cell membrane. In these cells, depolarization of the cell membrane leads to a small influx of Ca^{2+} through voltage-gated Ca^{2+}-channels (L-type). Binding of these Ca^{2+} ions to ryanodine sensitive Ca^{2+}-channels induces a rapid release of Ca^{2+} from the ER or other intracellular Ca^{2+} stores. Another Ca^{2+} release channel in the ER/SR is regulated by IP_3.

Specific Ca^{2+}-ATPases in the plasma membrane and in the SR (only in muscle cells) or ER pump Ca^{2+} into the SR/ER or into the extracellular space. The Na^+ and K^+ gradients are restored by Na^+-, K^+-, ATPase.

Whereas acetylcholine is degraded by a membrane-anchored acetylcholine esterase (ACE) in the synaptic cleft (choline is taken up presynaptically), the biogenic amines, adrenaline, noradrenaline, serotonin and dopamine, are taken up by the presynaptic membrane via transporters. These transporters have similar structure and contain 12 transmembrane regions. Once in the presynapse, the neurotransmitters are either degraded by MAO or catecholamine *O*-methyltransferase (COMT) or taken up by synaptic vesicles. This may occur via diffusion of the free bases into the vesicles, where they become protonated and concentrated by an 'ion trap' mechanism, or the amines are taken up and concentrated via specific proton-coupled antiporters.

2.2.2 Interactions of allelochemicals with individual steps of neurosignalling

In Figure 2.2, the various steps in neuronal signalling and signal transduction, which were discussed above, have been schematically summarized. The following targets are affected by several allelochemicals, some of which are listed in Tables 2.1 to 2.10:

- The neuroreceptor itself (Tables 2.1–2.7); agonists mimic the function of a signal compound by binding to its receptor and causing the normal response, whereas antagonists also bind to the receptor but do not activate the transmitter-induced effects. Thus, an antagonist acts as an inhibitor of the natural ligand by competing for binding sites on the receptor and, thereby, blocking the physiological response (antagonists are, therefore, often called 'blockers').

- Voltage-gated Na^+-, K^+- and Ca^{2+}- channels (Table 2.8).
- The enzymes which deactivate neurotransmitters after they have bound to a receptor (Table 2.9), such as acetylcholine esterase, monoamine oxidase and catechol-*O*-methyltransferase.
- Transport processes (Table 2.10), which are important for the uptake and release of the neurotransmitters in the presynapse or synaptic vesicles. Na^+-, K^+- and Ca^{2+}-ATPases, which restore the ion gradients, must also be considered in this category.
- Modulation of key enzymes of signal pathways (Table 2.11):

1. adenylyl cyclase (producing cAMP);
2. phosphodiesterase (inactivating cAMP);
3. phospholipase C (PLC) (releasing inositol phosphates, such as IP_3 and DAG); all phospholipases have been considered (not only PLC, which produces IP_3 and DAG);
4. several protein kinases, such as protein kinase C or tyrosine kinase (activating other regulatory proteins or ion channels).

2.2.2.1 Cholinergic receptors

Agonists at the nicotinic acetylcholine receptor (Table 2.1) include tetramethylammonium hydroxide which is of simple structure, and small alkaloids, such as anabasine, nicotine, cotinine, coniine, pseudopelletierine and epibatidine. These alkaloids carry a secondary or tertiary nitrogen atom in a pyrrolidine or piperidine ring, which becomes protonated under physiological hydrogen ion concentrations in animals (Fig. 2.3).

Often the N is methylated. If a second ring structure is present, it is not coupled via an ester or ether bond and does not carry bulky oxygen substituents, which could mimic an ester group as in acetylcholine. It is obviously not pure chance that an additional pyridine ring is often found in nAChR agonists, since a pyridine tetramer, nemertilline, exhibits agonistic properties alone. Several quinolizidine alkaloids, such as cytisine or *N*-methylcytisine, are strong nAChR agonists; although these molecules are much bigger, they still share the important characteristics described for the simpler nAChR agonists, i.e. a protonable quaternary N and an α-pyridone ring (which resembles a pyridine ring) at the correct distance.

Antagonist of nAChR also carry a tertiary nitrogen, which becomes protonated under physiological conditions. In a few powerful antago-

Table 2.1 Examples of alkaloids which bind to nicotinic acetylcholine receptors (nAChRs) (natural ligand: acetylcholine)

Alkaloid	Type	Occurrence	Activity	Reference
Acetylcholine	Amine	Endogenous neurotransmitter, several plants and in venom of Hymenoptera	nAChR agonist	Buckingham, 1996; Teuscher and Lindequist, 1994
Ammodendrine	Piperidine	*Lupinus*, *Cytisus*, *Genista* and other Leguminosae	nAChR binding	Schmeller *et al.*, 1994; Wink *et al.*, 1998b
Anabasine	Pyridine	*Anabasis aphylla* (Chenopodiaceae), *Nicotiana* spp., *Duboisia* spp. (Solanaceae)	nAChR agonist	Buckingham, 1996; Wink *et al.*, 1998b
Arborine	Quinazolinone	*Glycosmis arborea*, *Ruta graveolens*, *Zanthoxylum budrunga* (Rutaceae)	ACh blocker	Buckingham, 1996; Johne, 1983
Arecoline	Piperidine	*Areca catechu* (Palmae)	Binding to nAChR	Wink *et al.*, 1998b
Berberine	Protoberberine	*Berberis* spp., *Mahonia* spp. (Berberidaceae) and other families	Binding to nAChR	Schmeller *et al.*, 1997b
Boldine	Aporphine	*Peumus boldo*	nAChR antagonist	Hue *et al.*, 1994; Wink *et al.*, 1998b
Brucine	Indole	*Strychnos nux-vomica* and many other *Strychnos* spp. (Loganiaceae)	Binding to nAChR	Wink *et al.*, 1998b
C-Toxiferine I, II (calebassine, curarine)	Bis-indole	*Strychnos divaricans* and several other *Strychnos* spp. (Loganiaceae)	nAChR antagonist (muscle, neuronal)	Buckingham, 1996
Codeine	Morphinane	*Papaver somniferum* (Papaveraceae)	Non-competitive nACh agonist	Maelicke *et al.*, 1995; Storch *et al.*, 1995
Coniine	Piperidine	*Conium maculatum* (Apiaceae)	nAChR agonist	Forsyth *et al.*, 1996; Wink *et al.*, 1998b
Cytisine and related alpha-pyridone alkaloids	Quinolizidine	*Cytisus*, *Genista*, *Laburnum*, *Baptisia*, *Thermopsis* and several other Leguminosae	Neuronal nAChR agonist (muscle, neuronal)	Papke and Heinemann, 1994; Schmeller *et al.*, 1994

Table 2.1 (Continued)

Alkaloid	Type	Occurrence	Activity	Reference
Dihydro-β-erythroidine	Erythrina	*Erythrina* spp. (Leguminosae)	nAChR antagonist (muscle, neuronal)	Williams and Robinson, 1984
Epibatidine	Pyridine	*Epipedobates tricolor* (Dendrobatidae)	Potent agonist at neuronal nAChR	Badio and Daly, 1994; Daly *et al.*, 1993; Elguero *et al.*, 1996
Erysodine	Erythrina	*Erythrina* spp. (Leguminosae)	Very potent neuronal nAChR antagonist	Decker *et al.*, 1995
Galanthamine	Lycorine	*Galanthus* spp. and many other Amaryllidaceae	Non-competitive nACh agonist	Maelicke *et al.*, 1995; Storch *et al.*, 1995
Gramine	Indole	Gramineae, Leguminosae, Aceraceae	Binding to nAChR	Wink *et al.*, 1998b
Hirsuteine and related alkaloids	Indole	*Mitragyna* spp., *Uncaria* spp.	Antagonist of nicotine-evoked dopamine release	Watano *et al.*, 1993
Histrionicotoxin	Piperidine	Skin of *Dendrobates* spp. (Dendrobatidae) and other frogs	Non-competitive nAChR blocker	Daly *et al.*, 1993
Lobeline	Piperidine	*Lobelia* spp., Campanula *medium* (Campanulaceae)	nAChR binding	Buckingham, 1996
Lupanine and other quinolizidines	Quinolizidine	*Lupinus, Cytisus, Genista* and other Leguminosae	nAChR binding	Schmeller *et al.*, 1994; Wink *et al.*, 1998b
Methyllycaconitine	Norditerpenoid	*Delphinium* spp., *Consolida* spp. (Ranunculaceae)	Potent antagonist at neuronal, α-bungarotoxin-sensitive nAChR	Coates *et al.*, 1995; Hardick *et al.*, 1995
Nicotine and other Nicotiana alkaloids	Pyridine	*Nicotiana* spp. (Solanaceae), *Asclepias syriaca* (Asclepiadaceae) and many other families	Potent nAChR agonist (muscle, neuronal)	Buckingham, 1996

Physostigmine	Indole	*Physostigma venenosum* (Leguminosae)	Non-competitive nACh agonist	Maelicke *et al.*, 1995; Schrattenholz *et al.*, 1993; Storch *et al.*, 1995
Pseudopelletierine	Piperidine	*Punica granatum* (Punicaceae)	Strong binding to nAChR	Wink *et al.*, 1998b
Pumiliotoxin-C	Quinoline	*Dendrobates* spp., *Epipedobates* spp., *Phyllobates* spp., *Melanophryniscus* spp.	Non-competive nAChR blocker	Daly *et al.*, 1993
Sanguinarine	Benzophenan-thridine	Several Papaveraceae, Fumariaceae	Binding to nAChR	Schmeller *et al.*, 1997b; Wink *et al.*, 1998b
Strychnine	Indole	*Strychnos* spp. (Apocynaceae)	Binding to nAChR	Wink *et al.*, 1998b
D-Tubocurarine	Bis-isoquinoline	*Chondodendron tomentosum* (Menispermaceae)	nAChR antagonist (muscle, neuronal)	Buckingham, 1996

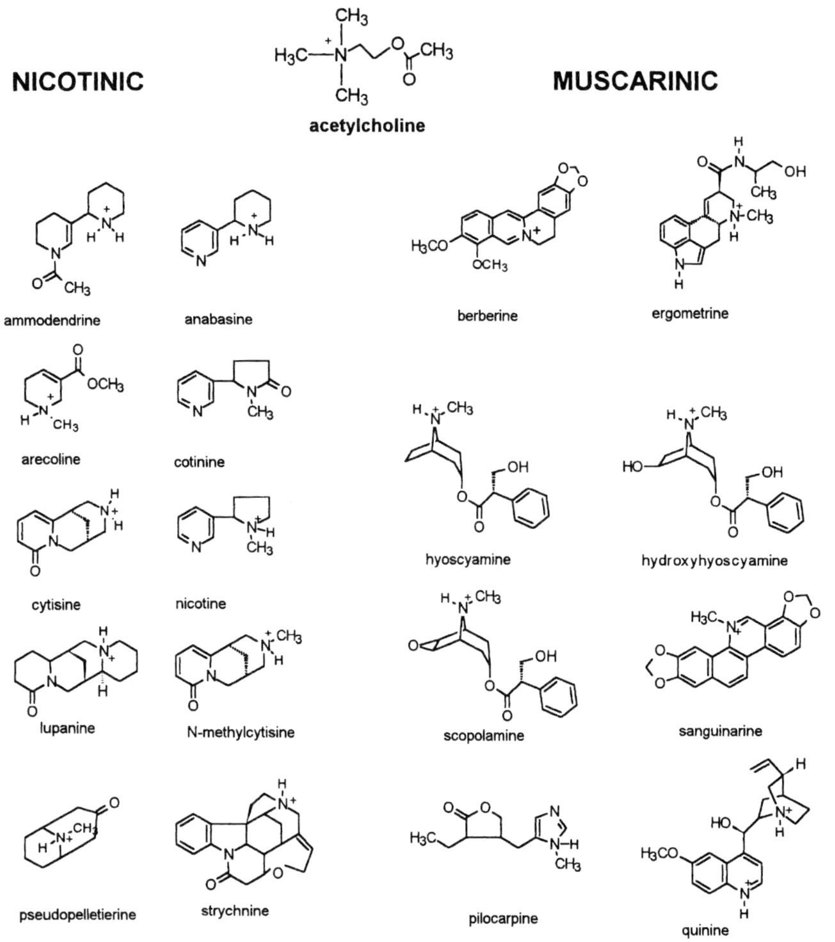

Figure 2.3 Structural similarities between acetylcholine and alkaloids that bind to nicotinic or muscarinic acetylcholine receptors. Under physiological conditions, alkaloids are protonated.

nists, the N is permanently quaternary, as in toxiferine or tubocurarine, which thus closely mimic the terminal quaternary N in acetylcholine. Because the dimeric alkaloids have two ligand sites, they can cross-link both alpha subunits of the nAChR. A much greater structural diversity is apparent in nAChR antagonists than in agonists. In addition to a tertiary or quaternary nitrogen in piperidine- or indolizidine-type rings, bulky ring structures, such as indoles or benzyls, are found in the vicinity (e.g. in boldine, toxiferine, tubocurarine, dihydro-β-erythroidine, erysodine, hirsutine or methyllycaconitine). Smaller molecules are esters with long alkyl substituents, as in pahutoxin, or smaller N- and O-containing rings, as in murexine. A few bulky peptides from animal venoms, such as α-bungarotoxin or α-conotoxin, also block the acetylcholine binding site. These compounds appear to interact with other functional groups of the receptor, so that a conformational change (to open the cation channel) is no longer possible.

At muscarinic acetylcholine receptors (Table 2.2), a similar situation is generally found. Agonists are usually small alkaloids with piperidine or imidazol rings (as in arecaidine, arecoline or pilocarpine), or they can be regarded as structural analogues of acetylcholine (such as tetramethylammonium hydroxide or muscarine). The agonists carry a quaternary N under physiological conditions but also an oxygen substituent at a distance similar to the ester oxygen in acetylcholine (Fig. 2.3).

Muscarinic AChR antagonists are much bigger alkaloids with a tertiary nitrogen, being present as *N*-methylpiperidine, *N*-methylpyrrolidine, indolizidine or quinolizidine skeletons as in all tropane alkaloids (anisodine, hyoscyamine, littorine, scopolamine or 3-tigloyltropine), in cryptolepine, gephyrotoxin, ginderine, imperialine or usambarensine (Fig. 2.3). Again, this nitrogen is quaternary under physiological conditions, like that of acetylcholine. In addition, these alkaloids have either an ester group, as in tropane alkaloids, or hexane, indole, benzyl or other bulky rings or alkane chains in the vicinity, as in cyclostellettamines, gephyrotoxins, ginderine, himandravine, himbacine or usambarensine. These compounds appear to bind to the ACh binding site but interact with other functional groups of the receptor, so that a conformational change (to activate the adjacent G protein) is no longer possible.

In conclusion, acetylcholine agonists are rather small alkaloids, which competetively bind to the AChR and induce conformational changes, whereas antagonist share the quaternary N for binding but inhibit the necessary conformational changes usually elicited by acetylcholine. Whether a molecule binds to nicotinic or muscarinic receptors, which exhibit a number of subtypes, depends on the fine structure of the respective bindings sites and their potential interactions with the alkaloid molecules carrying various substitutents.

2.2.2.2 Adrenergic receptors

Agonists at alpha and beta receptors (Table 2.3) clearly represent analogues of the natural ligands, adrenaline and noradrenaline. Ephedrine and its derivative, norephedrine, have an *R*-configurated hydroxyl function at position 1 of the side chain (as do the endogenous ligands; only cathinone has a keto-function instead) and either a free amino group, as in noradrenaline, or a methylated amino group, as in adrenaline (Fig. 2.4), both of which are protonated under physiological conditions. If the hydroxyl group at C1 is *S*-configurated, the activity is reduced. The phenolic hydroxy groups at position 3 and 4 of the aromatic ring are

Table 2.2 Examples of alkaloids which bind to muscarinic acetylcholine receptors (mAChR) (natural ligand: acetylcholine)

Alkaloid	Type	Occurrence	Activity	Reference
Acetylcholine	Amine	Endogenous neurotransmitter, several plants and in venom of Hymenoptera	mAChR agonist	Buckingham, 1996; Teuscher and Lindequist, 1994
Acetylheliosupine and related alkaloids	Pyrrolizidine	Boraginaceae	Binding to mAChR	Schmeller *et al.*, 1997a; Wink *et al.*, 1998b
Aconitine	Diterpene	*Aconitum* spp. (Ranunculaceae)	Binding to mAChR	Wink *et al.*, 1998b
Angustifoline and related alkaloids	Quinolizidine	*Lupinus* spp. (Leguminosae)	Binding to mAChR	Schmeller *et al.*, 1994; Wink *et al.*, 1998b
Arecoline/arecaidine	Piperidine	*Areca catechu* (Palmae)	mAChR agonist	Buckingham, 1996; Wink *et al.*, 1998b
Berbamine	Bis-isoquinoline	Berberidaceae, Menispermaceae, Ranunculaceae	Binding to mAChR	Hou and Liu, 1988
Berberine, palmatine and related alkaloids	Protoberberine	*Jateorhiza* spp. (Menispermaceae), *Berberis* spp., *Mahonia* spp. (Berberidaceae)	Binding to mAChR	Schmeller *et al.*, 1997b
Boldine	Aporphine	*Peumus boldo*	Binding to mAChR	Wink *et al.*, 1998b
Brucine/strychnine	Indole	*Strychnos nux-vomica* and many other *Strychnos* spp. (Loganiaceae)	Binding to mAChR	Wink *et al.*, 1998b
Cocaine	Tropane	*Erythroxylum* spp. (Erythroxylaceae)	Binding to mAChR	Schmeller *et al.*, 1995; Wink *et al.*, 1998b
Cryptolepine	Indole	*Cryptolepis sanguinolenta* (Periplocaceae)	M1, M2, M3 antagonist	Rauwald *et al.*, 1992
Dicentrine	Aporphine	Menispermaceae, Lauraceae, Fumariaceae, Papaveraceae	Binding to mAChR	Liu *et al.*, 1989b

Ebeinone	Homosteroidal	*Fritillaria imperialis* (Liliaceae)	M2 receptor antagonist	Gilani *et al.*, 1997
Echihumiline *N*-oxide	Pyrrolizidine	Boraginaceae	Binding to mAChR	Schmeller *et al.*, 1997a; Wink *et al.*, 1998b
Emetine	Isoquinoline	*Alangium* spp., *Psychotria* (Cephaelis) spp. (Alangiaceae, Rubiaceae)	Binding to mAChR	Wink *et al.*, 1998b
Harmaline and related alkaloids	β-Carboline	*Peganum harmala* (Zygophyllaceae), *Banisteriopsis* spp. (Malpighiaceae)	Binding to mAChR	Wink *et al.*, 1998b
Himandravine	Piperine	*Himantandra* spp. (Himantandraceae)	mAChR antagonist	Darroch *et al.*, 1990
Himbacine	Piperidine	*Himantandra* spp. (Himantandraceae)	mAChR antagonist	Kozikowski *et al.*, 1992; Wess *et al.*, 1992
13-Hydroxylupanine and related alkaloids	Quinolizidine	*Lupinus, Cytisus, Genista* and other Leguminosae	Weak binding to mAChR	Schmeller *et al.*, 1994; Wink *et al.*, 1998b
Ibogaine	Indole	*Tabernanthe iboga*, *Voacanga thouarsii*, *Tabernaemontana* spp. (Apocynaceae)	Binding to alpha 1 and 2 receptors	Sweetnam *et al.*, 1995
Imperialine	Homosteroidal	*Fritillaria* spp., *Petilium* spp., *Rhinopetalum* spp. (Liliaceae)	(Cardioselective) M2 antagonist	Eglen *et al.*, 1992
Laudanosine	Isoquinoline	*Papaver somniferum*, *Argemone grandiflora* (Papaveraceae)	Binding to mAChR	Wink *et al.*, 1998b
Lupinine	Quinolizidine	*Lupinus* spp. and other Leguminosae	Weak binding to mAChR	Schmeller *et al.*, 1994; Wink *et al.*, 1998b
Martinelline	Pyrroloquinoline	*Martinella iquitosensis* (Bignoniaceae)	Affinity for mAChR	Witherup *et al.*, 1995
Multiflorine and related alkaloids	Quinolizidine	*Lupinus* spp. and other Leguminosae	Binding to mAChR	Schmeller *et al.*, 1994; Wink *et al.*, 1998b

Table 2.2 (Continued)

Alkaloid	Type	Occurrence	Activity	Reference
Muscarine	Furan	*Amanita muscaria*, *Inocybe* spp., *Clitocybe* spp. and other fungi	mAChR agonist (M1-M5)	Buckingham, 1996
Palmatine	Protoberberine	*Jateorhiza palmata* (Menispermaceae), *Berberis* spp., *Mahonia* spp. (Berberidaceae), but also many families	Binding to mAChR	Wink *et al.*, 1998b
Physostigmine	Indole	*Physostigma venenosum* (Leguminosae)	Binding to mAChR	Wink *et al.*, 1998b
Pilocarpine	Imidazole	*Pilocarpus* spp. (Rutaceae)	mAChR agonist (M1-M5)	Wink *et al.*, 1998b
Quinine and related alkaloids	Quinoline	*Cinchona* spp. (Rubiaceae)	Binding to mAChR	Wink *et al.*, 1998b
Sanguinarine	Benzophenan-thridine	Several Papaveraceae, Fumariaceae	Binding to mAChR	Schmeller *et al.*, 1997b; Wink *et al.*, 1998b
Scopolamine, hyoscyamine and other tropane alkaloids	Tropane	*Atropa, Hyoscyamus, Datura, Mandragora, Scopolia, Duboisia, Brugmansia* (Solanaceae)	mAChR antagonist (M1-M5)	Kebabian and Neumeyer, 1994; Schmeller *et al.*, 1995
Seneciphylline/senecionine	Pyrrolizidine	*Senecio* spp. and other Asteraceae, *Crotalaria* spp. (Leguminosae)	Binding to mAChR	Schmeller *et al.*, 1997a; Wink *et al.*, 1998b
Sparteine and other tetracyclic alkaloids	Quinolizidine	*Lupinus, Cytisus, Genista* and other Leguminosae	Binding to mAChR	Schmeller *et al.*, 1994; Wink *et al.*, 1998b
Tetrandrine	Bis-isoquinoline	*Cocculus, Cyclea, Stephania* and other Menispermaceae	Binding to mACHR	Hou and Liu, 1988

Table 2.3 Examples of alkaloids which bind to adrenergic receptors (natural ligand: noradrenaline, adrenaline)

Alkaloid	Type	Occurrence	Activity	Reference
Acetylheliosupine	Pyrrolizidine	Boraginaceae	Binding α_{1+2} receptors	Schmeller *et al.*, 1997a; Wink *et al.*, 1998b
Adrenaline	Phenylalkyl amine	Banana (*Musa* spp.) and several other plants, venom of Hymenoptera	α_2, β-agonist	Buckingham, 1996; Teuscher and Lindequist, 1994
Ajmalicine	Indole	*Rauwolfia* spp., *Catharanthus roseus* and other Apocynaceae	α blocker, vasodilator	Buckingham, 1996; Nagase and Hagihara, 1986
Ammodendrine	Piperidine	*Lupinus*, *Cytisus*, *Genista* and other Leguminosae	α_2 receptor binding	Schmeller *et al.*, 1994; Wink *et al.*, 1998b
Anabasine	Pyridine	*Anabasis aphylla* (Chenopodiaceae), *Nicotiana* spp., *Duboisia myoporoides* (Solancaceae)	Binding to α_2 receptor	Buckingham, 1996; Wink *et al.*, 1998b
Arecoline	Piperidine	*Areca catechu* (Palmae)	Binding to α_2 receptor	Wink *et al.*, 1998b
Berbamine	Bis-isoquinoline	*Berberis* spp. and other Berberidaceae, Menispermaceae, Ranunculaceae	α_2 blocker	Han and Liu, 1988
Berberine and related alkaloids	Protoberberine	*Berberis* spp., *Mahonia* spp., *Jateorhiza* (Berberidaceae) and other families	α_{1+2} blocker	Dong *et al.*, 1997; Liu *et al.*, 1989a; Liu *et al.*, 1989b; Schmeller *et al.*, 1997b
Boldine	Aporphine	*Peumus boldo* (Monimiaceae) and several Annonaceae and Lauraceae	α_{1A} antagonist; strong binding to $\alpha_{1\ \mathrm{and}\ 2}$ receptors	Ivorra *et al.*, 1993a; Madrero *et al.*, 1996; Wink *et al.*, 1998b
Brucine/strychnine	Indole	*Strychnos nux-vomica* and many other *Strychnos* spp. (Loganiaceae)	Binding to α_1 receptor	Wink *et al.*, 1998b
Cinchonidine and related alkaloids	Quinoline	*Cinchona* spp. (Rubiaceae)	Strong binding to α_{1+2} receptors	Wink *et al.*, 1998b

Table 2.3 (Continued)

Alkaloid	Type	Occurrence	Activity	Reference
Cocaine	Tropane	*Erythroxylum* spp. (Erythroxylaceae)	Adrenergic antagonist	Buckingham, 1996
Colchicine	Tropolone	*Colchicum autumnale* and many other *Colchicum* spp., several *Merendera* spp., *Gloriosa superba* and other Liliaceae	Binding to α_2 receptor	Wink *et al.*, 1998b
Corynanthine	Indole	*Corynanthe*, *Rauwolfia*, *Pausinystalia*, (Rubiaceae, Apocynaceae)	α_1 blocker	Kebabian and Neumeyer, 1994
Crebanine	Aporphine	*Stephania* spp. (Menispermaceae)	α_1 blocker	Han and Liu, 1988; Liu *et al.*, 1989b
Daurisoline	Bis-isoquinoline	*Menispermum dauricum* (Menispermaceae)	Binding to α_{1+2} receptors	Waldmeier *et al.*, 1995
Dehydroevodiamine	Isaquinazolino-carboline	*Evodia rutaecarpa* (Rutaceae)	Endothelial α_1 blockade	Chiou *et al.*, 1994
Dispegatrine	Indole	*Rauwolfia verticillata* (Apocynaceae)	α blocker	Buckingham, 1996
Dopamine	Phenylalkylamine	*Musa sapientium* (Musaceae), *Cytisus* spp., (Leguminosae) and other families	Adrenergic agonist	Buckingham, 1996
Emetine	Isoquinoline	*Alangium*, *Psychotria* (*Cephaelis*) sp. (Alangiaceae, Rubiaceae)	Binding to α_1, and strongly to α_2 receptor	Wink *et al.*, 1998b
Ephedrine, cathinone and related alkaloids	Phenylalkylamine	*Ephedra* spp. (Ephedraceae) *Aconitum napellus* (Ranunculaceae), *Catha edulis* (Celastraceae), *Taxus baccata* (Taxaceae), *Sida cordifolia* (Malvaceae), *Roemeria refracta* (Papaveraceae)	α, β agonist; binding to α_1 receptor	Buckingham, 1996; Wink *et al.*, 1998b
Epiallocorynantheine	Indole	*Uncaria rhynchophylla*	Binding to β receptors	Zhu *et al.*, 1997

Ergometrine (ergonovine)	Ergot	*Claviceps* and several Convolvulaceae (*Argyreia*, *Stictocardia*, *Rivea*, *Ipomoea*)	Binding to α_2, strongly to α_1 receptor	Wink *et al.*, 1998b
Ergosine and related alkaloids	Ergot	*Claviceps purpurea* and several Convolvulaceae	α_1 antagonist	Kazic *et al.*, 1989
Ergotamine	Ergot	*Claviceps purpurea*	Peripheral α_1 antagonist, vasoconstrictor; partial α_2 agonist	Buckingham, 1996; Roquebert and Grenie, 1986
Glaucine	Aporphine	Many Monimiaceae, Berberidaceae, Annonaceae, Lauraceae, Ranunculaceae, Papaveraceae	$\alpha_{1A,2}$ antagonist	Madrero *et al.*, 1996; Orallo *et al.*, 1993; 1995
Gramine	Indole	Gramineae, Leguminosae, Aceraceae	Binding to $\alpha_{1\ \mathrm{and}\ 2}$ receptors	Wink *et al.*, 1998b
Harmaline and related alkaloids	β-Carboline	*Peganum harmala* (Zygophyllaceae), *Banisteriopsis* spp. (Malpighiaceae)	Binding to α_{1+2} receptors	Wink *et al.*, 1998b
Heliosupine	Pyrrolizidine	*Heliotropium* spp., *Cynoglossum* spp. (Boraginaceae)	Binding to α_1 (weak) and α_2 receptors	Schmeller *et al.*, 1997a; Wink *et al.*, 1998b
Higenamine	Isoquinoline	*Aconitum japonicum* (Ranunculaceae), *Gnetum parviflorum* (Gnetaceae)	β-receptor agonist	Yun-Choi and Kim, 1994
Hirsutine	Indole	*Uncaria rhynchophylla*	Binding to α_2, β receptor	Zhu *et al.*, 1997
Hyoscyamine (atropine)	Tropane	*Atropa* spp., *Hyoscyamus* spp., *Datura* ssp., *Mandragora* spp., *Scopolia* spp., *Duboisia* spp. (Solanaceae)	Binding to α_{1+2} receptors	Schmeller *et al.*, 1995; Wink *et al.*, 1998b
Ibogaine	Indole	*Tabernanthe iboga*, *Voacanga thouarsii*, *Tabernaemontana* spp. (Apocynaceae)	Binding to norepinephrine uptake sites	Sweetnam *et al.*, 1995
Laudanosine	Isoquinoline	*Papaver somniferum*, *Argemone grandiflora* (Papaveraceae)	Cardiovascular α_1 blocker; binding to $\alpha_{1\ \mathrm{and}\ 2}$ receptors	Chulia *et al.*, 1994; Wink *et al.*, 1998b
Lysergamid	Ergot	*Claviceps*, *Acremonium* and other fungi, various Convolvulaceae	Vasoconstrictor, adrenergic	Oliver *et al.*, 1993
Martinelline	Pyrroloquinoline	*Martinella iquitosensis* (Bignoniaceae)	Affinity for α_1	Witherup *et al.*, 1995

Table 2.3 (Continued)

Alkaloid	Type	Occurrence	Activity	Reference
N-methyldopamine (epinine)	Phenylalkylamine	*Cytisus scoparius*, *Vicia faba* (Leguminosae), *Lophophora williamsii* (Cactaceae)	Adrenergic agonist	Buckingham, 1996
Noradrenaline	Phenylalkylamine	Many plants, venom of many Hymenoptera	α agonist	Buckingham, 1996
Octopamine	Phenylalkylamine	*Capsisum frutescens* (Solanaceae), *Citrus* spp. (Rutaceae), *Cyperus* spp. (Cyperaceae), venom of Octobrachia and spiders	α agonist	Buckingham, 1996; Teuscher and Lindequist, 1994; Kebabian and Neumeyer, 1994
Predicentrine	Aporphine	Many Lauraceae, Magnoliaceae, Fumariaceae, Papaveraceae	α_{1A} antagonist	Madrero *et al.*, 1996
Reserpine	Indole	*Rauwolfia* spp. *Vinca minor*, *Alstonia constrictor*, *Tenduzia longifolia*, *Vallesia dichotoma*, *Excavatia coccinea* and other genera (Apocynaceae)	Adrenergic antagonist	Buckingham, 1996
Salsoline	Isoquinoline	*Salsola* spp. and many other Chenopodiaceae, Leguminosae, Cactaceae, Alangiaceae	Binding to α_2 receptor	Wink *et al.*, 1998b
Sanguinarine	Benzophenanthridine	Several Papaveraceae, Fumariaceae	Binding to α_{1+2} receptors	Schmeller *et al.*, 1997b; Wink *et al.*, 1998b
Stephanine and related alkaloids	Aporphine	*Stephania* spp. (Menispermaceae)	Modulation of central α_1 receptor	Han and Liu, 1988
Topsentin B2	bis-indole	Marine sponges (*Topsentia* spp., *Spongosorites* spp., *Hexadella* spp.)	Binding to $\alpha_{1A \text{ and } 1B}$ receptors	Buckingham, 1996
Xylopine	Aporphine	*Xylopia* spp. (Annonaceae)	α_1 antagonist	Han and Liu, 1988
Yohimbine	Indole	Several Apocynaceae and Rubiaceae	strong α_2 antagonist	Buckingham, 1996; Renouard *et al.*, 1994

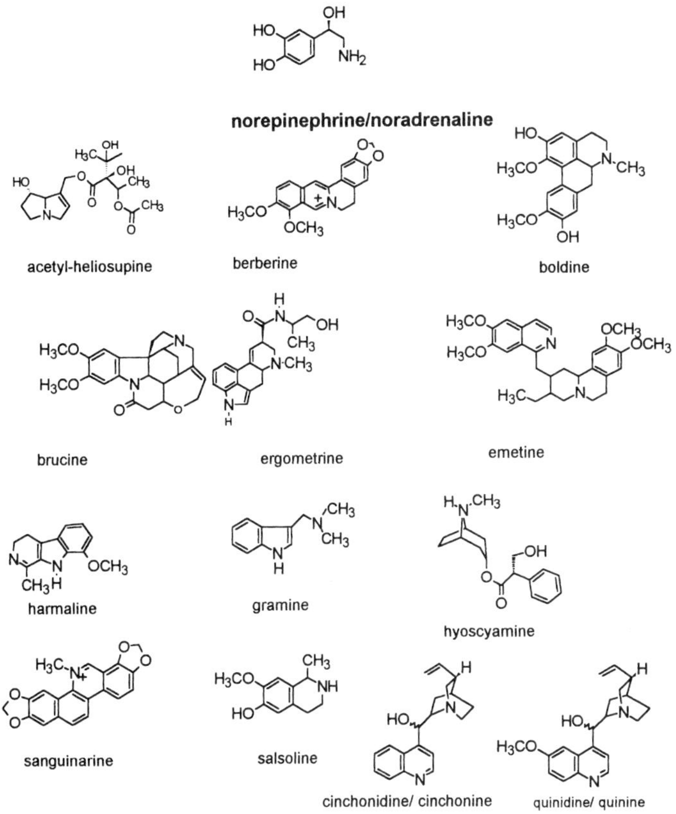

Figure 2.4 Structural similarities between noradrenaline and alkaloids which bind to adrenergic neuroreceptors.

apparently not essential for agonistic properties, although ephedrine is a weaker agonist than adrenaline. If the hydroxy groups are not present, this provides the advantage, firstly, that the molecules cannot be transformed by catecholamine-*O*-methyltransferases and, secondly, that they can cross biomembranes by simple diffusion. One or two phenolic hydroxyl groups are present in octopamine, dopamine and *N*-methyldopamine; the two latter compounds have no hydroxyl group at C1 of the side chain and are weaker agonists than the endogenous ligands.

Many of the alkaloids shown in Figure 2.4 are adrenergic antagonists. In alkaloids with an isoquinoline ring, such as aporphines (e.g. boldine, crebanine, glaucine or xylopine), the protoberberines (e.g. berberine, palmatine, govedine, gindarine or stylopine) or the simple or dimeric tetrahydroisoquinolines (e.g. higenamine, laudanosine, berbamine or oxyacanthine), the structure of the biogenic precursor dopamine can

Table 2.4 Examples of alkaloids which bind to dopamine receptors (DRs) (natural ligand: dopamine)

Alkaloid	Type	Occurrence	Activity	Reference
Agroclavine	Ergot	*Argyreia*, *Rivea*, *Cuscuta*, *Ipomoea* (Convolvulaceae) and *Penicillium* spp.	Dopamine receptor agonist	Buckingham, 1996
Anisocycline	Protoberberine	*Anisocycla cymosa* (Menispermaceae)	Binding to D_1, D_2	Markstein *et al.*, 1992
Bulbocapnine	Aporphine	*Corydalis cava* (Fumariaceae)	Dopamine antagonist	Buckingham, 1996; Kebabian and Neumeyer, 1994
Canadine and related alkaloids	Protoberberine	Several Fumariaceae, Papaveraceae	Binding to D_2 receptor	Chen *et al.*, 1987
3,4-Dihydroxy-phenylalanine	Dopa	Many Leguminosae, Aristolochiaceae	Dopamine precursor	Buckingham, 1996
Dopamine	Phenylalkylamine	Endogenous neurotransmitter, and in several higher plants, Leguminosae, *Musa* spp., in venoms of Hymenoptera	Endogenous ligand	Buckingham, 1996; Teuscher and Lindequist, 1994
Epinine	Phenylalkylamine	*Cytisus scoparius*, *Vicia faba* and *Lophophora williamsii* (Leguminosae, Cactaceae), animal skin secretions (Anura)	Dopamine agonist	Buckingham, 1996; Kebabian and Neumeyer, 1994
Ergocornine	Ergot	*Claviceps purpurea*	Dopamine receptor agonist	Buckingham, 1996
Ergovaline and related alkaloids	Ergot	*Claviceps purpurea* and other endophytic fungi	D_2 agonist	Larson *et al.*, 1995; Strickland *et al.*, 1994
Ioline	Pyrrolizidine	*Lolium* spp., *Festuca* spp. (Gramineae)	D_2 agonist	Strickland *et al.*, 1994
Salsolinol	Isoquinoline	*Annona reticulata* (Annonaceae), *Musa* spp. (Musaceae), *Aconitum carmichaeli* (Ranunculaceae)	Dopamine agonist	Nimit *et al.*, 1983

Stephanine	Aporphine	*Stephania* spp. (Menispermaceae)	High affinity for D_2 receptor	Chen *et al.*, 1987
Stylopine	Protoberberine	*Corydalis* spp. (Fumariaceae), *Chelidonium majus* and other Papaveraceae	Binding to D_2 receptor	Chen *et al.*, 1987
Tetrahydropalmatine	Protoberberine	*Corydalis* spp. (Fumariaceae), *Stephania glabra* (Menispermaceae), *Berberis tinctoria* (Berberidaceae) and *Coptis tecta* (Ranunculaceae)	Binding to D_1, D_2 receptor	Chen *et al.*, 1992; Chen *et al.*, 1987; Markstein *et al.*, 1992
Tyramine	Biogenic phenylalkyl amine	In several plant species, Magnoliaceae, Leguminosae, Gramineae, Cactaceae	Dopamine agonist	Kebabian and Neumeyer, 1994

Abbreviation: DOPA, 3,4-dihydroxy-phenylalanine.

easily be detected (Fig. 2.4): The aromatic ring usually contains either two free hydroxyl groups or corresponding methoxy groups, or even a methylenedioxy bridge, as in berberine. The secondary or tertiary amino group is still protonable under physiological conditions. Indole alkaloids, such as ajmalicine, corynanthine, reserpine, rauwolfscine, yohimbine and most ergot alkaloids are adrenergic antagonists. These alkaloids are usually of complex structure (see Fig. 2.4) and carry part of the adrenaline backbone: an aromatic ring (usually unsubstituted as in ephedrine) but a protonable tertiary N that is two carbon atoms adjacent to the aromatic ring. Whereas the protonable N and the aromatic ring A contribute to the binding of these alkaloids at adrenergic receptors, the other functional groups appear to interact with other amino acid residues in the receptor and inhibit a consequent conformational change.

2.2.2.3 Dopaminergic receptors

Table 2.4 lists several D1 and D2 agonists. *N*-Methyldopamine and salsolinol can be regarded as sharing most functional groups with dopamine, such as a catechol moiety and a protonable nitrogen. Tyramine, with a single hydroxyl group, is also accepted as an agonist. Surprisingly, the rather large ergot alkaloids, such as agroclavine, ergocornine and ergovaline, also function as dopamine agonists. The structure of dopamine can be superimposed on ergot alkaloids. The aporphine alkaloid, bulbocapnine, is a potent dopamine receptor antagonist. Two molecules of dopamine can be superimposed on this molecule.

2.2.2.4 Serotonergic receptors

Table 2.5 lists several serotonin receptor agonists and antagonists. Agonists include simple structural analogues, such as bufotenine, psilocine, *N*-methyltryptamine and *N*,*N*-dimethyltryptamine, differing in the presence/absence of a hydroxylfunction at C5 or C6 of ring A or whether the still protonable amino group bears one or two methyl groups (Fig. 2.5). In psilocybine, the hydroxyl group is phosphorylated, which does not interfere with the agonistic activity. β-Carboline alkaloids can be regarded as compounds that are closely related to serotonin; the structural similarity is close enough for harmaline and harmine to function as agonists. The simple ergot alkaloid lysergamide is also an agonist, in which the serotonin backbone can easily be detected. Mescaline also affects the serotonin receptor, although a structure-function relationship is less visible. Most of these serotonin agonists appear to provoke hallucinations in humans.

A few serotonin receptor antagonists share the serotonin backbone with the endogenous ligand, such as the indole alkaloid, akuammine, or

Table 2.5 Examples of alkaloids which bind to serotonin receptors (5-HTR) (natural ligand: serotonin)

Alkaloid	Type	Occurrence	Activity	Reference
Akuammine	Indole	*Picralima* spp., *Cabucala* spp. and *Vinca* spp. (Apocynaceae)	Serotonin antagonist	Lebanidze and Gedevanishvili, 1985
Annonaine	Aporphine	Annonaceae, Nelumbonaceae, also from the Lauraceae, Magnoliaceae, Monimiaceae, Papaveraceae, Rhamnaceae and Menispermaceae.	Binding to 5-HT_{1A}	Hasrat *et al.*, 1997
Asimilobine	Aporphine	Several Annonaceae, Nelumbonaceae, Magnoliaceae	Serotonergic antagonist	Hasrat *et al.*, 1997; Shoji *et al.*, 1987
Berberine and related alkaloids	Protoberberine	*Berberis* spp., *Mahonia* spp. (Berberidaceae) and other families	Strong binding to 5-HT_2 receptor	Schmeller *et al.*, 1997b
Boldine	Aporphine	*Peumus boldo* (Monimiaceae) and several Annonaceae and Lauraceae	Strong binding to 5-HT_2 receptor	Wink *et al.*, 1998b
Bufotenine and related alkaloids	Indolamine	Poisonous secretion of toads (*Bufo* spp.), of *Piptadenia peregrina*, many genera in the Leguminosae and Gramineae, and in some mushrooms (*Amanita* spp.), also in the gorgonian *Paramuricia chamaeleon*	Agonist at 5-HT receptors, hallucinogenic	Buckingham, 1996; Kebabian and Neumeyer, 1994
Cinchonidine and related alkaloids	Quinoline	*Cinchona* spp. (Rubiaceae)	Binding to 5-HT_2 receptor	Wink *et al.*, 1998b
Confusameline	Furoquinoline	*Evodia* spp., *Melicope* spp. (Rutaceae)	5-HT_2 antagonist	Cheng *et al.*, 1994
Emetine	Isoquinoline	*Alangium*, *Psychotria* (*Cephaelis*) spp. (Alangiaceae, Rubiaceae)	Binding to 5-HT_2 receptor	Wink *et al.*, 1998b
Ephedrine	Phenylalkylamine	*Ephedra* spp. (Ephedraceae) *Aconitum napellus* (Ranunculaceae), *Catha edulis* (Celastraceae), *Taxus baccata* (Taxaceae), *Sida cordifolia* (Malvaceae), *Roemeria refracta* (Papaveraceae)	Weak binding to 5-HT_2 receptor	Buckingham, 1996; Kebabian and Neumeyer, 1994; Wink *et al.*, 1998b

Table 2.5 (Continued)

Alkaloid	Type	Occurrence	Activity	Reference
Ergometrine and related alkaloids	Ergot	*Claviceps* and several Convolvulaceae (*Argyreia*, *Stictocardia*, *Rivea*, *Ipomoea*)	5-HT_2 antagonist	Buckingham, 1996; Wink *et al.*, 1998b
Ergotamine	Ergot	*Claviceps purpurea*	5-HT_{1C} antagonist	Brown *et al.*, 1992
Gramine	Indole	Gramineae, Leguminosae, Aceraceae	Binding to 5-HT_2 receptor	Wink *et al.*, 1998b
Harmaline and related alkaloids	β-Carboline	*Peganum harmala* (Zygophyllaceae), *Banisteriopsis* spp. (Malpighiaceae)	5-HT_{1A} agonist, binding to 5-HT_2	Abdel-Fattah *et al.*, 1995; Wink *et al.*, 1998b
Hirsutine	Indole	*Uncaria rhynchophylla*	Binding to 5-HT_{1A} and 5-HT_2 receptors	Zhu *et al.*, 1997
Hyoscyamine and related alkaloids	Tropane	*Atropa* spp., *Hyoscyamus* spp., *Datura* spp., *Mandragora* spp., *Scopolia* spp., *Duboisia* spp. (Solanaceae)	Binding to 5-HT_2 receptor	Schmeller *et al.*, 1995
Ibogaine	Indole	*Tabernanthe iboga*, *Voacanga thouarsii*, *Tabernaemontana* spp. (Apocynaceae)	Binding to 5-HT_2 and 5-HT_3 receptors, modulation of serotonin-mediated dopamine release	Sershen *et al.*, 1996, 1997; Sweetnam *et al.*, 1995
Kokusaginine	Furoquinoline	*Evodia* spp., *Orixa* spp. (Rutaceae)	5-HT_2 antagonist	Cheng *et al.*, 1994
Liridinine	Aporphine	*Liriodendron tulipifera* (Magnoliaceae)	Serotonergic antagonist	Shoji *et al.*, 1987
Mescaline	Phenylalkylamine	*Lophophora williamsii*, Trichocereus spp., *Gymnocalycium gibbosum*, *Opuntia cyclindrica* and other species, Cactaceae.	5-HT agonist	Kebabian and Neumeyer, 1994
Mitragynine	Indole	*Mitragyna speciosa*, *Uncaria* spp. (Rubiaceae)	Narcotic, serotonergic	Matsumoto *et al.*, 1996
N-methyltryptamine and related alkaloids	Indolamine	*Girgensohnia diptera*, *Acacia maidenii*, *Anadenanthera piptadena*, *Desmodium* spp., *Mimosa hostilis*, *Arthrophytum leptocladum*, *Arundo donax*, *Phalaris* spp., *Banisteriopsis argentea*, *Psychotria* spp., *Virola* spp., *Zanthoxylum* spp. (Myristicaceae, Chenopodiaceae, Leguminosae, Gramineae, Rutaceae)	Binding to 5-HT receptors;	Buckingham, 1996; McKenna *et al.*, 1984

Nornuciferine	Aporphine	Nelumbonaceae, Rhamnaceae, Annonaceae, Magnoliaceae, Menispermaceae	Binding to 5-HT_{1A} receptor	Hasrat *et al.*, 1997
Psilocybine/Psilocine	Indole	*Psilocybe* spp., *Conocybe* spp., *Stropharia* spp. and other fungi	Hallucinogen, 5-HT affinity	Buckingham, 1996; Teuscher and Lindequist, 1994
Serotonin	Indoleamine	Many plants (stinging hairs), also in venoms of Cnidaria, molluscs, and arthropods and skin secretions of amphibia	Endogenous 5-HT ligand	Buckingham, 1996; Teuscher and Lindequist, 1994
Stephanine	Aporphine	*Stephania* spp. (Menispermaceae)	High affinity for 5-HT_1, 5-HT_2	Chen *et al.*, 1987

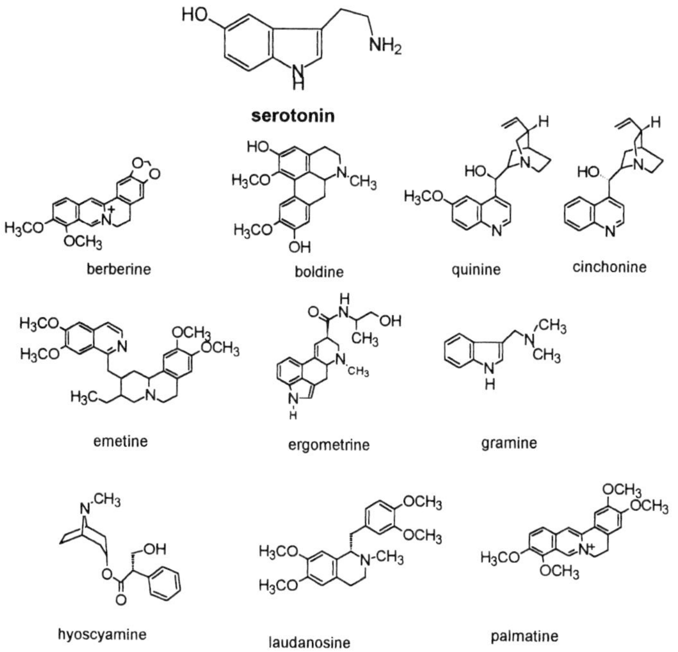

Figure 2.5 Structural similarities between serotonin (5-HT) and alkaloids which bind to 5-HT receptors.

the more bulky ergot alkaloids (e.g. ergometrine, ergosine or ergotamine) (Fig. 2.5). Such a structural relationship is not apparent for other antagonists, such as aporphine alkaloids (e.g. asimilobine, liridinine), quinoline alkaloids (e.g. confusameline, kokusaginine), or complex imidazole alkaloids (e.g. hymenine or keramadine).

2.2.2.5 GABA-ergic receptors

As compared to cholinergic and adrenergic alkaloids, only few GABA receptor agonists or antagonists have been detected; these are listed in Table 2.6. The small fungal metabolite, muscimol, functions as a GABA agonist. Like the endogenous ligand GABA, muscimol contains a protonable amino group. In GABA, there are three methylene groups between the amino group and the bulky carboxyl group. In muscimol, two or four carbon atoms are found between the amino group and the next oxygen function. GABA antagonists include the phthalideisoquinoline alkaloids, bicuculline, corlumine and hydrastine, and the securidan alkaloids, securinine, dihydrosecurinine and virosecurinine. Common to these alkaloids is a tertiary nitrogen, which becomes protonated under physiological conditions, and bulky oxygen substituents, which are three

Table 2.6 Examples of alkaloids which bind to gamma aminobutyric acid (GABA) receptors (natural ligand: GABA)

Alkaloid	Type	Occurrence	Activity	Reference
Bicuculline	Isoquinoline	*Dicentra* spp., *Corydalis*, *Fumaria* spp. (Fumariaceae)	$GABA_A$ blocker	Ameri, 1997a,b; Kardos *et al.*, 1984; Simonyi, 1987
Chelerythrine	Benzophenanthridine	Several Papaveraceae, Fumariaceae	Binding to $GABA_A$	Haeberlein *et al.*, 1996
Cryptopine	Isoquinoline	Several Papaveraceae, Fumariaceae, Ranunculaceae	Binding to GABA receptors	Kardos *et al.*, 1984
Corlumine	Isoquinoline	*Corydalis* spp., *Dicentra cucullaria* (Fumariaceae)	GABA antagonist	Buckingham, 1996
Dihydroergosine	Ergot	Sclerotia of *Sphacelia sorghi*	Binding to $GABA_A$ receptor	Pelassy and Aussel, 1993
Harmaline and related alkaloids	β-Carboline	*Peganum harmala* (Zygophyllaceae), *Passiflora* spp. (Passifloraceae), *Banisteriopsis* spp. (Malpighiaceae), *Picrasma quassioides* (Simaroubaceae), *Lolium perenne* and *Festuca arundinacea*	Binding to benzodiazepine receptor ($GABA_A$)	Rommelspacher *et al.*, 1980
β-Hydrastine and related alkaloids	Isoquinoline	*Corydalis fimbrillifera* (Fumariaceae), *Stylomecon heterophylla* (Papaveraceae), *Berberis laurina*, *Hydrastis canadensis* (Berberidaceae)	Competitive $GABA_A$ antagonist	Huang and Johnston, 1990
Muscimol		*Amanita muscaria*	$GABA_A$ agonist	Buckingham, 1996
Protopine	Protoberberine	Several Papaveraceae, Fumariaceae, Berberidaceae	Allosteric modulation of GABA receptors	Haeberlein *et al.*, 1996; Kardos *et al.*, 1984
Sanguinarine	Benzophenanthridine	Several Papaveraceae, Fumariaceae	$GABA_A$ binding	Haeberlein *et al.*, 1996
Securinine and related alkaloids	Indolixidine	*Phyllanthus* spp., *Securinega* spp. (Euphorbiaceae)	Potent central antagonist, inhibits GABA stimulated benzodiazepine binding	Beutler *et al.*, 1985

Table 2.7 Examples of alkaloids and NPAAs which bind to other receptors: glutamate (NMDA, AMPA, kainate), opiate, somatostatin, bradykinin and others

Alkaloid	Type	Occurrence	Activity	Reference
Acromelic acid	NPAA	Fungal metabolite, *Clitocybe acromelalga*	Glutamate receptor agonist	Teuscher and Lindequist, 1994
Akuammine	Indole	*Picralima nitida* (*P. klaineana*), *Cabucala erythrocarpa* and *Vinca* spp. (Apocynaceae)	Binding to mu and kappa opiate receptors	Lewin *et al.*, 1992
Baikiain	NPAA	*Baikiaea plurijuga*, *Caesalpinia tinctoria* (Leguminosae), fungus, *Russula subnigricans*	Glutamate receptor antagonist	Teuscher and Lindequist, 1994
Caffeine	Purine	*Coffea arabica*, many other *Coffea* spp., *Theobroma cacao*, *Camellia thea*, *Cola acuminata* and several other *Cola* spp., and several other plants (Rubiaceae, Sterculiaceae, Theaceae)	Binding to adenosine receptor	Buckingham, 1996
Corymine	Indole	*Hunteria* spp. (Apocynaceae)	Inhibition of glycine receptor	Leewanich *et al.*, 1997
Daurisoline	Bis-isoquinoline	*Menispermum dauricum* (Menispermaceae)	Antagonistic binding	Waldmeier *et al.*, 1995
Domoic acid	Pyrrolidine	From blue mussels (*Mytilus edulis*), red algae, *Chondria armata*, *Alsidium corallinum*	Glutamate/kainate receptor agonist	Buckingham, 1996
Ergocryptine and other ergot alkaloids	Ergot	*Claviceps purpurea*	Inhibition of cAMP production stimulated by vasoactive intestinal peptide	Larson *et al.*, 1995
Histrionicotoxin	Piperidine	Skin of *Dendrobates* spp. (Dendrobatidae) *Mantella madagascariensis* (Ranidae, subfamily Mantellinae)	NMDA receptor blocker	Daly *et al.*, 1993
Ibogaine and related alkaloids	Indole	*Tabernanthe iboga*, *Voacanga thouarsii*, *Tabernaemontana* spp. (Apocynaceae)	NMDA receptor antagonist, interaction with NMDA associated Na^+-channels, binding to mu, kappa and delta opioid receptors, hallucinogenic	Chen *et al.*, 1996; Sweetnam *et al.*, 1995

Ibotenic acid	NPAA	Fungal metabolite, *Amanita* spp.	Glutamate (aspartate) receptor agonist	Buckingham, 1996
Kainic acid		Red algae, *Digenea simplex* and *Centroceras clavulatum*	Glutamate receptor agonist (kainate site)	Buckingham, 1996
Martinellic acid	Pyrroloquinoline	*Martinella iquitosensis* (Bignoniaceae)	Bradykinin receptor antagonist	Witherup *et al.*, 1995
Martinelline	Pyrroloquinoline	*Martinella iquitosensis* (Bignoniaceae)	Affinity for histaminergic receptors	Witherup *et al.*, 1995
Mitragynine	Indole	*Mitragyna speciosa*, *Uncaria* spp. (Rubiaceae)	Narcotic, binding to opioid receptors	Watanabe *et al.*, 1997
Morphine	Morphinane	*Papaver somniferum* (Papaveraceae)	Stimulation of NO release in endothelial cells; binding to μ3 receptors, inhibition of presynaptic dopamine release	Liu *et al.*, 1996; Stefano *et al.*, 1995, 1997
Nuciferine	Aporphine	*Nelumbo lutea* (Nelumbonaceae), *Colubrina faralaotra* (Rhamnceae)	Glutamate receptor blocker	Buckingham, 1996; Kettenes *et al.*, 1981
Psycholeine	Tris-indole	*Psychotria oleoides* (Rubiaceae)	Somatostatin receptor antagonist	Rasolonjanahary *et al.*, 1995
Quisqualic acid	NPAA	*Quisqualis indica* (Combretaceae)	Glutamate receptor agonist (AMPA)	Buckingham, 1996
Rutaecarpine	Indole	*Evodia rutaecarpa*, *Hortia* spp., *Zanthoxylum* spp. (Rutaceae)	Interaction with endothelial nitric oxide and guanylyl cyclase	Chiou *et al.*, 1994
Stizolobic acid	NPAA	*Amanita pantherina*, *Stizolobium* spp. and *Mucuna irukande* (Leguminosae)	Glutamate receptor antagonist	Teuscher and Lindequist, 1994
Strychnine	Indole	*Strychnos* spp. (Apocynaceae)	Antagonist of glycine-gated Cl^--channels	Leewanich *et al.*, 1997; Perez Leon and Salceda Sacanelles, 1996
Willardiine	NPAA	*Acacia willardiana* and other *Acacia* spp. (Leguminosae)	AMPA/kainate receptor agonist	Buckingham, 1996

Abbreviations, cAMP, cyclic adenosine monophosphate; NMDA, *N*-methyl-D-aspartate; AMPA, alpha-amino-3-hydroxy-5-methyl-isoxazol-proprionic acid; NPAA, nonprotein amino acid.

carbon atoms apart, thus mimicking the structure of GABA to some degree. The GABA receptor binding alkaloids with a benzophenanthridine skeleton (e.g. cheleythrine, sanguinarine), the protopine alkaloids (e.g. fagarine I, cryptopine, protopine) and the β-carboline alkaloids also share this structural element. Analogous to the situation at other receptors, the agonist is a small molecule, whereas the antagonist shares a binding core but is usually a larger and bulkier compound that is able to interact with other parts or subunits of the receptor.

2.2.2.6 Glutamate/NMDA receptor

Table 2.7 lists several alkaloids and NPAAs that bind to the glutamate *N*-methyl-D-aspartate (NMDA) receptor. Several NPAAs, such as acromelic acid, domoic acid and kainic acid, are NMDA receptor agonists, which exhibit the following structural similarities with glutamate. Firstly, they contain an alpha nitrogen atom (sometimes in a heterocyclic ring) adjacent to a carboxyl group, common in amino acids. Secondly, they have a second carboxyl group, which is three carbon atoms apart. In the NPAAs, quisqualic acid and willardiine, there are two carbon and one N between the amino group and the next oxygen substitutent. In the fungal agonist, ibotenic acid, four carbon atoms are found between the amino group and the next hydroxyl group. It is more difficult to detect a common structural theme in glutamate/NMDA receptor antagonist (e.g. histrionicotoxins, ibogaine, nuciferine or philanthotoxin 433). Only in kynurenin and stizolobic acid are there structural elements similar to those found in the NPAA-type agonists.

2.2.2.7 Other neuroreceptors

Although several more neuroreceptors exist, rather few alkaloids have been identified as agonists or antagonists so far (Table 2.7). The indole alkaloids, brucine, corymine and strychnine, act as glycine receptor antagonists. Morphine, akuammine, daurisoline, 12-hydroxyibogaine, ibogaine and mitragynine affect opiate receptors, which normally bind endorphins and other peptides. The purino receptor is strongly affected by purine alkaloids, such as caffeine. Psycholeine and gelliusine A and B interact with the somatostatin receptor, and the latter compounds also interact with the receptor of neuropeptide Y. Martinellic acid binds to bradykinin receptors. This list is incomplete and it is certain that many more interactions with alkaloids will be discovered as soon as research directed towards these targets is carried out.

2.2.2.8 Ion channels

Alkaloids which modulate ion channels are tabulated in Table 2.8. Voltage-gated Na^+-channels are a target for several steroidal alkaloids,

Table 2.8 Alkaloids which affect ion channels

Alkaloid	Type	Occurrence	Activity	Reference
Aconitine and related alkaloids	Diterpene	*Aconitum* spp. (Ranunculaceae)	Activation of Na^+-channel (site 2)	Friese *et al.*, 1997; Hardick *et al.*, 1995
Ajmaline	Indole	*Rauwolfia* spp., *Melodinus balansae*, *Tonduzia logifolia* (Apocynaceae)	Inhibition of Na^+- and K^+-channel, antiarrhythmic (class Ia)	Körper *et al.*, 1998
Antioquine	Bis-isoquinoline	*Pseudoxandra lucida* (Annonanceae)	Ca^{2+} entry blocker, via benzothiazepine recognition site	Ivorra *et al.*, 1993b
Berbamine	Bis-isoquinoline	*Berberis* spp. and other Berberidaceae, Menispermaceae, Ranunculaceae	Ca^{2+} channel blocker	Pan *et al.*, 1989
Bisnordehydro-toxiferine	Bis-indole	*Strychnos* spp. (Loganiaceae)	Inhibition of voltage-dependent Ca^{2+} channels	da Silva *et al.*, 1993
Boldine	Aporphine	*Peumus boldo* (Monimiaceae), several Annonaceae, Lauraceae	Ca^{2+} blocker via benzothiazepine receptor site	Ivorra *et al.*, 1993a
Caffeine	Purine	*Coffea* spp., *Theobroma cacao*, *Camellia thea*, *Cola* spp. and several other plants (Rubiaceae, Sterculiaceae, Theaceae)	Inhibition of sarcoplasmic Ca^{2+} IP_3 sensitive release channels, activation of ryanodine sensitive SR Ca^{2+} release channels	Berridge and Bootman, 1996; Teuscher and Lindequist, 1994
Capsaicin	Amide	*Capsicum* spp. (Solanaceae)	Inhibition of type I K^+ currents	Baker and Ritchie, 1994
Cocaine	Tropane	*Erythroxylum* spp. (Erythroxylaceae)	Inhibition of Ca^{2+} release channel from SR	Gawin, 1991
Cordobimine	Bis-isoquinoline	*Crematosperma* spp. (Annonaceae)	Ca^{2+} entry blocker	Ivorra *et al.*, 1992a
Corlumidine	Isoquinoline	*Corydalis scouleri* (Fumariaceae)	Increase of Ca^{2+} currents	Kadota *et al.*, 1996
Daurisoline	Bis-isoquinoline	*Menispermum dauricum* (Menispermaceae)	Inhibition of P-type Ca^{2+}-channels	Lu *et al.*, 1994; Waldmeier *et al.*, 1995
Dehydroevodiamine	Isaquinazolino-carboline	*Evodia rutaecarpa* (Rutaceae)	Endothelial K^+-channel activation; Ca^{2+} blockade	Chiou *et al.*, 1996

Table 2.8 (Continued)

Alkaloid	Type	Occurrence	Activity	Reference
Dicentrine	Aporphine	Menispermaceae, Lauraceae, Fumariaceae, Papaveraceae	Inhibition of Na^{+}-, K^{+}-channels	Su *et al.*, 1994
Dopamine	Phenylalkyl amine	Many plants	Activation of K^{+}-channels in snail neurons	Nesic and Pasic, 1992
Egenine	Isoquinoline	*Corydalis* spp., *Fumaria vaillantii* (Fumariaceae)	Inhibition of Ca^{2+} currents	Kadota *et al.*, 1996
Ervatamine	Indole	*Ervatamia* spp. (Apocynaceae)	Na^{+}-channel blocker	Buckingham, 1996
Glaucine	Aporphine	Many Monimiaceae, Berberidaceae, Annonaceae, Lauraceae, Ranunculaceae, Papaveraceae	Inhibition of voltage-dependent Ca^{2+}-channels	Ivorra *et al.*, 1993c; Loza *et al.*, 1993
Gonyautoxins	Purine	*Gonyaulax* and *Protogonyaulax* spp. and isolated from marine animals, mussels, crabs, fish	Inhibition of Na^{+}-channels	Buckingham, 1996; Teuscher and Lindequist, 1994
Granjine	Bis-isoquinoline	*Crematosperma* spp. (Annonaceae)	Ca^{2+} entry blocker	Ivorra *et al.*, 1992a
Hernandezine	Bis-isoquinoline	*Thalictrum* spp. (Ranunculaceae)	Inhibition of Ca^{2+}-channels	Low *et al.*, 1996
Hirsutine	Indole	*Mitragyna* spp., *Uncaria* spp., *Cephalanthus* spp.	Inhibition of voltage-gated Ca^{2+}-channels	Horie *et al.*, 1992; Nakazawa *et al.*, 1991; Yano *et al.*, 1991
Isotetrandrine	Bis-isoquinoline	Several Atherospermataceae, Berberidaceae, Menispermaceae, Ranunculaceae	Ca^{2+}-channel blocker	D'Ocon *et al.*, 1992
Liensinine	Bis-isoquinoline	*Nelumbo nucifera* (Nelumbonaceae)	Ca^{2+}-channel antagonist	Wang *et al.*, 1993b
Liriodenine	Aporphine	Many genera in the Annonaceae, Araceae, Eupomatiaceae, Lauraceae, Magnoliaceae, Menispermaceae, Nelumbonaceae, Papaveraceae, Rhamnaceae, Rutaceae, Atherospermataceae	Inhibition of Na^{+}-channels, L-type Ca^{2+} currents, and 4-AP sensitive transient outward currents	Chang *et al.*, 1996

Mitragynine	Indole	*Mitragyna speciosa*, *Uncaria* spp. (Rubiaceae)	Neuronal Ca^{2+}-channel blocker	Horie *et al.*, 1995
Monterine	Bis-isoquinoline	*Crematosperma* spp. (Annonaceae)	Ca^{2+} entry blocker	Ivorra *et al.*, 1992a
Norushinsunine	Aporphine	In a wide variety of genera in the Annonaceae, Magnoliaceae Menispermaceae, Eupomatiaceae and Monimiaceae	Blocker of L-type Ca^{2+}-channels	Chulia *et al.*, 1995
Paragracine	Imidazol	*Dentitheca habereri* (coral)	Selective Na^{+}-channel blocker	Buckingham, 1996
Paspalinine	Indole	Fungal toxins, *Claviceps paspali*, *Aspergillus flavus*	Inhibition of Ca^{2+}-sensitive K^{+}-channels	Knaus *et al.*, 1994
Paspalitrem A, C	Indole	Fungus, *Claviceps paspali*	Inhibition of Ca^{2+}-sensitive K^{+}-channels	DeFarias *et al.*, 1996
Paxilline	Indole	*Penicillium paxilli*, *Acremonium lorii*, *Emericella foveolata*, *Emericella desertorum* and *Emericella striata*	Inhibition of Ca^{2+} sensitive K^{+}-channels	DeFarias *et al.*, 1996
Penitrem A	Indole	Fungus, *Penicillium crustosum*	Inhibition of Ca^{2+} sensitive K^{+}-channels	Edwards and Weston, 1996
Phalloidin	Cyclic peptide	Fungal metabolite, *Aminata phalloides* and *Amanita* spp.	Voltage-gated K^{+}-channel blocker	Buckingham, 1996
Quinidine/quinine	Quinoline	*Cinchona* spp. (Rubiaceae)	Inhibition of voltage-gated Na^{+}-channel, opening of hemigap junctional channel	Körper *et al.*, 1998; Malchow *et al.*, 1994
Rhynchophylline	Indole	*Uncaria rhynchophylla*, *Mitragyna* spp. (Rubiaceae)	Inhibitor of voltage-gated Ca^{2+}-channels	Huang *et al.*, 1993
Ryanodine and derivatives	Pyrrole	*Ryania speciosa* (Flacourtiaceae)	Activation of ER Ca^{2+}-channels at nano-micromolar concn., inhibition of SR Ca^{2+} release channels at higher concn.	Allouche *et al.* 1996; Bidasee *et al.*, 1995; Inui, 1992; Schmitt *et al.*, 1996; Tinker *et al.*, 1996; Vais and Usherwood, 1995

Table 2.8 (Continued)

Alkaloid	Type	Occurrence	Activity	Reference
Saxitoxin	Purine	*Protogonyaulax tamarensis*, *Gonyaulax cantenella* (dinoflagellates), accumulate in clams (*Saxidomus giganteus*) and mussels (*Mytilus californianus*)	Inhibition of Na^+-channels (I, II, III, h1)	Buckingham, 1996; Kebabian and Neumeyer, 1994
Sparteine and related alkaloids	Quinolizidine	*Cytisus*, *Lupinus*, *Genista* and many other Leguminosae	Inhibition of ATP-regulated K^+-channel in insulin-secreting beta-cells, in muscle, stronger inhibition of K^+-than of Na^+-channels	Ashcroft *et al.*, 1991; Körper *et al.*, 1998
Strychnine	Indole	*Strychnos* spp. (Loganiaceae)	Inhibition of muscle Na^+-channels	Körper *et al.*, 1998
Tetrandrine	Bis-isoquinoline	*Cocculus*, *Cyclea*, *Stephania* and other Menispermaceae	Inhibits voltage-dependent L- and T-type Ca^{2+}-channels in excitable cells	Bickmeyer *et al.*, 1996; Bickmeyer and Wiegand, 1993; Dworetzky *et al.*, 1996; Gribkoff *et al.*, 1996; Wang *et al.*, 1993a, 1994; Wang and Lemos, 1992; Wu *et al.*, 1997
Tetrodotoxin	Guanidinium	Potent neurotoxin isolated from the ovaries and liver of fish, many amphibians and marine organisms, probably a metabolic product of a *Alteromonas* sp.	Sodium channel (I, II, III, mu,1, h1) blocker	Buckingham, 1996; Lu and De Clerck, 1993; Teuscher and Lindequist, 1994
Thalidasine	Bis-isoquinoline	*Thalictrum* spp. (Ranunculaceae)	Interaction with voltage- and receptor-dependent Ca^{2+}-channels	Wu *et al.*, 1977

Veratrine/ veratridine	Steroidal	*Schoenocaulon officinale*, *Veratrum lobelianum* and *Veratrum viride* (Liliaceae), *Helleborus viridis* (Ranuculaceae)	Activation of Na^+ channel, no effect on inward rectifier K^+ current	Honerjaeger *et al.*, 1992; Matsumoto and Shimizu, 1995; Nanasi *et al.*, 1990; Sheldon *et al.*, 1990; Van Huizen *et al.*, 1988
Vincamine	Indole	*Vinca* spp., *Tabaernaemontana rigida* (Apocynaceae)	Inhibitor of voltage-gated Na^+-channels	Erdo *et al.*, 1996
Vincanol	Indole	*Kopsia* spp., *Melodinus celastroides*	Inhibitor of voltage-gated Na^+-channels	Erdo *et al.*, 1996
Warifteine	Bis-isoquinoline	*Cissampelos* spp. (Menispermaceae)	Inhibition of voltage-gated Ca^{2+}-channels	De Freitas *et al.*, 1996
Yohimbine	Indole	Several Apocynaceae and Rubiaceae	Inhibition of muscle Na^+-channel	Körper *et al.*, 1998
Zygadenine	Steroidal	*Zygadenus* spp. (Liliaceae)	Activation of voltage-gated Na^+-channels	Badria *et al.*, 1995

Abbreviations: SR, sarcoplasmic reticulum; IP_3, inositol-1,4,5-triphosphate; ER, endoplasmic reticulum; ATP, adenosine triphosphate.

(e.g. batrachotoxinin, samandarine, veratrine, veratridine and zygadenine), indole alkaloids (e.g. ajmaline), aporphines (e.g. dicentrine, liriodenine), quinoline alkaloids (quinine, quinidine), quinolizidine alkaloids (e.g. sparteine, lupanine), alkaloids present in animals (e.g. chiriquitoxin, μ-conotoxins, dibromosceptrine, gonyautoxins, histrionicotoxins, pumiliotoxins, saxitoxin and tetrodotoxin). Activation of Na^+-channels inhibits a subsequent repolarization and, thus, leads to a total blockage of neuronal and neuromuscular signalling. This might explain why most Na^+ agonists are highly potent poisons in animals and humans (e.g. aconitine, veratrine, tetrodotoxin).

Ca^{2+}-channels are inhibited by several bis-isoquinoline alkaloids (e.g. berbamine, hernandezine, liensinine, monterine, tetrandrine), aporphines (e.g. glaucine, norushinsunine), complex indole alkaloids (bis-nortoxiferine, hirsutine, mitragynine, paspaline, paspalitrem, paxilline, penitrem), or other bulky alkaloids (agelasine, contotoxins, crambescidin, ryanodine).

An apparent common theme is barely visible in these ion channel blockers but most of these alkaloids have tertiary Ns, which are protonated under physiological conditions, and are relatively large and bulky molecules that contain many functional groups, so that interactions are mediated with amino acid residues of the channel proteins.

2.2.2.9 Cell membranes

The integrity of biomembranes and the maintenance of membrane potential is of ultimate importance for the functioning of cells and of all neuronal activities. Compounds which disturb biomembranes, and thus make cells leaky, are usually strong cell poisons and interfere with membrane potential. Natural products that exhibit these properties are either highly lipophilic or amphiphilic. Several secondary metabolites, such as mono-, sesqui- and diterpenes, or triterpene and steroid saponins, respectively, fall into this category.

Steroidal alkaloids, such as solanine and tomatine, which are present in many members of the Solanaceae, can form complexes with the cholesterol and other lipids present in biomembranes. Important for this interaction is the presence of a lipophilic portion of the molecule (given by the steroidal moiety) and a hydrophilic portion (provided by the sugar side chain). Whereas the lipophilic moiety 'dives' into the lipophilic interior of the membrane and interacts with the structurally similar cholesterol, the hydrophilic side chain remains outside and binds to external sugar receptors. Since phospholipids are in a continuous motion (spin around their axis and horizontal movements), a tension easily builds up, which leads to membrane disruption; i.e. transient 'holes' occur in the biomembrane rendering the cell leaky.

Since particular steroidal alkaloids can specifically interact with receptors, ion channels or transmitter transforming enzymes (see veratrine, solanine; Tables 2.3–2.12), specific effects must be distinguished from more nonspecific membrane pertubations. The effects of the steroidal alkaloids of Solanaceae, α-chaconine, α-solanine and tomatine, on the intracellular free Ca^{2+} concentration were studied in various cell lines (Toyoda *et al.*, 1991). In all cultured cells treated with the alkaloids, the intracellular Ca^{2+} concentrations were raised in a dose-dependent manner. The Ca^{2+} influx evoked by α-chaconine could not be prevented by metal ions or by inhibitors of Ca^{2+} transport across membranes, such as voltage-operated channel antagonists, muscarinic and nicotinic antagonists, or Na^+- and K^+-channel blockers. These findings confirm that the ion flux across biomembranes caused by steroidal alkaloids is due to destabilization of the cell membrane. A similar mechanism is plausible for monodesmosidic saponins, a widely-distributed group of natural products, to which the steroidal alkaloids may be assigned according to their physicochemical properties.

The alkaloids, tetrandrine, 3β-hydroxylupanine and cepharanthine, have also been reported to interfere with membrane integrity (Wink, 1993a). Weak haemolytic properties were detected for berbamine, harmin, narcotine, norharman and sanguinarine (Wink *et al.*, 1998b). These membrane perturbances will also affect neuronal and neuromuscular signaling, since both processes require intact membranes. Whereas interactions of alkaloids with neuroreceptors, ion channels and corresponding enzymes of the signal pathways show a high degree of specificity, membrane interactions, as shown here for steroidal alkaloids and saponins, are non-specific but, nevertheless, powerful defence strategies of many plants and animals.

The inhibition of Na^+-, K^+-ATPase (Table 2.10) by cardiac glycosides, anthraquinones, some proanthocyanidins and a few alkaloids might also be discussed in this context, since this inhibition will prevent the generation of ion gradients and, thus, of the membrane potential.

2.2.2.10 Other elements of neuronal signalling

Alkaloids which inhibit acetylcholine esterase (ACE), monoamine oxidase (MAO) and catecholamine-*O*-methyltransferase are tabulated in Table 2.9. Potent ACE blockers include indole alkaloids of the physostigmine type (e.g. eseramine, geneserine, physovenine, eserine), protoberberine alkaloids (berberine, columbamine, coptisine, jatrorrhizine, palmatine), steroidal alkaloids (leptine I, solanine, solamargine and tomatidine), and others (e.g. galanthamine). No plausible structure-function relationship is apparent, except that all these alkaloids are

Table 2.9 Examples of alkaloids as inhibitors of neurotransmitter degrading enzymes

Enzyme	Transmitter	Alkaloid	Type	Occurrence	Reference
Acetylcholine esterase	Acetylcholine	Anatoxin A	Imidazole	*Anabaena flos-aquae* (Cyanobacterium)	Buckingham, 1996
		Berberine and related alkaloids	Protoberberine	Many *Berberis* spp. and *Mahonia* spp. (Berberidaceae) and in several different families	Buckingham, 1996; Schmeller *et al.*, 1997b
		Cimiciphytine and related alkaloids	Indole	*Haplophyton cimicidum* (Apocynaceae)	Buckingham, 1996
		Demissine and related alkaloids	Steroidal	*Solanum* spp., *Lycopersicon* spp. (Solanaceae)	Buckingham, 1996
		Decarbomethoxy-tetrahydrosecodine	Indole	*Tabernaemontana cumminsii*, *Haplophyton crooksii* (Apocynaceae)	Buckingham, 1996
		Galanthamine		*Galanthus* spp. and many other Amaryllidaceae	Thomsen and Kewitz, 1990
		Harmaline and related alkaloids	β-Carboline	*Peganum harmala* (Zygophyllaceae), *Passiflora* spp. (Passifloraceae), *Banisteriopsis* spp. (Malpighiaceae)	Wink *et al.*, 1998b
		Huperzine and related alkaloids	Lycopodium	*Lycopodium* (Huperzia) *serrata* and other *Lycopodium* spp. (Lycopodiaceae)	Kozikowski *et al.*, 1996; Xu and Tang, 1987
		Physostigmine and related alkaloids	Indole	*Physostigma venenosum* (Leguminosae), also metabolite of *Streptomyces pseudogriseolus*	Buckingham, 1996; Marta and Pomponi, 1987; Yu *et al.*, 1988
		Sanguinarine	Benzophen-anthridine	Several Papaveraceae, Fumariaceae	Schmeller *et al.*, 1997b
		Strychnine	Indole	*Strychnos* spp. (Loganiaceae)	Buckingham, 1996
		Thebaine	Morphinane	*Papaver bracteatum*, *P. somniferum* (Papaveraceae)	Buckingham, 1996
		Vasicinol	Quinazoline	*Adhatoda vasica*, *Sida cordifolia* and from other *Sida* spp. (Acanthaceae, Malvaceae).	Buckingham, 1996

Monoamine oxidase (MAO)	NA, dopamine, serotonin, histamine	Alstovenine	Indole	*Alstonia venenata* (Apocynaceae)	Buckingham, 1996
		Carnegine (MAO A/B)	Isoquinoline	*Haloxylon* spp. and other Chenopodiaceae and Cactaceae	Bembenek *et al.*, 1990
		N,*N*-dimethyltryptamine and related alkaloids (MAO I)	Phenylalkyl amine	*Mimosa hostilis*, *Acacia* spp., *Arundo donax*, *Desmodium* spp., *Phalaris* spp., *Banisteriopsis argentea*, *Girgensohnia* spp., *Psychotria* spp., *Virola* spp., *Zanthoxylum* spp. and others (Leguminosae, Gramineae, Malphigiaceae, Rubiaceae, Myristicaceae, Rutaceae	McKenna *et al.*, 1984
		Harmaline and related alkaloids (MAO A)	β-Carboline	*Peganum harmala* (Zygophyllaceae), *Passiflora* spp. (Passifloraceae), *Banisteriopsis* spp. (Malpighiaceae)	McKenna *et al.*, 1984
		O-methylcorypalline (MAO B)	Isoquinoline	Cactaceae, Papaveraceae, Ranunculaceae, Nelumbonaceae	Bembenek *et al.*, 1990
		Quinine and related alkaloids	Quinoline	*Cinchona* spp. (Rubiaceae)	Mitsui *et al.*, 1989
		Saracodine	Steroidal	*Sarcococca pruniformis* (Buxaceae)	Buckingham, 1996
		Salsolidine and related alkaloids (MAO A/B)	Isoquinoline	*Salsola* spp. and many other Chenopodiaceae, Leguminosae, Cactaceae, Annonaceae	Bembenek *et al.*, 1990
		Vinblastine and related alkaloids (MAO B)	Indole	*Cantharanthus roseus* (Apocynaceae)	Son *et al.*, 1990

Abbreviations: BSE, butylcholine esterase; ACE, acetylcholine esterase; NA, noradrenaline.

Table 2.10 Examples of alkaloids as inhibitors of neurotransmitter uptake (transport via presynaptic membrane and into vesicles), and of Na^{+}-, K^{+}- or Ca^{2+}- ATPases

Alkaloid	Type	Occurrence	Activity	Reference
Annonaine	Aporphine	Annonaceae, Nelumbonaceae, Lauraceae, Magnoliaceae, Monimiaceae, Papaveraceae, Rhamnaceae and Menispermaceae	Inhibition of dopamine uptake	Bermejo *et al.*, 1995; Protais *et al.*, 1995
Arecaidine and related alkaloids	Piperidine	*Areca catechu* (Palmae)	Inhibits GABA reuptake	Teuscher and Lindequist, 1994
Cocaine	Tropane	*Erythroxylum* spp. (Erythroxylaceae)	Inhibition of dopamine transporter	Berger *et al.*, 1990; Kilty *et al.*, 1991; Lever *et al.*, 1993
Ephedrine and related alkaloids	Phenylalkyl amine	Ephedraceae, *Aconitum napellus* (Ranunculaceae), *Catha edulis* (Celastraceae), *Taxus baccata* (Taxaceae), *Sida cordifolia* (Malvaceae), *Roemeria refracta* (Papaveraceae)	Induces release of NA and inhibits reuptake	Mutschler, 1996; Teuscher and Lindequist, 1994
Ibogaine and related alkaloids	Indole	*Tabernanthe iboga*, *Voacanga thouarsii*, *Tabernaemontana* spp. (Apocynaceae)	Binding to vesicular dopamine and monoamine transporter, binding to 5-HT transporter	Staley *et al.*, 1996
Norharman	β-Carboline	*Chrysophyllum lacourtianum*, *Nocardia* spp. and *Streptomyces* spp., *Catharanthus roseus*, *Lolium perenne* and *Festuca arundinacea* (Sapotaceae, Apocynaceae, Gramineae)	Inhibition of dopamine and tryptamine uptake	Brossi, 1993; Rommelspacher *et al.*, 1980

Reserpine and related alkaloids	Indole	*Rauwolfia* spp., *Vinca minor*, *Alstonia constrictor*, *Tenduzia longifolia*, *Vallesia dichotoma*, *Excavatia coccinea* and other genera (Apocynaceae)	Depletes stores of NA and 5-HT in vesicles, H^+-ATPase inhibitor	Buckingham, 1996; Eiden *et al.*, 1984; Mutschler, 1996
Salsolinol	Isoquinoline	*Annona reticulata* (Annonaceae), *Musa paradisiaca* (Musaceae), *Theobroma cacao* (Sterculiaceae) and *Aconitum carmichaeli* (Ranunculaceae)	Inhibition of uptake of biogenic amines	Melchior and Collins, 1982
Tyramine	Phenylalkyl amine	In several plants of Magnoliaceae, Leguminosae, Gramineae, Cactaceae	Induces release of NA and inhibits reuptake	Mutschler, 1996
Veratramine	Steroidal	*Veratrum* spp. (Liliaceae)	Both releaser and uptake inhibitor of 5-HT	Nagata *et al.*, 1991

Abbreviations: GABA, gamma-aminobutyric acid; NA, noradrenaline; 5-HT, 5-hydroxytryptamine (serotonin); ATPase, adenosine triphosphatase.

quaternary under physiological conditions and that an oxygen can be traced 2-4 carbons adjacent to the N.

MAO inhibitors are in the group of simple indole alkaloids (e.g. β-carbolines, *N,N*-dimethyltryptamine; *N*-methyltryptamine), simple isoquinolines (carnegine, salsolidine, salsolinol), quinoline alkaloids (e.g. quinine) and even complex indole alkaloids (e.g. vinblastine and vincristine). A structural similarity can be seen between the MAO blocker and endogenous substrates: the indole alkaloids and serotonin; simple isoquinoline and dopamine; noradrenaline and adrenaline, which implies that these compounds bind to the active site of the enzyme.

Uptake blockers (Table 2.10) display a certain degree of structural relatedness to the endogenous neurotransmitters whose transport is inhibited: dopamine uptake by annonaine, cocaine, ibogaine and salsolinol; serotonin by 12-hydroxyibogaine, ibogaine and norharman; noradrenaline and adrenaline by cathinone, ephedrine, salsolinol; GABA by arecaidine and guvacine. Reserpine and deserpidine inhibit a H^+-ATPase at the synaptic vesicle, which builds up a proton gradient, necessary to drive the transport of biogenic amines into the vesicles against a concentration gradient.

A few alkaloids have been recognized as adenylyl cyclase blockers (Table 2.11), as inhibitors of phosphodiesterase, protein kinases and phospholipases. The higher representation of fungal and animal alkaloids in the last two groups might be biased, due to the fact that the latter targets have been included in screening programmes only during the last two decades, when extensive programmes were directed towards metabolites from fungi and marine animals. At present, no apparent structure-function relationship can be seen between alkaloids that inhibit the same molecular target.

2.3 Interference of secondary metabolites with DNA and related targets

2.3.1 Intercalation

The genetic information of most organisms is encrypted in DNA (with the exception of some viruses that have RNA in their genome). Since DNA encodes all RNAs and, via messenger ribonucleic acid (mRNA), proteins and enzymes that are important for the metabolism and development of an organism, DNA is a highly vulnerable target. It is not surprising that, during evolution, a number of secondary metabolites have been selected which interact with DNA, DNA-processing enzymes and other DNA-related targets.

Table 2.11 Examples of alkaloids modulating enzymes involved in signal transduction

Enzyme	Alkaloid	Type	Occurrence	Activity	Reference
Adenylyl cyclase (AC)	Ergometrine and related alkaloids	Ergot	Claviceps and several Convolvulaceae	AC agonist	Rosenfeld *et al.*, 1980
	Nuciferine	Aporphine	*Nelumbo lutea* (Nelumbonaceae), *Colubrina faralaotra* (Rhamnceae)	AC inhibition	Buckingham, 1996
Phosphodiesterase (PDE)	Allocryptopine	Protoberberine	*Glaucium arabicum*	PDE inhibition	Abu-Ghalyun *et al.*, 1997
	Caffeine and related alkaloids	Purine	Rubiaceae, Aquifoliaceae, Sapindaceae, Sterculiaceae, Theaceae	PDE inhibition	Buckingham, 1996
	Chelerythrine	Benzoanthridine	Several Papaveraceae, Fumariaceae	PDE inhibition	Moriyasu *et al.*, 1990
	Colchicine	Tropolone	*Colchicum autumnale*, many other *Colchicum* spp., several *Merendera* spp., *Gloriosa superba* and other Liliaceae	PDE inhibition	Ewart and Bradford, 1988; Teuscher and Lindequist, 1994
	Glaucine	Aporphine	Many Monimiaceae, Berberidaceae, Annonaceae, Lauraceae, Ranunculaceae, Papaveraceae	Inhibition of a Ca^{2+}-independent PDE	Ivorra *et al.*, 1992b
	Griseolic acid	Purine	Fungal metabolite, *Streptomyces*	PDE inhibition	Buckingham, 1996

Table 2.11 (Continued)

Enzyme	Alkaloid	Type	Occurrence	Activity	Reference
	Infractine	Indole	Fungus, Cortinarius infractus, *Picrasma quassioides* (Simaroubiaceae)	PDE inhibition	Bracher and Hildebrand, 1995
	Laccarin		Fungus, *Laccaria vinaceoavellanea*	PDE inhibition	Matsuda *et al.*, 1996
	Papaverine	Isoquinoline	*Papaver somniferum* (Papaveraceae); *Rauwolfia serpentina* (Apocynaceae)	Unselective PDE inhibition	Buckingham, 1996; Ivorra *et al.*, 1992b
	Persicanidine A	Steroidal	*Fritillaria persica* (Liliaceae)	PDE inhibition	Ori *et al.*, 1992
	Pseudodistomin	Piperidine	Tunicate *Pseudodistoma kanoko*	PDE inhibition	Ishibashi *et al.*, 1987
	Sanguinarine	Benzophenanthridine	Several Papaveraceae, Fumariaceae	PDE inhibition	Moriyasu *et al.*, 1990
Protein kinase C (PKC)	Balanol		Fungal metabolite, *Cordyceps ophioglossoides*, *Fusarium merismoides*, *Verticillium balanoides*	PKC inhibition	Buckingham, 1996
	Cepharanthine	Bis-isoquinoline	*Stephania* spp. (Menispermaceae)	PKC inhibition	Edashige *et al.*, 1991
	Chelerythrine	Benzophenanthridine	*Chelidonium majus*	Selective PKC inhibition	Herbert *et al.*, 1990
	Erbstatin	Amine	*Streptomyces* spp.	EGF tyrosine kinase inhibitor	Buckingham, 1996

Ellipticine	Indole	*Ochrosia elliptica* and several other *Ochrosia* spp., *Aspidosperma subincanum*, *Bleekeria vitiensis* (Apocynaceae)	Selective inhibition of p53 phosphorylation	Ohashi and Matsuoka, 1985
9-Hydroxyellipticine	Indole	*Strychnos dinklagei* (Loganiaceae)	Selective inhibition of p53 phosphorylation, suppression of CDK2 kinase	Ohashi and Matsuoka, 1985
Lavendustin A		*Streptomyces griseolavendus*	Tyrosine kinase inhibitor	Buckingham, 1996
Lyngbyatoxin A	Indole	Marine blue-green alga, *Lyngbya majuscula*, also Streptomyces mediocidicus	PKC activator	Buckingham, 1996
Melittin	Peptide	Bee venom (*Apis* spp.)	Inhibitor of PKC and cAMP protein kinases	Buckingham, 1996
16-Methylpendolmycin	Indole	*Nocardiopsis* sp.	PKC inhibition	Sun *et al.*, 1991
Michellamine B	Isoquinoline	*Ancistrocladus* spp. (Ancistocladraceae)	PKC inhibition	Upender *et al.*, 1996
Polymyxin B	Cyclic peptide	*Bacillus polymixa*	PKC inhibition	Buckingham, 1996
Swainsonine	Indolizidine	*Swainsonia canescens*, *Astragalus* spp. (Leguminosae)	Indirect PKC activation	Breton *et al.*, 1990

Table 2.11 (Continued)

Enzyme	Alkaloid	Type	Occurrence	Activity	Reference
Phospholipase	Aristolochic acid	'aporphine'	*Aristolochia* spp. (Aristolochiaceae)	PLA_2 blocker	Moreno, 1993; Teh *et al.*, 1990
	Berbamine and related alkaloids	Bis-isoquinoline	*Berberis* spp. and other Berberidaceae, Menispermaceae, Ranunculaceae	Inhibition of cytosolic PLA_2, uncoupling of G-protein from PL	Akiba *et al.*, 1992, 1995; Hashizume *et al.*, 1991
	Topsentin B2	Bis-indole	Marine sponges (*Topsentia* spp., *Spongosorites* spp., *Hexadella* spp.)	Inactivation of PLA_2	McConnell *et al.*, 1993

Abbreviations: cAMP, cyclic adenosine monophosphate; PLA_2, phospholipid A_2; EGF, epidermal growth factor.

Some alkaloids are known to bind or to intercalate with DNA (Krey and Hahn, 1969; Maiti *et al.*, 1982; Nandi and Maiti, 1985; Schmeller *et al.*, 1997b; Wink and Latz-Brüning, 1995; reviews in Wink 1993a; Wink *et al.*, 1998a,b). Many of these molecules are planar, hydrophobic molecules, which fit between the planar stacks of adenine-thymine (AT) and guanine-cytosine (GC) base pairs. Important alkaloids that fall into this group include; sanguinarine, harmin, berberine, berbamine, ergometrine, harmaline, emetine, quinidine, quinine, cinchonidine, cinchonine, boldine, norharman, solanine, canadine, chelidonine, lobeline, ajmalicine and, possibly, ajmaline (Wink and Latz-Brüning, 1995; Wink *et al.*, 1998a,b). The degree of DNA intercalation is strongly and positively correlated with inhibition of DNA and RNA processing enzymes, such as DNA polymerase I and reverse transcriptase (Wink and Latz-Brüning, 1995; Wink *et al.*, 1998a,b). Other alkaloids act at the level of DNA and RNA polymerases, DNA topoisomerase, thus impairing the processes of replication, DNA repair and transcription.

The effects of DNA-binding or intercalating compounds can be mutations, which may result in malformation of newborn animals, in the initiation of cancer or in defective proteins, with negative consequences on metabolism. This topic will be reviewed in more detail in the following paragraphs.

2.3.2 Secondary plant metabolites with mutagenic and carcinogenic properties

2.3.2.1 Significance and distribution in plants

Mutagenic and carcinogenic properties of secondary plant metabolites have, for a long time, been considered to be merely a curious exception in the spectrum of biological activities of naturally-occurring plant products. However, since the discovery of the mutagenicity of some pyrrolizidine alkaloids in *Drosophila melanogaster* (Clark, 1960), our knowledge in the field of plant mutagens and carcinogens capable of displaying DNA-damaging activity in prokaryotic and eukaryotic cells and organisms has vastly increased. It is now certain that compounds with mutagenic potential are widely distributed throughout the plant kingdom. They have been isolated from bacteria, fungi, algae and lichens, but also from ferns and from many members of families belonging to the spermatophytes.

In the past, such compounds were discovered mainly as a result of outbreaks of disease in agricultural livestock. During the last two decades, however, screening programmes have been developed with the specific aim of identifying natural mutagens in our environment. This was initiated by the understanding that mutagenicity is associated with carcinogenicity. It was also established that mutagenicity is characteristically connected with considerable cytotoxicity. It was, therefore, assumed that mutagenic properties reflect particular aspects of the intrinsic toxicity and have not been evolved as a direct basis upon which selection might act.

The term 'mutagenicity' is often used in a more strict sense in order to distinguish it from the term 'clastogenicity', which means the ability to induce structural or numerical chromosomal aberrations, e.g. breaks, exchanges and gaps. Loss of chromosomal material is also an indication of genetic damage and can be detected by special methods, e.g. the micronucleus test and the comet assay. In addition, DNA damage is indicated by the induction of sister chromatid exchange (SCE) and, indirectly, by lethality in recombination repair deficient bacterial strains. But the most widely-employed *in vitro* mutagenicity test is the Ames assay. It is performed with *Salmonella typhimurium* strains, which are constructed to detect base pair and frameshift mutagens. Independently of the genetic endpoint tested, DNA-damaging agents are generally termed 'genotoxic agents'.

The purpose of the present review is to discuss the current status of our knowledge of genotoxic plant metabolites, their interaction with DNA and the evidence of their association with tumour formation in experimental animals and with human cancer.

Table 2.12 is not a complete list of the presently known natural mutagens but it gives an overall impression of the great variability in the chemical structures involved and the abundance of such compounds. However, in the present review, it is not possible to discuss all of the mutagens listed. It should also be pointed out that the mutagenic potential of certain genotoxic agents has not yet been evaluated accurately or has been associated with *in vitro* conditions. Although genotoxic carcinogens are apparently the main cause of tumour induction in animals, tumour incidence can also be forced by tumour promoters, which occur in plants and microorganisms. Tumour promoters, such as phorbol esters of many Euphorbiaceae, exert their action through non-genotoxic mechanisms by activating protein C kinases or affecting cell proliferation. It has been shown that some genotoxic agents also possess tumour-promoting properties.

2.3.2.2 Mode of action

Genotoxic agents can damage DNA and chromosomes through very different mechanisms. Many naturally-occurring mutagens do not show genotoxic properties *per se*. They need metabolic activation by mammalian enzymes. *In vitro*, the exogenous metabolization is achieved with rat liver enzyme preparations used as an S9 fraction or S9 mix. If the bacterial enzymes are not capable of activating a compound to

Table 2.12 Secondary plant metabolites with genotoxic properties

Producing organisms	Type of compounds	Examples
Bacteria	**'Antibiotics'**	Streptozotocin, azaserin, daunomycin, adriamycin, mitomycin C, bleomycin
Fungi	**Mycotoxins**	Aflatoxins, sterigmatocystins, ochratoxin A, luteoskyrin, patulin, sporidesmin, griseofulvin, citrinin, fusarin C, alternariolmethylether, gyromitrin, nivalenol, isovelleral, illudin S, necatorin, ergotamine and other ergot alkaloids
Lichens		Physodalic acid
Algae		Plocamenon
Ferns		Ptaquiloside, hypoloside A and C
Spermatophytes	**Alkaloids**	
	Acridone alkaloids	Rutacridone, rutacridone epoxide, isogravacridonchlorine
	Quinolizidine alkaloids	Cryptopleurine
	Furoquinoline alkaloids	Dictamnine, γ-agarine, maculine, evolitrine, kokusaginine, pteleine
	Indole alkaloids	Ellipticine, vincristine, voacristine, strychnine
	Quinoline alkaloids	Camptothecin
	Isoquinoline alkaloids	Liriodenine, roemerine, lysicamine, noscapine, tetrandrine
	Cephalotaxine alkaloids	Harringtonine, homoharringtonine
	Piperidine alkaloids	Arecoline, arecaidine
	Pyrrolizidine alkaloids	Clivorine, heliotrine, monocrotaline, senecionine, senkirkine, seneciphylline, retrorsine, echimidine, fulvine, jacobine, ligularidine, lycopsamine, intermedine, petasitenine

Table 2.12 (Continued)

Producing organisms	Type of compounds	Examples
	Phenylalkylamines	Cathinone
	Purines	Caffeine, theobromine, theophylline
	β-Carboline alkaloids	Harman, harmine, brevicolline, harmalol, harmol
	Further N-containing plant metabolites	Capsaicine, benzoxazinones
	Nitro aromatic compounds and related lactams	Aristolochic acid I, II, IV, methoxytariacuripyrone, aristolactam I, II
	N-containing glycosides	Cycasin, neocycasins, macrozamin
	Glucosinolates and mustard oils	Sinigrin, allylisothiocyanate, thiourea, phenethylisothiocyanate
	Anthranoids	Aloe-emodin, emodin, physcion, lucidin, purpurin, rubiadin
	Flavonoids	Quercetin, kaempferol, galangin, wogonin, norwogonin
	Furocoumarins	Bergapten, heraclenin, imperatorin, xanthotoxin
	Phenols	Gossypol, hydroquinone
	Phenylpropanoids	Estragole, safrole, isosafrole, β-asarone, (6)-gingerole, (6)-shogaol, caffeic acid, cinnamaldehyde
	Terpenoids	Citronellal, menthone, catalpin, costunolide, hymenoxon, steviol, valepotriates
	Xanthones	Gentisin, isogentisin, swertianin, bellidifolin, methylbellidifolin, desmethylbellidifolin.

become a mutagen and mammalian enzymes are necessary for induction of mutation, the genotoxin is characterized as a promutagen or an indirect-acting mutagen.

Promutagens are metabolized into the ultimate mutagens from which the electrophilic intermediates are formed. These are detoxified or react with nucleophilic biopolymers, e.g. nucleic acids, forming DNA adducts. Safrole and aflatoxin B_1 are known to be promutagens. Aflatoxins need microsomal monooxygenases (cytochrome P_{450} enzymes) to be activated, safrole is converted into the ultimate mutagen through cytochrome P_{450} enzymes and a sulfotransferase.

Naturally-occurring glycosides, e.g. cycasin and rutin, are also inactive *per se* in mutagenicity tests. But when they are hydrolyzed by the action of bacterial glycosidases, stable or unstable aglycones with mutagenic properties are formed. In mammals, hydrolysis is mediated by the intestinal microflora.

A considerable number of plant mutagens have been found to be frameshift mutagens. This is suggested for molecules with a planar structure, e.g. quercetin. Planarity facilitates the intercalation within the DNA. After intercalation, the compound can interact further with the DNA. The nature of the binding and the binding properties are important for the consequences of intercalation and DNA complexation.

More potent mutagens result from irreversible bonds with DNA. Very strong effects are expected when a mutagen is capable of forming cross-links, i.e. irreversible bonds with both DNA strands. The pyrrolizidine alkaloids and some furocoumarins belong to this mutagen-type. Several plant metabolites interfere in the process of mitosis (e.g. vinblastine, colchicine, taxol, podophyllotoxin), damaging the function and the structure of the spindle apparatus. This can lead to aneuploidy, as evidenced from experiments with noscapine.

Finally, some genotoxic compounds can affect the function of the topoisomerases. The importance of this mechanism of action for genotoxicity has now been established for several mutagens with cytostatic properties, e.g. ellipticine and camptothecin (Anderson and Berger, 1994). Detailed references for section 2.3.2 can be found in the review of Clark (1982), Hirono (1987), Lai and Woo (1987), Ishidate and co-workers (1988), Stich (1991), and in the corresponding paragraphs.

2.3.2.3 Significant endogenous mutagens/carcinogens

Mycotoxins. Aflatoxins (AFs) are chemically-related compounds with a common difuranocoumarin structural element. They are produced by several strains of *Aspergillus flavus*. AF B_1 (Fig. 2.6) is the most potent carcinogen of the group. AF B_1 shows mutagenic, clastogenic and recombinogenic activity in most test systems. In the Ames assay,

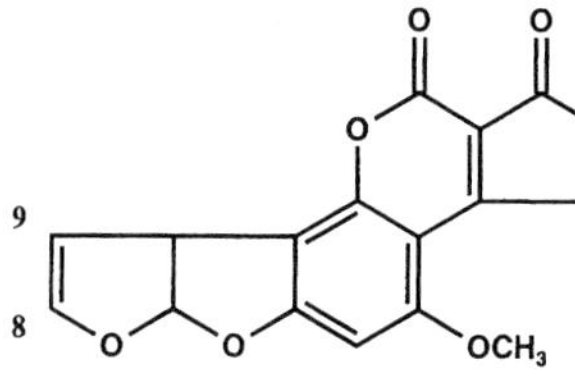

1

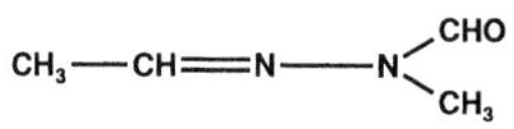

2

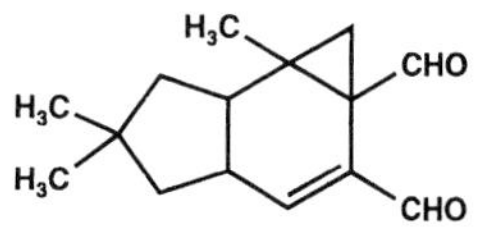

3

Figure 2.6 Structure of the mycotoxins aflatoxin B1 (**1**), gyromitrin (**2**) and isovelleral (**3**).

metabolic activation is needed for mutagenicity. This leads to the formation of an unstable epoxide, which is further converted into an electrophile intermediate. The guanosine moieties are the reactive sites, where the intermediate is covalently bound. The major DNA adduct *in vivo* was the 8,9-dihydro-8-(N^7-guanyl)-9-hydroxyaflatoxin B_1 (Groopman and Cain, 1990). Formation of DNA adducts correlated to mutagenicity, SCE induction, clastogenicity and recombinogenic effects. Recent data support the hypothesis that oncogenes may have been mutated as a result of AF-DNA adduct formation. The most important results of AF research can be found in an excellent review by Chu (1991).

Experiments with rats, fish and non-human primates have produced evidence that AF B_1 is a potent hepatocarcinogen. But neoplastic formations in other organs were also induced. Details, such as the sensitivity of animals, dose, route of administration and the role of the factors, can be found in reviews by Tazima (1982) and Groopman and Cain (1990). Circumstantial evidence from epidemiological studies indicates a causal relationship between the incidence of primary liver cancer in humans and the mean daily intake of AF B_1 (Tazima, 1982). Shen and Ong (1996) examined more than 1500 human hepatocellular carcinoma samples and found evidence that oncogenes are critical molecular targets for AF B_1.

Brief mention can be made of several other mycotoxins that might have genetic significance to herbivores and man. Whereas extensive studies have been made of AF, little is known about the mode of action of other mycotoxins produced by *Aspergillus* and *Penicillium* species. Most of them have carcinogenic properties *in vivo* but show different responses in mutagenicity tests *in vitro* (Table 2.13). Fusarin C, a secondary metabolite of *Fusarium moniliforme*, showed mutagenicity comparable to that of AF B_1 in the Ames assay. It remains to be determined whether the presence of fusarin C is associated with the known carcinogenicity of *Fusarium moniliforme* isolates. Very recently, additional *Fusarium* mycotoxins were investigated with respect to their genotoxicity in bacteria and rat hepatocytes (Knasmüller *et al.*, 1997). Nivalenol and mycotoxins with trichothecene structure also showed mutagenic effects but their carcinogenic potential has not been tested (Chu, 1991).

Table 2.13 Mutagenicity and carcinogenicity of selected mycotoxins (combined data from Tazima, 1982)

Compound	Carcinogenicity	Mutagenicity *in vitro* Ames assay	Bacterial recombinant assay	Mammalian cell cultures
Aflatoxin B_1	+ (rat, trout, mouse)	+	+	+
Sterigmatocystin	+ (rat, mouse)	+	+	+
Patulin	+ (rat)	-	+	+
Citrinin	+ (rat)	-	+	?
Luteoskyrin	+ (mouse)	-	+	-
Ochratoxin A	+ (mouse)	-	-	-

Gyromitrin. In 1967, gyromitrin (Fig. 2.6) was isolated from *Gyromitra esculenta*, false morel (Pezizales, Ascomycetes), and was found to be the main toxic principle of this mushroom. Under acidic conditions, gyromitrin forms *N*-methyl-*N*-formylhydrazine. In the stomach it is converted to *N*-methylhydrazine. *N*-methyl-*N*-formylhydrazine can be oxidized to the *N*-nitroso derivative by liver cytochrome P_{450} enzymes. Positive results were obtained with *N*-methyl-*N*-formylhydrazine in the Ames assay. A stronger effect was found after metabolic activation (von der Hude and Braun, 1983). The authors postulated the formation of the *N*-nitroso derivative to explain the genotoxic effect. *N*-methylhydrazine showed a positive effect in *Escherichia coli*. The high bactericidal activity made it difficult to evaluate the genotoxic potential of *N*-methylhydrazine accurately (IARC, 1983).

A summary of data reported for carcinogenicity in experimental animals and evaluation of such data is presented in the review by Natori (1987) and the IARC (1983). The results provide sufficient evidence for the carcinogenicity of gyromitrin and its metabolites in animals.

Mutagens in larger fungi. Screening experiments have resulted in the discovery of some compounds present in *Lactarius* species that showed weak but significant mutagenicity in the Ames assay. Isovelleral (Fig. 2.6) and hydroxy-isovelleral were mutagenic after metabolic activation. The activity is possibly connected with the unsaturated dialdehyde structure of the molecules (Sterner *et al.*, 1987). Their role in chemical defence was discussed by Sterner and co-workers (1985).

Fern toxins: ptaquiloside. Ptaquiloside (Fig. 2.7) is an *O*-glycoside with a norsesquiterpene aglycone. It has a planar structure and is relatively

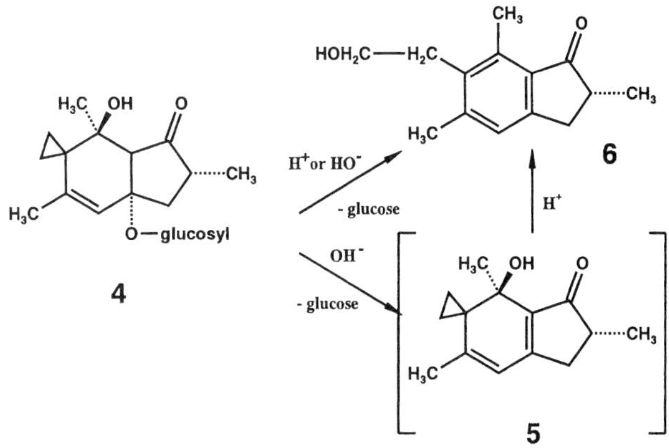

Figure 2.7 Ptaquiloside (**4**) and its conversion into an unstable dienone (**5**) and pterosin B (**6**) as end-product (Redrawn from Hirono and Yamada, 1987).

unstable under heat, light and acidic or alkaline conditions. In alkaline aqueous solution, it is converted into pterosin B via an unstable conjugated dienone (Fig. 2.7) (Hirono and Yamada, 1987). Ptaquilosid was first isolated from *Pteridium aquilinum* (L.) Kuhn (syn. *Pteris aquilina*) but more recently it was also detected in *Pteris cretica* and other ferns (Saito *et al.*, 1990). *Pteridium aquilinum* (bracken fern) is widely-distributed in many parts of the world. The hypolosides are related compounds and were isolated from other pteridophyta (Saito *et al.*, 1990). Ptaquiloside showed marked mutagenicity in a modified Ames assay after preincubation at pH 8.5 (Nagao *et al.*, 1989). It proved to be a direct-acting mutagen, with an activity comparable to that of the illudins, metabolites of certain basidiomycetes. Ptaquiloside induced chromosomal aberrations in cultured Chinese hamster lung cells (Matsuoka *et al.*, 1989), as did the illudins and the hypolosides. The clastogenic effect was pH-dependent.

Two possibilities have been discussed for the mechanism of action: an electrophilic intermediate may be formed either via the highly reactive cyclopropane ring or via the cyclopropylcarbinol structure (Hirono and

Yamada, 1987). In 1965, the carcinogenicity of bracken fern was clearly demonstrated by experiments with rats. The tumour-inducing effect of a bracken fern diet was confirmed by many research groups in the following decades. After being fed a diet containing bracken powder, tumours were observed in mice, rats, hamsters, guinea-pigs and cattle. Target organs were the urinary bladder and the intestinal tract. The carcinogenicity of ptaquiloside was shown in female CD rats. For details of the carcinogenicity experiments see the review by Hirono and Yamada (1987).

Cycasin and related azoxyglycosides. Cycasin was first isolated from *Cycas revoluta* and *C. circinalis*. Subsequently, macrozamin and the neocycasins were described as metabolites of other Cycads. They differ in their sugar moiety from cycasin. Cycads, which represent a group of ancient gymnosperms, occur in tropical and subtropical zones. Azoxyglycosides are present in higher amounts in the seeds, which are used as a source of food starch. The glycosides are hydrolyzed by glycosidases in herbivores and humans, forming methylazoxymethanol (MAM), an aglycone that causes acute intoxications as well as mutations and tumour initiation.

Cycasin was inactive in the Ames assay because the standard S9 mix from rat liver lacks the appropiate β-glucosidase. MAM, however, induced genetic alterations in *Salmonella typhimurium* and *Bacillus subtilis*. Gene mutations and chromosomal aberrations were observed in yeast, plant cells, *Drosophila melanogaster* and mammalian cells. The genotoxicity was possibly due to the spontaneous decomposition of MAM into alkylating intermediates, presumably diazomethane or, more likely, into another methyl donor. After administration of cycasin, N^7-methylguanine was discovered in rat DNA *in vitro* and *in vivo*. The carcinogenicity of cycasin and the other azoxyglycosides has been demonstrated in mice, rats, hamsters, guinea-pigs, rabbits, fish and monkeys (Hoffman and Morgan, 1984; Hirono, 1987).

Aristolochic acids and related compounds. Aristolochic acids (AAs) occur in the roots and in the aerial parts of many members of the genus *Aristolochia*, e.g. *Aristolochia clematitis*. AAs are often accompanied by aristolactams or related nitro aromatic compounds (Mix *et al.*, 1982; Achenbach *et al.*, 1992). In *Aristolochia* species, AAs are normally present as a mixture of at least six compounds. The main ingredients are the AAs I and II (Fig. 2.8). AAs have been employed in a number of *in vitro* mutagenicity tests. A mixture of AA I and II proved to be mutagenic in the Ames test with *Salmonella typhimurium* strains (Robisch *et al.*, 1983; Schmeiser *et al.*, 1984). Subsequently, it was established that AA II was

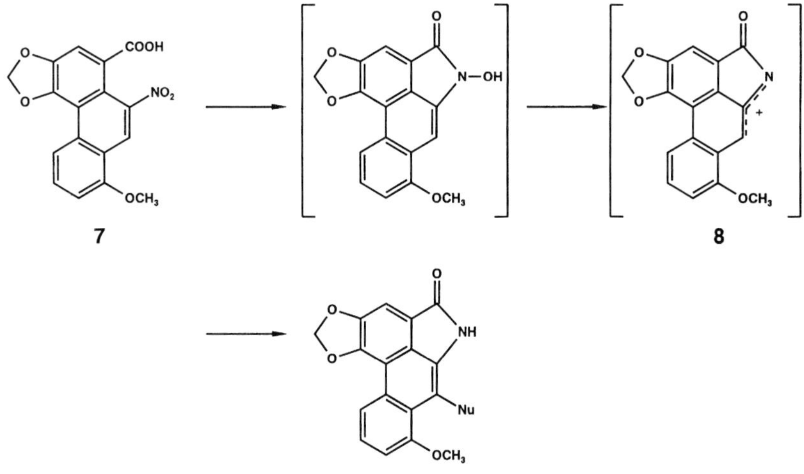

Figure 2.8 Proposed mechanism of activation of aristolochic acid I (**7**) via a pentacyclic nitrenium ion (**8**). The two DNA adducts identified were deoxyguanosine-N^2-yl-aristolactam and deoxyadenosine-N^6-yl-aristolactam (**9**,**10**) (redrawn from Pfau *et al.*, 1990).

more active in *Salmonella typhimurium* than the acids I and IV (Götzl and Schimmer, 1993; Pistelli *et al.*, 1993). The aristolactams, which occur naturally but are also formed metabolically after oral ingestion of the AAs (Krumbiegel *et al.*, 1987), showed weak mutagenic activities after metabolic activation by liver enzymes in the Ames assay (Schmeiser *et al.*, 1986). Aristolactams were also produced *in vitro* when AAs were incubated with S9 mix (Schmeiser *et al.*, 1986).

AAs are direct-acting mutagens in bacteria (Robisch *et al.*, 1983; Schmeiser *et al.*, 1984; Pezzuto *et al.*, 1988). They were activated by bacterial nitroreductases analogous to the activation of other nitro aromatic compounds. A cyclic nitrenium ion is formed via hydroxylamine (Fig. 2.8). Delocalization of the positive charge finally leads to DNA binding via the C7-position. The intermediates are covalently bound at the exocyclic amino groups of the purines (Pfau *et al.*, 1990a). AA II can also be activated by the cytosolic fraction of liver homogenates from Wistar rats (Schimmer and Drewello, 1994). In *Salmonella typhimurium*, the microsomal fraction was capable of detoxification and reduced the mutagenic effect. It is therefore suggested that, in the presence of mammalian liver enzymes, activation and deactivation processes take place concurrently and at a comparable level.

The important role of the nitro group for activation could be established using nitroreductase overproducing strains (Götzl and

Schimmer, 1993). This was also deduced from results obtained with nitrophenanthrene derivatives (Pfau *et al.*, 1990c). The lower activity of AA I compared to acid II was attributed to the methoxy group, which possibly produced steric hindrance for binding of the genetically active intermediate to DNA or for binding of the substrate to the active site of the enzyme(s).

Many positive results are available concerning the mutagenicity of AAs in eukaryotic test systems. They induced point mutations and recombinations in *Drosophila melanogaster* (Frei *et al.*, 1985) and caused gaps and chromosome breaks and induced SCE in human lymphocytes *in vitro* (Abel and Schimmer, 1983). AAs also induced point mutations in V79 Chinese hamster cells at the HPGRT locus (Manolache *et al.*, 1985). Positive results were obtained in the point mutation test on L5178Y/TK$^{+/-}$ mouse lymphoma cells, in the DNA repair test on rat hepatocytes, and in the cell transformation test with BALB 3T3 cells (Puri and Müller, 1985). In the Granuloma Pouch Assay, which detects gene mutations *in vivo*, AAs were more potent at equimolar doses than *N*-methyl-*N*-nitro-*N*-nitrosoguanidine, a mutagen that has been known for its strong alkylating potency (Maier *et al.*, 1985). AA was found to be mutagenic in further *in vivo* mutagenicity tests. Genotoxic effects on bone marrow cells of mice were reported by Mengs and Klein (1988), who used the micronucleus test system.

During the last decade, experiments *in vitro* and *in vivo* have been performed to elucidate the nature of DNA adducts and their relevance for carcinogenicity. Using the ^{32}P-postlabelling assay, it could be shown that AA I forms covalent DNA adducts upon metabolic activation *in vitro* (Pfau *et al.*, 1990b). Incubation of AA I or II with rat liver S9 and calf thymus DNA also gave rise to DNA adduct formation. The aristolactams I and II (corresponding to the acids I and II) did not form DNA adducts in the presence of S9 (Schmeiser *et al.*, 1988). However, evidence exists that aristolactams can bind to natural and synthetic DNA by a mechanism of intercalation and exhibit considerable specificity towards alternating GC polymers (Nandi *et al.*, 1991). When AA I or II was administered orally to male Wistar rats and the DNA from different target and non-target tissues was analyzed for DNA adducts, the patterns were similar to those obtained from *in vitro* incubations (Schmeiser *et al.*, 1988).

Several papers have been published during the last 5 yrs on the chemical nature of the adducts and the molecular consequences. The two main adducts were identified as 7-(deoxyguanosin-N^2-yl)-aristolactam and 7-(deoxyadenosin-N^6-yl)-aristolactam (Pfau *et al.*, 1991; Fernando *et al.*, 1992) (Fig. 2.8). Further experiments showed that irrespective of the AA used to induce DNA adducts, deoxyadenosine was the major target for chemical carcinogenesis of these compounds (Stiborova *et al.*,

1994). In a study with synthetic oligonucleotides, it was demonstrated that all purine adducts provided severe blocks to DNA replication but the guanine adducts may not be very efficient mutagenic lesions. The adenine adducts, however, exhibited a distinct mutagenic potential, resulting from deoxyadenylate adenosine monophosphate (dAMP) incorporation by polymerase.

AT→TA transversions would be the mutagenic consequences of adenine adducts (Broschard *et al.*, 1994). This is consistent with the detection of transversion mutations in *c-ras* genes in the analysis of AA-induced tumours in rodents (Schmeiser *et al.*, 1990, 1991). In this context, it is noteworthy that DNA adducts formed by AAs were also identified in renal tissues from patients with Chinese herbs nephropathy (Schmeiser *et al.*, 1996).

The carcinogenicity of AAs was first recognized in rodents even before mutagenicity was proved. Carcinogenicity in rats was first reported by Mengs and co-workers (1982), who found that AAs caused gastric carcinomas with a short latency period. Further reports on carcinogenicity with respect to organ or tissue specificity followed (Mengs, 1988; Schmeiser *et al.*, 1990). The animals developed papillomas or squamous cell carcinomas in the forestomach, renal pelvis and urinary bladder. Liver cell carcinogenesis was initiated after partial hepatectomy (Rosiello *et al.*, 1993). AAs are potent inhibitors of seed germination (Watanabe *et al.*, 1988). They also have insect chemo-sterilant activity (Mathur *et al.*, 1980; Watanabe *et al.*, 1988). This may be discussed as a particular evolutionary aspect of these secondary plant products.

9-Methoxytariacuripyrone, another naturally-occurring and closely-related nitro aromatic compound showed much stronger mutagenicity than AA II in the Ames assay (Schimmer and Drewello, 1994).

Pyrrolizidine alkaloids. Pyrrolizidine alkaloids (PAs) occur mainly in members of the families Boraginaceae (e.g. *Symphytum, Echium, Anchusa, Heliotropium*), Asteraceae (e.g. *Senecio, Adenostyles*) and Leguminosae (mainly tribe Crotalarieae). Many representatives of these families are used for medicinal purposes (Röder, 1995). More than 6000 species contain alkaloids with this basic structure.

The hepatotoxicity and the mutagenicity of PAs has been known for more than 30 yrs. The toxicity and carcinogenicity are connected with certain structural elements: a 1,2-double-bond in the pyrrolizidine ring and branched chain acids, esterifying a 9-hydroxyl and preferably also the 7-hydroxyl substituent. PAs with a saturated ring system are inactive. The structures of three active PAs are presented in Figure 2.9. Other active compounds are listed in Table 2.12. The alkaloids occur as free bases and *N*-oxides. *N*-oxides can be reduced to the free bases by the intestinal

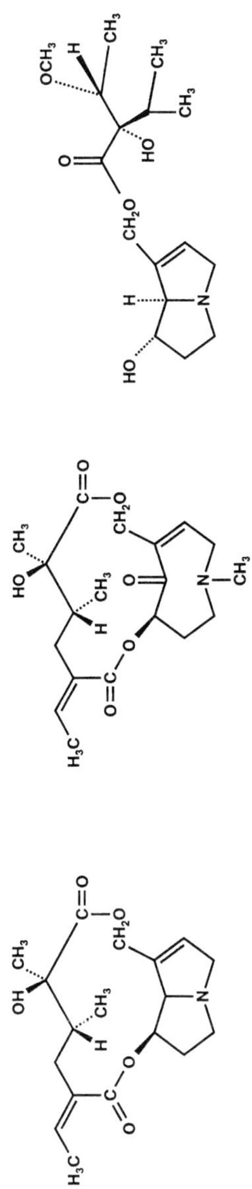

11 12 13

Figure 2.9 Pyrrolizidine alkaloids with cyclic diester and monoester structures, senecionine (**11**), senkirkine (**12**) and heliotrine (**13**).

microflora and then exhibit activities similar to those of the free bases in animals. In the liver, the bases are metabolized into pyrrole derivatives (dehydroalkaloids) by cytochrome P_{450} enzymes (monooxygenases). They can produce electrophilic intermediates and then act as monofunctional or as bifunctional alkylating agents (Fig. 2.10). The pyrrole derivatives are considered to be the ultimate genotoxic compounds. Their rate of formation varies greatly from PA to PA and may, in part, explain the variation in activity. The pyrrolic alcohol metabolites show antimitotic effects, which are relevant for hepatotoxicity. PAs cross the placental barrier and are transferred into the milk of mammals fed with plants containing PAs.

Most PAs with a 1,2-double-bond in their pyrrolizidine ring exhibit only weak or insignificant mutagenicity in *Salmonella typhimurium*. A slight increase in mutants was observed under preincubation conditions and in the presence of S9 mix (Rubiolo *et al.*, 1992). More pronounced effects were obtained in *Drosophila melanogaster*. Chromosomal aberrations were also induced in plant cells, fungi and mammalian cells. Positive effects were reported with the SCE assay, the micronucleus test and DNA repair test. Earlier results from genotoxicity experiments were reviewed by Mattocks (1986) and Furuya and co-workers (1987). The activity of the individual PAs varied greatly from assay to assay. In human lymphocytes *in vitro*, heliotrine was capable of inducing chromosomal aberrations but senkirkine was not (Kraus *et al.*, 1985). SCE and chromosomal breaks were also induced by crude extracts from the roots of *Symphytum officinale*, which contained the monoester lycopsamine and intermedine (Behninger *et al.*, 1989). *In vitro*, both the dehydroalkaloids and the dehydroaminoalcohols interacted with DNA. In *E. coli* and rat liver, DNA cross-links were formed, indicating the bifunctional nature of these metabolites. The DNA adducts formed with dehydroretronecine were identified as derivatives with a covalent bond between the C-7 position of the PA and the N^2-position of deoxyguanosine.

The carcinogenicity of PAs appears to parallel their mutagenicity. The most active PAs belong to the macrocyclic diester and the open diester type, in which the amino alcohol part is retronecine, heliotridine or otonecine. Crude plant extracts and numerous PAs and metabolites have been tested for carcinogenicity. The studies were primarily carried out on rats. Lasiocarpine produced the largest yield of tumours. Hepato-cellular carcinomas and haemangiosarcoma were the most common tumour types. The carcinogenicity studies of PAs in experimental animals were extensively reviewed by Mattocks (1986), Furuya and co-workers (1987) and the World Health Organization (WHO) (1988).

Isoquinoline alkaloids. Isoquinoline alkaloids with an aporphine structure (Fig. 2.11) occur mainly in genera belonging to the families,

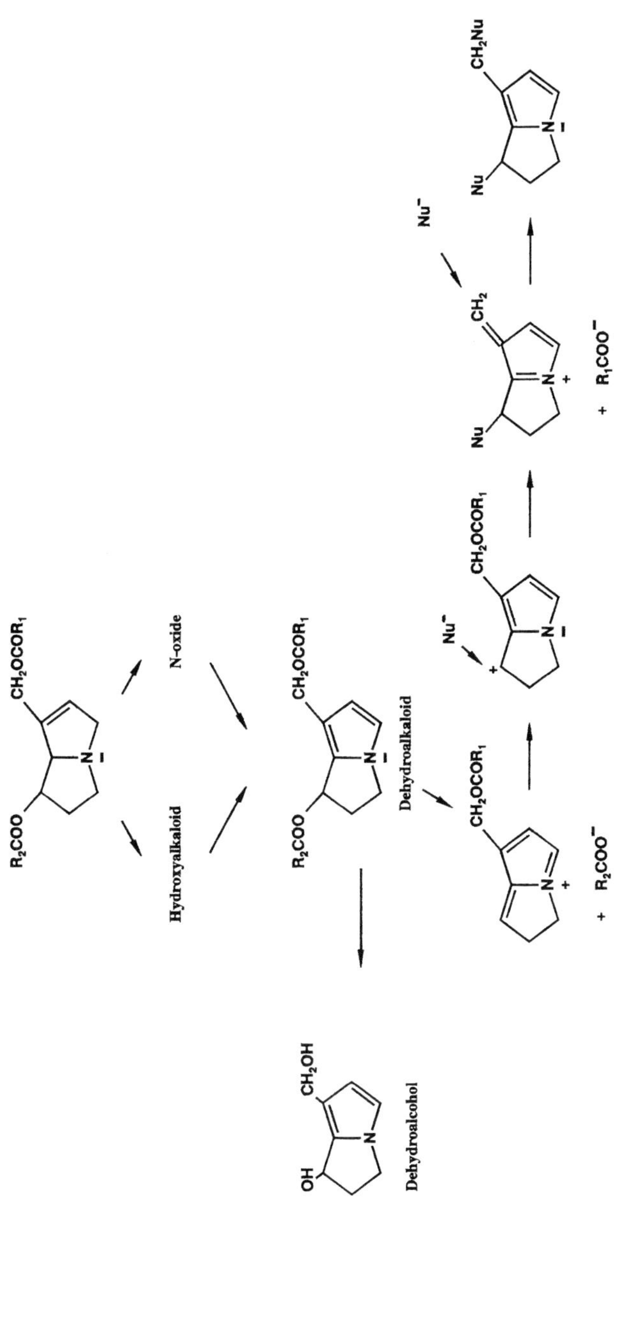

Figure 2.10 Metabolism of 1,2-unsaturated pyrrolizidine alkaloid (PA). Formation of monofunctional and bifunctional pyrrolic intermediates.

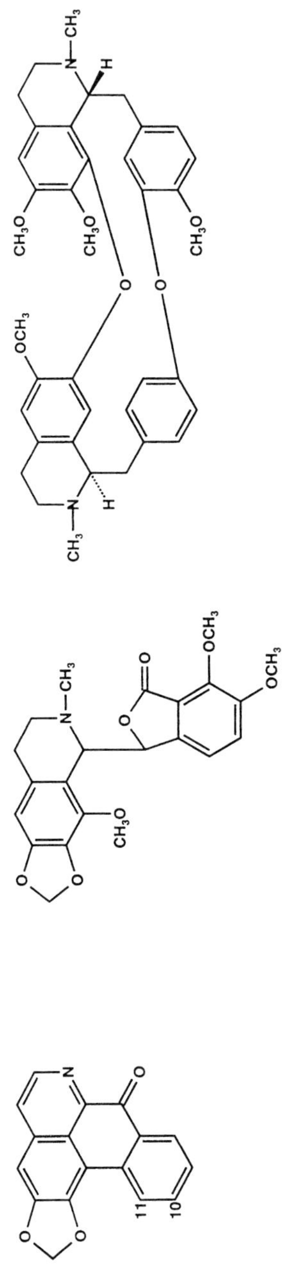

14 15 16

Figure 2.11 Structure of the isoquinoline alkaloids, liriodenine (**14**), noscapine (**15**) and tetrandrine (**16**).

Menispermaceae and Annonaceae. Nozaka and co-workers (1987) researched the mutagenic principles in *Sinomenum acutum* (*Sinomeni caulis et rhizoma*). They isolated *N*-demethyl-*N*-formyl-dehydronuciferine as the compound responsible for mutagenicity in *Salmonella typhimurium*. Subsequently, 44 alkaloids of this type were screened for mutagenicity in *Salmonella typhimurium* (Nozaka *et al.*, 1990). Most aporphine-type alkaloids, e.g. dicentrine, nornuciferine, roemerine, liriodenine and lysicamine, were reported to be positive. Liriodenine was found to be the most active alkaloid. The same research group investigated the clastogenicity of 18 related aporphines in the chromosomal aberration test *in vitro*, using a Chinese hamster lung cell line (Tadaki *et al.*, 1991). Although the alkaloids showed mutagenicity to *Salmonella* strains only in the presence of S9 mix, many of the compounds tested induced chromosomal aberrations in the absence as well as in the presence of rat liver enzymes. Among these compounds, liriodenine was the most potent clastogen. Liriodenine and roemerine also induced polyploidy in the Chinese hamster lung cells. The authors postulated that the mutagenicity may be due to an epoxide that is formed at the C10–C11 position (see Fig. 2.11). Boldine, however, did not exhibit mutagenicity in the Ames assay but showed weak activity in inducing mitochondrial mutants in *Saccharomyces cerevisiae* (Moreno *et al.*, 1991).

Among the biosynthetically-related phthalide isoquinoline alkaloids, noscapine (Fig. 2.11), a centrally acting antitussive agent, has been shown to induce polyploidy in Chinese hamster lung cells, in V79 Chinese hamster cells and in human lymphocytes *in vitro* (Gatehouse *et al.*, 1991). In addition, spindle damage was observed in V79 cells and human skin fibroblasts using concentrations of 30 and 60 mg/ml. From these studies and further preliminary results, it appears that noscapine might induce an increase in chromosome loss and polyploidy through effects upon spindle structure and function.

The investigation of the bisbenzyl isoquinoline alkaloid, tetrandrine (Fig. 2.11) resulted in particularly interesting results. From the molecular structure, it has been predicted that the alkaloid would be a genotoxic carcinogen (Rosenkranz and Klopman, 1990). Using the ^{32}P-postlabelling assay, it was recently shown that tetrandrine is capable of adduct formation with DNA (Schmeiser, DKFZ Heidelberg, personal communication). From *in vitro* experiments, however, only weak effects were observed in Salmonella and in Chinese hamster lung cells and tetrandrine was found to be a weak indirect-acting genotoxicant (Whong *et al.*, 1989; Xing *et al.*, 1989). Nonetheless, the alkaloid was a potent enhancer of the mutagenicity of benzo(a)pyrene, 2-aminoanthracene, mitomycin C and cigarette smoke condensate (Whong *et al.*, 1989; Xing *et al.*, 1989).

Furoquinoline alkaloids. Furoquinoline alkaloids are characteristic secondary metabolites of the Rutaceae, in which they co-occur with the chemically-related furocoumarins. Dictamnine (Fig. 2.12) was isolated from the roots of *Dictamnus albus* and was detected in the herb of *Ruta graveolens*. Both plants have been used in the past for medical purposes.

Several reports have been published on the biological activity of dictamnine. The activities of the mono- and dimethoxy derivatives, e.g. γ-fagarine and skimmianine, were investigated to a lesser extent (Table 2.14). The mutagenicity of dictamnine after metabolic activation with rat liver microsomes was first reported by Mizuta and Kanamori (1985) using *Salmonella typhimurium* as the indicator organism. Paulini and co-workers (1987) and Häfele and Schimmer (1988) confirmed its activity in the Ames assay and investigated further naturally-occurring furoquinolines. Structure-mutagenicity relationships were tested with a series of 11 furoquinolines using base pair and frameshift indicator strains of *Salmonella typhimurium* (Paulini *et al.*, 1989). Klier and co-workers (1990) suggested that dictamnine is metabolized by the liver microsomes via an unstable 2,3-epoxide to an electrophilic intermediate which is covalently bound to DNA, analogous to the mechanism of aflatoxin B_1 (Fig. 2.12).

In *E. coli*, dictamnine was reported to be a direct-acting mutagen producing bacterial frameshift mutations in the dark (Ashwood-Smith *et al.*, 1982). Schimmer and Leimeister (1989), who studied the SCE-inducing potency of γ-fagarine in human lymphocytes *in vitro*, also showed a direct effect in this system, which could not be enhanced after addition of liver microsomes.

Besides exhibiting genotoxic activity in the dark, furoquinolines were also capable of inducing genetic damage in the light. When prokaryotic and eukaryotic cells were irradiated with long-wave ultraviolet (UV-A) in the presence of dictamnine, the furoquinoline was covalently bound to DNA. Monoadducts were formed with the DNA bases, particularly with the thymine moieties (Pfyffer and Towers, 1982a; Pfyffer *et al.*, 1982b). Monoadducts were formed *in vitro* as well as *in vivo*. The photobinding of dictamnine to DNA is thought to be the reason for its phototoxicity and photomutagenicity. The sites in the DNA for the photobinding of dictamnine are probably identical with those for monoadducts of furocoumarins (Pfyffer *et al.*, 1982b). Since furocoumarins react with their furan 2,3-double-bond, the furan ring appears to be the crucial reactive site for the induction of light-dependent mutations. If this were confirmed, it would mean that both mechanisms of mutagenic activation take place at the same site of the furoquinoline molecule (Fig. 2.12).

Photomutagenic effects were found in *E. coli* (Ashwood-Smith *et al.*, 1982; Fujita and Kakishima, 1989) and in the green alga, *Chlamydomonas reinhardtii* (Schimmer and Kühne, 1991). In the alga, dictamnine had the

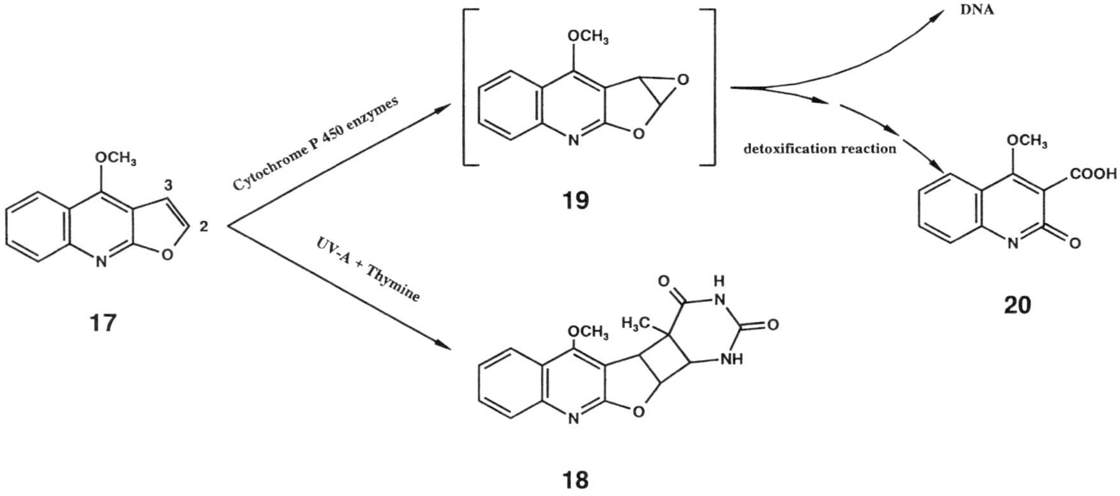

Figure 2.12 Dictamnine (**17**) and its enzymatic and UV-A light activation. Cyclobutane adduct with thymine (**18**), (*syn-cis*-configuration not shown), dictamnine oxide (**19**) and the non-active dictamnic acid (**20**).

Table 2.14 Summary of published mutagenicity data for dictamnine and gamma-fagarine

Compounds	Organism tested	Assay	Reference
Dictamnine	*Escherichia coli*	Reversion (WP 2 lac^-)	Ashwood-Smith *et al.*, 1982
	Salmonella typhimurium	Ames assay	Mizuta and Kanamori, 1985
			Paulini *et al.*, 1987
			Häfele and Schimmer, 1988
	Human cell cultures	SCE/chromosomal aberrations	Schimmer (unpublished)
Dictamnine plus UV-A	*E. coli*	Reversion	Ashwood-Smith *et al.*, 1982
	CHO cell cultures	SCE	Ashwood-Smith *et al.*, 1982
	CHO cell cultures	Chromosomal aberrations	Towers and Abramowski, 1983
	Chlamydomonas reinhardtii	Reversion (arg -1)	Schimmer and Kühne, 1991
Gamma-fagarine	*Salmonella typhimurium*	Ames assay	Mizuta and Kanamori, 1985
			Paulini *et al.*, 1987
	Human cell cultures	SCE	Schimmer and Leimeister, 1989

Abbreviations: SCE, sister chromatid exchange; UV-A, long wave ultraviolet; CHO, Chinese hamster ovary.

strongest effect among the furoquinolines tested. But the activity was lower than that of bergapten, the most active furocoumarin. This may be due to the lower photobinding to DNA (Pfyffer *et al.*, 1982b). In Chinese hamster ovary cells, chromosome aberrations were induced by dictamnine and skimmianine (Towers and Abramowski, 1983).

Recently, it was established with newly synthesized furoquinolines that, after intercalation and subsequent UV-A irradiation, only monoadducts with thymine with cis-syn configurations were formed; and in the lower energy conformation the furan ring was turned toward the minor groove of the polynucleotide, in such a way that photoreaction of this ring with thymine was favoured (Rodighiero *et al.*, 1996). No information was found in the literature concerning the carcinogenic potential of this group of alkaloids. However, it appears to be significant that dictamnine has ovicidal properties against spider mites (Tanaka *et al.*, 1985) and possibly participates in the antifertility effect of the roots of *Dictamnus albus* (Woo *et al.*, 1987).

Furocoumarins. Furocoumarins are present in many members of the family, Apiaceae, and are also accumulated in some representatives of the families, Rutaceae and Leguminosae. They are found in roots and are more concentrated in fruits and leaves, where they are usually stored in resin ducts as components of the essential oil.

The most outstanding property of furocoumarins is their great ability to sensitize cells to visible light, sunlight and, especially, near ultraviolet light. This results in strong toxicity, mutagenicity and possibly carcinogenicity. The mechanism of action is well known. After intercalation into the double helix of the DNA and molecular complexing, the light-activated furocoumarins react with the pyrimidine bases, especially with thymine. The reactive sites for the covalent photobinding are the 2′3′-furan double-bond and the 3,4-pyrone double-bond. This leads to the formation of monoadducts and cross-links. Both reactions contribute to mutagenicity. The linear furocoumarins, e.g. bergapten (5-methoxypsoralen) and xanthotoxin (8-methoxypsoralen) were more effective in mutation induction than the angular compounds or those linear furocoumarins, which have bulky side chains. Bergapten and xanthotoxin were capable of inducing mutations in bacteria, yeasts, algae and mammalian cells *in vitro*. Positive reactions were obtained in the SCE test, the cell transformation assay and in most other conventional test systems.

Without UV light, furocoumarins showed only weak effects in *E. coli* (Clarke and Wade, 1975) and in human lymphocytes *in vitro* (Abel *et al.*, 1985). Heraclenin, a linear furocoumarin with an epoxidic structure in its side chain, was a weak mutagen in bacteria (Ivie *et al.*, 1980) but a strong clastogen in human lymphocytes (Abel and Schimmer, 1986). The photobiology and the genetic basis of furocoumarin reactions were reviewed by Scott and co-workers (1976), Song and Tapley (1979), Averbeck (1989), and in an IARC report (1986).

Two recently published observations have received special attention. It has long been known that treatment with furocoumarins plus UV-A (PUVA) induces skin cancer in mice. It has now been shown that PUVA-induced mouse skin cancers display carcinogen-specific mutations in the p53 tumour suppressor gene (Nataraj *et al.*, 1996). Calsou and co-workers (1996) investigated the role of mono- and biadducts for mutagenicity. They found further evidence that the monoadducts persist for a longer time, whereas the cross-links are rapidly repaired via excision. Since furocoumarins are strong phototoxic compounds, their presence in a plant may be indicative and has been demonstrated to be a protective mechanism against phytopathogenic microorganisms and herbivores.

Flavonoids. The flavonoids constitute one of the largest groups of secondary plant metabolites universally distributed among vascular plants. Although some occur naturally in a sugar-free form, the majority exist in glycosidic form. The glycosides themselves are probably not mutagenic but they can be hydrolyzed enzymatically to liberate the aglycones, which may show mutagenicity. Whereas the flavone and the flavonol glycosides were inactive in short-term mutagenicity tests, several aglycones, e.g. quercetin, kaempferol and galangin, showed moderate mutagenic effects in bacteria and yeasts. Among the flavone group, wogonin and norwogonin showed mutagenicity comparable to that of quercetin. Their mutagenicity spectrum, however, differed from that of quercetin by a difference in mechanism of activation. A few compounds belonging to other flavonoid groups also showed weak activities in bacteria. The genotoxic effect of the various flavonoids in microorganisms with and without metabolic activation was reviewed by Brown (1980), Nagao and co-workers (1981) and Elliger and co-workers (1984).

The mutagenicity of quercetin has been established in many experiments using prokaryotic and eukaryotic organisms. In the Ames assay, quercetin was the most active flavonoid. Mutagenicity was shown without metabolic activation but was strongly increased when microsomes from rat liver were added. Quercetin possibly acts via intercalation into DNA, inducing frameshift mutations. However, using biophysical methods, Solimani (1996, 1997) suggested that a frameshift mutagenicity of quercetin is highly improbable.

After metabolic activation, quercetin may be enzymatically oxidized to quinoidic intermediates, which could be responsible for the stronger effect observed in the experiment with rat liver enzymes. However, the possible role of such oxidation products for mutation induction is not yet clear.

From further results, it has been suggested that quercetin might exert DNA damage via more than one mechanism (Rueff *et al.*, 1986).

Quercetin induced chromosomal aberrations, SCEs and micronuclei in mammalian cells *in vitro* without using metabolic activation systems. Quercetin was also capable of inducing single DNA strand breaks and of transforming hamster embryo cells. Micronuclei and polyploidy were induced in human lymphocytes (Popp and Schimmer, 1991). It seems quite clear that quercetin is mutagenic *in vitro*. *In vivo*, quercetin was inactive in most assays. The published mutagenicity data from *in vivo* and *in vitro* experiments have been summarized by Müller and co-workers (1991). In further papers, the divergent reactivity of quercetin *in vitro* and *in vivo* was confirmed, even when the same model organism or cell line was used (Caria *et al.*, 1995). Müller and co-workers (1992) explained the lack of genotoxicity of quercetin *in vivo* with detoxification reactions or low absorption capacity. Two papers recently published reported studies in which the role of cytochrome P_{450} enzymes in the bioactivation of kaempferol and galangin and its relevance to genotoxicity were investigated (Silva *et al.*, 1997a,b).

Most carcinogenicity tests with experimental animals have produced negative results (Natori and Ueno, 1987; Ito *et al.*, 1989). The lack of mutagenicity effects *in vivo* correlates with the negative carcinogenicity tests and is an indication that the metabolic competence of the organisms may play an important role. Flavonoids are scarcely absorbed from the intestinal tract of mammals. Furthermore, the microflora of the large intestine is capable of decomposing flavonols very rapidly. However, it must be pointed out that quercetin can occasionally act as enhancer of the activity of mutagens/carcinogens (Ogawa *et al.*, 1986).

Anthranoids. Anthranoids are secondary plant metabolites with a basic anthrone or anthraquinone structure. In the intact plant, they normally occur as glycosides. The 1,8-dihydroxy derivatives are of primary ecological and pharmaceutical interest owing to their laxative properties. They occur in many members of the families, Liliaceae, Polygonaceae, Rhamnaceae and Caesalpinioideae. A second type of anthranoid without laxative activity is present in the family Rubiaceae. Only free anthraquinones, e.g. luteoskyrin (Table 2.12) were synthesized in *Penicillium* and *Aspergillus* species. Free aglycones are formed from the glycosides in higher plants by the action of enzymes during drying and storage. The genotoxic data of 80 phenolic anthraquinones were summarized in an early review by Brown (1980). In the *Salmonella typhimurium* assay, lucidin showed the strongest mutagenicity; aloe-emodin, emodin and physcion were less active and needed metabolic activation. A frameshift mechanism was postulated from these results and it was suggested that a nonspecific free radical mechanism may be involved in the DNA damage elicited by these agents. A great deal of information on *in vitro* genotoxicity of anthranoids in bacterial and mammalian systems was presented in a review by Westendorf (1993). Positive results were obtained with aloe-emodin, emodin and with the rubiadin-type anthraquinones, lucidin, purpurin and rubiadin, in the Ames assay. Various anthraquinones induced gene mutations in Chinese hamster V79 cells, in the DNA repair test and in the transformation assay with mouse fibroblasts, but controversial results in V79 cells were also reported (Bruggeman and van der Hoeven, 1984; Heidemann *et al.*, 1996). Heidemann and co-workers discussed the genotoxicity data of aloe-emodin in detail and referred to the negative *in vivo* results. However, *in vitro*, the DNA-damaging activity of several anthraquinones was again confirmed in the micronucleus assay and the newly developed comet assay (Müller *et al.*, 1996). From these results, it was concluded that the genotoxicity of aloe-emodin is caused by interaction with the topoisomerase II activity.

No evidence was found that emodin or emodin metabolites were covalently bound to rat liver DNA or *Salmonella* DNA (Bösch *et al.*, 1987). 2-Hydroxyemodin formed *in vitro* with hepatic microsomes was the most active mutagen in *Salmonella* (Masuda and Ueno, 1984). The metabolic activation of emodin was shown to proceed through the cytochrome P_{450} system (Tanaka *et al.*, 1987). Emodin and chrysophanol were not capable of producing oxidative damage in DNA, in comparison to luteoskyrin, which enhanced the number of 8-hydroxyguanosine sites in the DNA of hepatoma cells (Akuzawa *et al.*, 1992). In contrast, the clastogenicity of the anthraquinones was not mediated by cytochrome P_{450} enzymes (Simi *et al.*, 1995). It was, therefore, concluded that anthraquinones are active as intercalative agents due to their planar structure and do not need to be metabolized in mammalian cells *in vitro*.

Lucidin, however, was shown to exhibit its genotoxicity through a different mechanism. As indicated from results obtained with the ^{32}P-postlabelling method, lucidin forms adducts with DNA (Poginsky *et al.*, 1991). After feeding mice and rats with lucidin, DNA adducts were observed in the intestinal tissue, liver and kidney.

The carcinogenic effects of 1,8-dihydroxyanthraquinone and 1-hydroxyanthraquinone, which was identified as a metabolite of alizarin primeveroside (Blömeke *et al.*, 1992), were demonstrated after rats and mice were fed with these compounds over a longer period of time (Mori *et al.*, 1985, 1986, 1990). The carcinogenic risk of anthranoids for humans is the subject of controversy. It should be mentioned, however, that hydroxyanthraquinones and anthrones have tumour-promoting properties *in vitro* (di Giovanni *et al.*, 1985; Wölfle *et al.*, 1991) and lead to a considerable increase in the proliferation of intestinal cells *in vivo*, after oral administration (Kleibeuker *et al.*, 1995). It should also be noted that 1-hydroxyanthraquinone acts synergistically on methylazoxymethanol-induced carcinogenesis in rats (Mori *et al.*, 1991).

Phenylpropanoids. Among the genotoxic phenylpropanoids, safrole and the chemically-related estragole and β-asarone, are the most important DNA-damaging agents. Safrole (Fig. 2.13) and estragole are

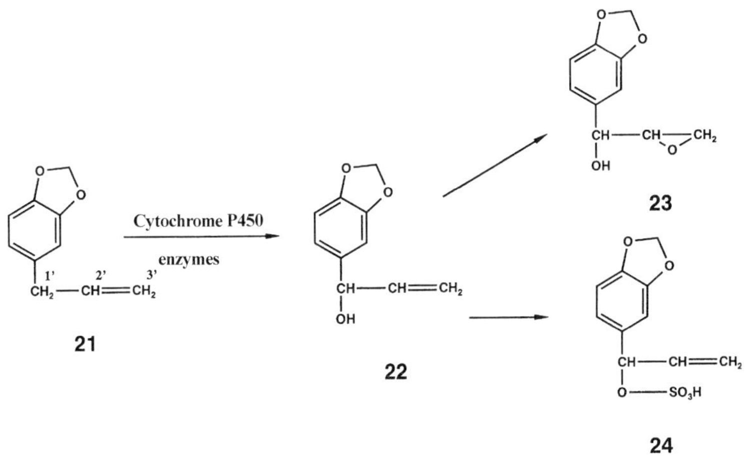

Figure 2.13 Metabolic activation of safrole (**21**). Metabolites with carcinogenic properties are 1'-hydroxy-safrole (**22**), 1'-hydroxy-2',3'-safrole oxide (**23**) and 1'-sulfoxy safrole (**24**).

characterized by an allylic side chain, in contrast to β-asarone, which bears a propenylic substituent. The allylic structure is apparently important for the metabolic activation and the formation of ultimate mutagens. Safrole is the main component in the essential oil of *Sassafras officinale* and occurs in small amounts in some species of the genera *Illicium*, *Asarum* and *Cinnamomum*. Estragole is a component of the essential oil of *Artemisia dracunculus* and β-asarone is present in the oil of *Acorus calamus*.

In the standard *Salmonella typhimurium* test, safrole and estragole were inactive with or without added rat liver preparations. The known or possible metabolites 1′-hydroxysafrole, 1′-hydroxyestragole, the 2′, 3′-oxides and the esters were mutagenic (Enomoto, 1987) (Fig. 2.13). Beta-asarone showed dose-dependent mutagenicity only with rat liver enzymes (Göggelmann and Schimmer, 1983), and was also active in inducing chromosomal aberrations and SCE in human lymphocytes *in vitro*

(Abel, 1987). Safrole and estragole showed low clastogenic effects *in vitro* and *in vivo*. In Chinese hamster cells, a slight increase in SCE was observed by safrole. The effect was enhanced with rat liver preparations (Tayama, 1996). Recently, the cytogenetic effect and DNA adduct formation induced by safrole in Chinese hamster lung cells were analyzed (Daimon *et al.*, 1997). Schiestl and co-workers (1989) reported that safrole induced intrachromosomal recombination in yeasts in a dose-related response.

Safrole and estragole were found to produce liver tumours in rats. The 1′-hydroxy derivatives, the 2′,3′-oxides and the 1′-acetoxy derivatives were also tumorigenic. *In vivo*, the sulfoxy derivative may be the ultimate carcinogen. Sulfotransferase catalyzes the formation of the sulfuric ester of 1′-hydroxysafrole. The enzyme was detected in rat liver cytosol. Inhibition of the enzyme represses hepatic tumour formation.

DNA adduct formation and DNA binding properties have been extensively investigated. Safrole and estragole exhibited strong binding to mouse liver DNA. Several DNA adducts were isolated and confirmed by nuclear magnetic resonance (NMR) and circular dichroism (CD) spectroscopy. The adducts persist for a long time. However, the possible carcinogenic relevance of the long-term presence of such unrepaired adducts is unknown. They have been detected in target as well as in non-target tissues. Abundant information on metabolic activation, DNA adduct formation and tumour incidence in experimental animals is available from reviews by Groopman and Cain (1990) and Enomoto (1987).

Beta-asarone was also active in inducing liver tumours, but the hepato-carcinogenicity was not inhibited by sulfotransferase inhibitors. Nothing is known at present of DNA binding and DNA adduct formation. Although the carcinogenic potential of safrole, estragole and β-asarone does not appear to be high, since their effects in animals are relatively weak, we need more information to calculate the health risk of these compounds for humans.

2.3.2.4 Other endogenous mutagens in plants

Numerous plant metabolites have been assayed for mutagenicity in only a few *in vitro* test systems. Thus, their genotoxic potential cannot be critically assessed at present.

Xanthones. Free and glycosidic xanthone derivatives have been identified as secondary metabolites of several members of the Gentianaceae (genera *Gentiana, Swertia*). Some of them, listed in Table 2.12, showed mutagenic effects after metabolic activation in certain *Salmonella typhimurium* strains. The corresponding glycosides were inactive when β-glucosidase was absent in the system (Morimoto *et al.*, 1983; Kanamori *et al.*, 1984; Matsushima *et al.*, 1985).

Hymenoxon. The sesquiterpene lactone, hymenoxon (isolated from *Hymenoxis odorata*, Asteraceae), was found to be a direct-acting mutagen in the Salmonella Ames assay but did not produce lethal DNA damage in recombination repair deficient *E. coli* and *Bacillus subtilis* strains (Jones and Kim, 1981). Alkylation of deoxyguanosine was reported (Sylvia *et al.*, 1985). Evidence was obtained that hymenoxon is capable of cross-linking the DNA strands (Sylvia *et al.*, 1987).

Cathinone. The mutagenicity of (−)cathinone has been evaluated in the germ cells of male albino mice using the dominant lethal test. Reduction in fertility and dose-dependent postimplantation loss but no dominant lethality were observed (Qureshi *et al.*, 1988). However, the alkaloid produced chromosomal aberrations in rats (de Hondt *et al.*, 1984) and in somatic cells of mice (Tariq *et al.*, 1987). The mechanism of action is unknown at present.

Sanguinarine. Sanguinarine is the main alkaloid in the seeds of *Argemone mexicana* and *Sanguinaria canadensis* but is also present in other members of the Papaveraceae. Chromosomal breaks in plant cells induced by the seed oil were attributed to the presence of sanguinarine (Subramanyam *et al.*, 1974).

Nandi and Maiti (1985) showed that sanguinarine has a high specificity to bind on GC-rich DNA *in vitro*. They concluded that the alkaloid preferentially binds to the GC pairs in the DNA template. Further data confirmed the intercalation model for sanguinarine (Faddejeva *et al.*, 1984; Schmeller *et al.*, 1997b). Single doses of 10 mg/kg increased the activity of certain liver enzymes in rats and caused a significant loss of microsomal cytochrome P_{450}. It was, therefore, suggested that sanguinarine was a potential hepatotoxic compound (Dalvi, 1985) but its role as a potential mutagen remains to be clarified. Farnsworth and co-workers (1976) reported that sanguinarine was carcinogenic in rats, guinea-pigs and hamsters but this result has not yet been confirmed.

Ellipticine. Ellipticine is an indole alkaloid with intercalating properties. It induced mutations in *Salmonella typhimurium*, especially in the frameshift indicator strains (Ashby *et al.*, 1980), mitochondrial mutations in *Saccharomyces cerevisiae* (Pinto *et al.*, 1982) and gene mutations in mouse lymphoma cells (Moore *et al.*, 1987). In addition, it showed SCE-inducing capacity and chromosome-breaking activity in human lymphocytes *in vitro* and clastogenic potency on bone marrow cells of Wistar rats *in vivo* (Sakamoto-Hojo *et al.*, 1988). The authors assumed that the chromosomal damage and the SCE formations may have been the result of interaction with topoisomerase enzymes. Additional information is available from Anderson and Berger (1994).

Acridone alkaloids. Only a few reports were found in the literature concerning the evaluation of genotoxicity of acridone alkaloids. Rutacridone epoxide was characterized as a direct-acting mutagen with high activity in *Salmonella typhimurium*. Rutacridone was less active in this assay and needed metabolic activation (Paulini and Schimmer, 1989). Isogravacridonchlorine was shown to induce frameshift mutations via a reactive intercalation mechanism (Paulini *et al.*, 1991). The occurrence of acridone alkaloids and their biological properties were described in a review by Gröger (1988).

Hydroquinone. Hydroquinone was detected in higher amounts in the basidiomycete, *Agaricus hondensis*, where it was assumed to be responsible for the toxicity of this species (Jovel *et al.*, 1996). In higher plants, it occurs in glycosidic forms, e.g. as arbutoside, which is a characteristic metabolite of the Ericaceae family.

Endogenous glycosidases or bacterial enzymes produced by the intestinal microflora are capable of hydrolyzing the glycosides and of releasing free hydroquinone. The first report concerning the genotoxicity of hydroquinone was by Boyland and co-workers (1964). The authors described the appearance of carcinomas after implantation of the compound into the urinary bladder of mice. Walles (1990) studied the effect of hydroquinone in primary hepatocyte cultures. DNA single-strand breaks were detected with the alkaline elution technique. Evidence was found that OH radicals were involved in the DNA-damaging effect of hydroquinone. Marrazzini and co-workers (1991) tested the genotoxicity of hydroquinone in bone marrow cells of mice. Micronucleated erythrocytes, structural chromosomal aberrations and aneuploidy were induced in this system. This indicates that hydroquinone may represent a risk both to somatic cells (carcinogenic) and/or germ cells (aneuploidy). Recently, Tsutsui and co-workers (1997) studied the genotoxic activity of hydroquinone in Syrian hamster embryo cells, using methods with different genetic endpoints.

Isothiocyanates and related compounds. Isothiocyanates are characteristic secondary metabolites of Brassicaceae, Tropaeolaceae and other members of the order Capparales (Wink and Waterman, 1999). In intact plants, they occur in their *S*-glycosidic form, as glucosinolates. Hydrolysis by the endogenous thioglucosidases leads to an unstable aglycone, which is rapidly converted into isothiocyanate and several byproducts, e.g. thiocyanate, and nitrile. Allylisothiocyanate is formed in this way from sinigrin. Allylisothiocyanate was one of the first natural mutagens to be studied (Auerbach and Robson, 1944). The compound is highly toxic to bacteria and fungi. After local application, it causes severe skin irritation in humans.

Allylisothiocyanate showed weak genotoxic effects in *Drosophila melanogaster* and fungi. Chromosomal damage was induced in onion root tip cell (reviewed by Clark, 1982). Neudecker and Henschler (1985) found that allylisothiocyanate was an indirect-acting mutagen in *Salmonella typhimurium*. The effect could be reduced when higher amounts of mammalian liver enzymes were present. Yamaguchi (1980) tested various isothiocyanates, thiourea compounds and the glycoside sinigrin under preincubation conditions. Most substances showed positive effects in the absence of any exogenous metabolic activation. In studies by Neudecker and Henschler (1985) and by Yamaguchi (1980), long preincubation times were required to express mutagenicity in *Salmonella typhimurium*.

The clastogenicity of allylisothiocyanate was investigated in Chinese hamster cells by Kamasaki and co-workers (1982) and later by Musk and co-workers (1995). Whereas, Kamasaki and co-workers (1982) observed chromosomal aberrations in a B241 cell line, Musk and co-workers (1995) found that the compound was inactive in Chinese hamster ovary cells. Phenethylisothiocyanate and – at higher doses – sinigrin and gluconasturtiin, the parent compounds of the isothiocyanates tested, were active in the chromosomal aberration test and in the SCE test.

Carcinogenicity experiments with allylisothiocyanate were performed by Dunnick and co-workers (1982). After long-term administration of the substance, transitional cell papillomas and epithelial hyperplasia were induced in the urinary blatter of male rats. An important property of glucosinolates and several of their decomposition products is the ability to modulate chemically-induced carcinogenesis (Morse *et al.*, 1988). There is some evidence that administration of these compounds during the initiation period of carcinogenesis may cause the inhibition of tumour production; administration during the promotion, however, may enhance the carcinogenic effect (McDanell *et al.*, 1988).

2.3.3 Outlook

There can be little doubt that further screening of secondary plant metabolites will bring to light many more examples of endogenous mutagens. Considering the numerous mutagens on which insufficient preliminary studies were performed, it will be necessary to fill the gaps in our knowledge first before being able to assess their genotoxic potential. This will include study of genetic and biochemical aspects as well as tissue specificities and deactivation mechanisms. We can expect that the role of intercalation in the process of mutagenicity/carcinogenicity will become clearer. The consequences of frameshift mutations for the eukaryotic cell, in particular, have to be elucidated. This would enable us to evaluate the health risk of intercalators, such as sanguinarine and berberine.

Fruitful future research might be the study of the interaction between mutagens co-occurring in the same plant or mutagens with different mechanisms of action. Under certain conditions, the first mutagen can act as an antimutagen, inhibiting the metabolic activation of the second mutagen or scavenging its mutagenic intermediate. Such interactions have already been demonstrated (Schimmer *et al.*, 1991).

2.4 Interference with molecular targets other than those related to neurotransmission or DNA

2.4.1 Cytoskeleton

Many cellular activities, such as motility, endo- and exocytosis and cell division, are mediated through elements of the cytoskeleton, including microfilaments and microtubules (for an overview see Alberts *et al.*, 1994; Lodish *et al.*, 1995). In plants and fungi, a number of allelochemicals have been identified that can interfere with tubulin and microtubules (e.g., colchicine, *Vinca* alkaloids, maytansine, maytansinine, taxol, and the lignan podophyllotoxin).

Cell stability, phagocytosis, cell-cell interactions and cell movements are also controlled by actin filaments, which are rapidly assembled or disassembled from actin monomers. Cytochalasin B, an alkaloid produced by a number of moulds, binds to the plus end of a growing actin filament, preventing the addition of actin monomers at that point. Latrunculin B from *Latrunculia magnifica* (a marine organism) is 10 to 100-fold more potent than cytochalasins in the inhibition of microfilament organisation. Phalloidin, produced by the fatally poisonous toadstool, *Amanita phalloides*, stabilizes actin filaments and inhibits their depolymerisation. Any allelochemical that impairs the function of microtubules or microfilaments is likely to be toxic and, from the point of view of defence, a well-designed molecule (reviews in Wink, 1993a, 1998).

2.4.2 Protein biosynthesis

Protein biosynthesis is essential for all cells and, thus, provides another important target. Indeed, a number of allelochemicals have been detected (although not many have been studied in this context) that inhibit protein biosynthesis *in vitro*. Emetine, from *Cephaelis ipecacuanha* (Rubiaceae), is the most potent plant constituent. Other alkaloids with the same ability include harringtonine, homoharringtonine, cryptopleurine, tubulosine, hemanthamine, lycorine, narciclasine, pretazettine, pseudolycorine, tylocrepine and tylopherine (Wink, 1993a). A weaker inhibition was observed for ajmaline, berberine, boldine, cinchonine, cinchonidine, harmaline, 'harmine', lobeline, norharman, papaverine, quinidine, quinine, salsoline, sanguinarine, solanine and yohimbine (Wink and Latz-Brüning, 1995; Wink *et al.*, 1998a,b). Tannins inhibit protein biosynthesis effectively *in vitro*; if they are resorbed, activity is also likely *in vivo*.

Most compounds that substantially affect DNA, DNA polymerase I and reverse transcriptase (RT) are also active as translation inhibitors; a significant ($p < 0.01$) correlation can be established between the degree of intercalation and inhibition of translation (Wink *et al.*, 1998a,b). Interaction of these alkaloids with ribosomal nucleic acids, for example ribosomal ribonucleic acid (rRNA), transfer ribonucleic acid (tRNA) or mRNA, is likely, in addition to interactions with ribosomal proteins (Wink and Twardowski, 1992). On the other hand, most compounds (such as aconitine, caffeine, colchicine, cytisine, gramine, hyoscyamine, lupanine, narcotine, scopolamine, sparteine or strychnine) that do not intercalate do not have a substantial influence on translation (Wink *et al.*, 1998a,b). Inhibition of protein biosynthesis shows a significant correlation with inhibition of bacterial and radicle growth in *Lepidium sativum* and insect and vertebrate toxicity, indicating the importance of this molecular target (Wink *et al.*, 1998a,b).

Quinolizidine alkaloids, such as sparteine, lupanine and cytisine, are relatively weak inhibitors of this target (they strongly affect ACh receptors and Na^+-channels) (see Tables 2.3–2.14). The stages that are inhibited are the loading of aminoacyl-tRNA with amino acids and the elongation step. The inhibitory activity was visible in heterologous systems but protein biosynthesis in the producing plants (in this case lupins) was not affected (Wink and Twardowski, 1992).

A number of antibiotics (from *Streptomyces* and other bacteria or fungi) inhibit protein biosynthesis at specific steps, such as initiation, peptidyl transferase or elongation. Depending on their affinity for prokaryotic or eukaryotic ribosomes, some of the antibiotics selectively inhibit microbial systems. Since mitochondria also contain ribosomes of prokaryotic origin, effects can also occur in animals. These compounds have apparently evolved as chemical defence compounds of fungi or bacteria against other microorganisms.

2.4.3 *Electron chains*

The respiratory chain and ATP-synthesis in mitochondria or photophosphorylation in chloroplasts demand the controlled flux of electrons and protons. These targets seem to be attacked by sanguinarine, ellipticine, gramine, alpinigenine, capsaicine, hydrocyanic acid, rotenone, nicotine and a few other allelochemicals (Wink, 1993a).

2.4.4 *Interaction with multiple targets*

In general, the interactions of a particular allelochemical with a molecular target (as described in section 2.2 and 2.3) suggest a high degree of specificity. A closer look, however, shows that many substances interfere with more than one target (compare Tables 2.1–2.12). Since the data included in these tables were often produced in studies that did not aim to assess all of the activities of an isolated compound (and also the tables do not indicate when no interaction was recognized), the picture could be misleading or incomplete. In order to have a more solid basis for discussion, the results of a comparative study performed on more than 70 alkaloids (representing most structural types) will be discussed (Wink *et al.*, 1998a,b).

For this purpose, we have determined whether an alkaloid can displace a specifically-bound ligand from a neuroreceptor, such as α_1 and α_2 adrenergic receptors, serotonin (5-HT_2) receptor, and nicotinic and muscarinic acetylcholine receptors (Table 2.15) obtained from porcine brains (Schmeller *et al.*, 1994, 1995, 1997a,b). In addition, we have determined whether the same alkaloids inhibit acetylcholine esterase, whether they intercalate DNA, inhibit DNA polymerase I, RT, protein biosynthesis and membrane stability (Table 2.16) (Wink and Latz-Brüning, 1995; Wink *et al.*, 1998a). The results are summarized below.

Most alkaloids displace radioligands at more than one receptor (Table 2.15); for a number of known alkaloids, specific interactions were discovered that (to our knowledge) had not been reported in the literature. Affinities to a particular receptor are often significantly correlated with effects at other neuroreceptors. For example, alkaloids that affect α_1 receptors also bind to α_2 and serotonin receptors, and those that affect α_2 are active at serotonin receptors. Activities between muscarinic and nicotinic acetylcholine receptors and between serotonin and adrenergic receptors are much weaker or lack a clear correlation.

For a few alkaloid groups, the hybrid character is well known and well documented, for example ergot and quinolizidine alkaloids. Ergot alkaloids, such as ergotamine, ergometrine or ergoclavine, are produced by fungi of the genus *Claviceps* and other endophytic fungi, which live in

Table 2.15 Interaction of alkaloids with neuroreceptors and acetylcholine-related enzymes (after Wink *et al*., 1998a,b)

	Adrenergic receptors		Serotonin receptor	Acetylcholine receptors				
	α_1	α_2	(5-HT_2)	mAChR	nAChR	AChE	BChE	ChAT
Imidazole alkaloids								
Pilocarpine	n.a.	n.a.	n.a.	11.0	1656	n.a.	n.a.	n.a.
Indole alkaloids								
Brucine	4.7	347.3	n.a.	51.3	13.6	n.a.	n.a.	n.a.
Ergometrine	4.4	0.9	1.5	2.0	178.2	n.a.	161.3	n.a.
Gramine	26.3	8.7	8.5	677.1	30.7	1019	232.3	n.a.
Harmaline	34.0	7.5	14.6	33.5	n.a.	173.2	90.4	n.a.
Harmalol	56.8	16.9	30.6	60.4	623.7	93.1	37.1	n.a.
Harman	n.d.	n.d.	n.d.	n.d.	n.d.	n.a.	72.0	n.d.
Harmine	n.d.	n.d.	n.d.	n.d.	n.d.	1005	175.3	n.d.
Harmol	n.d.	n.d.	n.d.	n.d.	n.d.	n.a.	26.1	n.d.
Norharman	n.d.	n.d.	n.d.	n.d.	n.d.	n.a.	63.5	n.d.
Physostigmine	n.a.	73.6	394.2	66.2	1992	0.03	16.2	n.a.
Strychnine	25.1	172.3	51.6	32.8	10.2	n.a.	130.8	n.a.
Isoquinoline alkaloids								
Berberine	3.2	0.48	1.9	1.0	35.5	167.4	55.8	n.a.
Boldine	0.53	0.09	0.67	118.1	11.1	n.a.	n.a.	n.a.
Emetine	6.2	0.07	7.6	58.2	n.a.	n.a.	n.a.	n.a.
Laudanosine	18.4	0.82	8.2	67.1	1313	n.a.	415.7	n.a.
Morphine	n.a.	n.a.	n.a.	n.a.	n.a.	n.a.	n.a.	n.a.
Palmatine	5.8	0.96	2.9	4.1	> 100	124.5	425.6	> 100
Salsoline	115.1	9.8	146.2	n.a.	n.a.	n.a.	n.a.	n.a.
Sanguinarine	33.6	6.4	91.7	2.4	11.8	10.9	17.4	0.3
Phenylalkylamines								
L-Ephedrine	186.4	14.6	60.0	2649	n.a.	n.a.	n.a.	n.a.
Piperidine/pyridine alkaloids								
Ammodendrine	257.5	11.6	109.0	523.6	9.1	n.a.	275.4	n.a.
Anabasine	2374	47.6	n.a.	n.a.	0.58	n.a.	n.a.	n.a.
Arecoline	n.a.	29.7	n.a.	32.1	5.7	n.a.	n.a.	n.a.
Coniine	n.a.	260.0	492.7	2071	19.0	n.a.	327.5	n.a.

Table 2.15 (Continued)

	Adrenergic receptors α_1	α_2	Serotonin receptor ($5\text{-}HT_2$)	Acetylcholine receptors mAChR	nAChR	AChE	BChE	ChAT
Cotinine	n.a.	n.a.	n.a.	1776	2.5	n.a.	n.a.	n.a.
Nicotine	n.a.	n.a.	n.a.	882.8	0.008	n.a.	n.a.	n.a.
Pseudopelletierine	n.a.	n.a.	n.a.	386.1	0.7	n.a.	n.a.	n.a.
Purine alkaloids								
Caffeine	n.a.	n.a.	n.a.	464.8	n.a.	n.a.	n.a.	n.a.
Pyrrolizidine alkaloids								
Acetylheliosupine	39.1	2.9	23.2	71.3	159.7	n.a.	n.a.	n.a.
Echihumiline	n.a.	358.8	549.0	89.2	n.a.	n.a.	314.4	n.a.
Echihumuline *N*-oxide	n.a.	> 50	182.0	8.7	n.a.	n.a.	n.a.	n.a.
Echimidine	n.a.	900	257.6	512.5	n.a.	n.a.	n.a.	n.a.
Heliosupine	148.1	15.0	77.1	392.0	n.a.	n.a.	n.a.	n.a.
Heliosupine *N*-oxide	n.a.	n.a.	n.a.	350.0	n.a.	n.a.	n.a.	n.a.
Heliotrine	n.a.	n.a.	535.4	52.2	n.a.	n.a.	n.a.	n.a.
Monocrotaline	n.a.	n.a.	203.4	n.a.	n.a.	n.a.	n.a.	n.a.
Pycnanthine	n.a.	n.a.	407.6	177.2	n.a.	n.a.	462.6	n.a.
Retronecine	n.a.	n.a.	n.a.	127.9	n.a.	n.a.	n.a.	n.a.
Riddelline	n.a.	n.a.	n.a.	208.7	n.a.	n.a.	n.a.	n.a.
Senecionine	n.a.	n.a.	249.4	43.0	n.a.	n.a.	n.a.	n.a.
Seneciphylline	n.a.	341.4	608.6	52.6	n.a.	n.a.	n.a.	n.a.
Quinoline alkaloids								
Cinchonine	5.2	1.4	5.5	19.2	n.a.	n.a.	22.9	n.a.
Cinchonidine	1.1	1.3	10.7	19.7	n.a.	n.a.	n.a.	n.a.
Quinine	5.7	2.5	6.4	4.5	n.a.	n.a.	109.0	n.a.
Quinidine	29.7	1.3	14.4	18.4	n.a.	n.a.	n.a.	n.a.
Quinolizidine alkaloids								
Albine	n.a.	n.a.	n.a.	32.9	237.7	n.a.	n.a.	n.a.
Anagyrine	n.a.	n.a.	n.a.	132.1	2096	n.a.	n.a.	n.a.
Angustifoline	n.a.	n.a.	n.a.	25.3	n.a.	n.a.	n.a.	n.a.

Annotinine	n.a.	n.a.	n.a.	260.5	n.a.	n.a.	n.a.	n.a.
Cytisine	n.a.	n.a.	n.a.	398.2	0.14	n.a.	n.a.	n.a.
13-hydroxylupanine	n.a.	n.a.	n.a.	139.7	467.2	n.a.	n.a.	n.a.
13-tigloyloxylupanine	n.a.	n.a.	508	11.1	99.8	n.a.	648.3	n.a.
Lupanine	n.a.	n.a.	n.a.	118.0	5.3	n.a.	n.a.	n.a.
Lupinine	n.a.	207.8	n.a.	189.9	n.a.	n.a.	n.a.	n.a.
N-methylcytisine	n.a.	n.a.	n.a.	416.7	0.05	n.a.	n.a.	n.a.
17-oxosparteine	n.a.	180.7	n.a.	117.9	155.0	n.a.	n.a.	n.a.
3β-hydroxylupanine	n.a.	n.a.	n.a.	74.1	192.4	n.a.	n.a.	n.a.
Multiflorine	n.a.	n.a.	n.a.	49.4	n.a.	n.a.	n.a.	n.a.
Sparteine	n.a.	127.7	n.a.	21.3	330.8	n.a.	165.5	n.a.
Tetrahydrorhombifoline	n.a.	n.a.	n.a.	128.8	347.6	n.a.	n.a.	n.a.
Terpene alkaloids								
Aconitine	n.a.	331.6	n.a.	1.3	n.a.	n.a.	n.a.	n.a.
Tropane alkaloids								
Atropine	6.1	10.1	6.0	0.005	284.4	n.a.	n.a.	n.a.
Cocaine	n.a.	506.7	317.3	56.7	371.4	n.a.	274.3	n.a.
6β-hydroxyhyoscyamine	12.6	42.0	n.d.	0.039	n.a.	n.a.	n.a.	n.a.
7β-hydroxyhyoscyamine	n.d.	37.6	n.d.	0.008	n.a.	n.a.	n.a.	n.a.
Littorine	66.3	72.3	68.9	0.003	909.8	n.a.	n.a.	n.a.
Noratropine	16.4	30.4	22.7	0.2	494.4	n.a.	n.a.	n.a.
Scopolamine	113.0	359.7	168.0	0.002	928.4	n.a.	n.a.	n.a.
Tropine	n.a.	n.a.	n.a.	2631	n.a.	n.a.	n.a.	n.a.
Tropolone alkaloids								
Colchicine	n.a.	23.8	133.2	347.3	30.0	n.a.	n.a.	n.a.

The data presented are the concentrations (in μM) which replace 50% of the specifically bound ligand or which inhibit the enzymes by 50% (IC_{50}). Abbreviations: 5-HT_2, serotonin receptor, mAChR, muscarinic acetylcholine receptor; nAChR, nicotinic acetylcholine receptor; AChE, acetylcholine esterase; BChE, butylcholine esterase; ChAT, choline acetyltransferase; n.a., not active at 500 μM; n.d., not determined.

Table 2.16 Interaction of alkaloids with basic molecular targets (after Wink and Latz-Brüning, 1995; Wink *et al.*, 1998a,b)

	DNA melting temperature increase[1] °C	DNA methylgreen release IC_{50}[2]	DNA DNA Pol I inhibition IC_{50}[3]	RNA RT inhibition IC_{50}[3]	Protein biosynthesis inhibition %[4]	Membrane permeability haemolysis %[5]
Imidazole alkaloids						
Pilocarpine	0	n.a.	n.a.	n.a.	n.a.	n.a.
Indole alkaloids						
Ajmalicine	1.0*	> 1 mM	n.a. (0.5 mM)	n.d.	n.d.	n.a.
Ajmaline	2.3*	> 5 mM	> 10 mM	7 mM	46 (4 mM)	n.a. (1 mM)
Brucine	n.d.	> 5 mM	n.a.	5 mM	36	n.a.
Ergometrine	n.d.	> 5 mM	n.a.	> 10 mM	n.a.	n.a.
Ergotamine	13.7*	n.d.	n.d.	n.d.	20 (0.25 mM)	n.a.
Gramine	n.d.	n.a.	n.a.	> 10 mM	n.a.	n.a.
Harmaline	8.6*	0.3 mM	3.2 mM	2.4 mM	70	n.a.
Harmine	16.1*	0.4 mM	0.9 mM	0.5 mM	95	2.3
Norharman	6.2*	< 1 mM	8 mM	< 0.2 mM	50 (0.2 mM)	3.3
Physostigmine	n.d.	n.a.	n.a.	> 10 mM	15	n.a.
Strychnine	n.d.	n.a.	n.a.	7 mM	20	n.a.
Vincamine	n.d.	n.d.	n.a.	n.a.	54	2 (1 mM)
Yohimbine	n.d.	n.a.	n.a.	> 10 mM	62	n.a.
Isoquinoline alkaloids						
Berbamine	13.2*	0.5 mM	0.7 mM	< 0.25 mM	n.d.	2.3 (2 mM)
Berberine	15*	0.1 mM	0.4 mM	0.2 mM	100	n.a.
Boldine	6*	< 0.7 mM	5 mM	< 1.5 mM	30	n.a.
Canadine	1.8*	n.d.	n.d.	n.d.	15 (0.1 mM)	1.5 (0.2 mM)
Chelidonine	1.5*	(> 2 mM)	(> 2 mM)	(> 2 mM)	28 (0.14 mM)	n.a.
Emetine	7.4*	0.7 mM	5 mM	< 0.5 mM	100 (0.01 mM)	n.a.
Laudanosine	n.d.	n.a.	n.a.	4 mM	15	n.a.
Narcotine	0	n.a.	1 mM	n.a. (0.5 mM)	n.a. (0.1 mM)	4.5 (0.5 mM)
Papaverine	0	n.a.	n.a.	n.d.	55 (0.8 mM)	1 (0.5 mM)
Salsoline	n.d.	n.a.	n.a.	n.a.	30	n.a.
Sanguinarine	24*	0.02 mM	< 0.02 mM	0.03 mM	n.d.	5.2 (0.1 mM)

Phenylalkylamines						
L-ephedrine	n.d.	n.a.	n.a.	n.a.	n.a.	n.a.
Piperidine/pyridine alkaloids						
Anabasine	n.d.	n.a.	n.a.	n.a.	n.a.	n.a.
Arecoline	n.d.	n.a.	n.a.	> 10 mM	n.a.	n.a.
Cycloheximide	n.d.	n.a.	n.a.	n.a.	90 (0.01 mM)	n.a.
Lobeline	1.5*	> 5 mM	> 10 mM	< 2 mM	45	n.a.
Nicotine	n.d.	n.a.	n.a.	n.a.	20	n.a.
Piperine	n.d.	n.a.	n.a.	n.d.	35 (0.3 mM)	n.a.
Pseudopelletierine	n.d.	n.a.	n.a.	n.a.	n.a.	n.a.
Purine alkaloids						
Caffeine	n.d.	n.a.	n.a.	n.a.	n.a.	n.a.
Pyrrolizidine alkaloids						
Heliotrine	n.d.	n.a.	n.a.	n.a.	20	1
Monocrotaline	n.d.	n.a.	n.a.	n.a.	25	n.a.
Retronecine	n.d.	n.a.	n.a.	n.a.	20	n.a.
Riddelline	n.d.	n.a.	n.a.	n.a.	20	n.a.
Senecionine	n.d.	n.a.	n.a.	n.a.	50 (5 mM)	1.2
Quinoline alkaloids						
Cinchonine	5*	n.a.	8 mM	6 mM	58 (5 mM)	n.a.
Cinchonidine	6*	< 5 mM	10 mM	< 1 mM	90 (5 mM)	n.a.
Quinidine	8*	1 mM	2.4 mM	< 1 mM	63	0.6 (1 mM)
Quinine	6*	2 mM	3.2 mM	< 2 mM	80	n.a.
Quinolizidine alkaloids						
Anagyrine	n.d.	n.a.	n.a.	n.a.	n.a.	n.a.
Angustifoline	n.d.	n.a.	n.a.	n.a.	15 (5 mM)	n.a.
Cytisine	n.d.	n.a.	n.a.	n.a.	n.a.	n.a.
3β-hydroxylupanine	n.a.	n.a.	n.a.	n.a.	n.a.	91
13-hydroxylupanine	n.a.	n.a.	n.a.	n.a.	38	3.5
Lupanine	n.a.	n.a.	n.a.	n.a.	15 (5 mM)	0.8
Lupinine	n.d.	n.a.	n.a.	n.a.	31 (5 mM)	n.a.

Table 2.16 (Continued)

	DNA melting temperature increase[1] °C	DNA methylgreen release IC_{50}[2]	DNA DNA Pol I inhibition IC_{50}[3]	RNA RT inhibition IC_{50}[3]	Protein biosynthesis inhibition %[4]	Membrane permeability haemolysis %[5]
17-oxosparteine	n.a.	n.a.	n.a.	n.a.	37	n.a.
Sparteine	n.d.	n.a.	n.a.	n.a.	n.a.	n.a.
Tetrahydrorhombifoline	n.d.	n.a.	n.a.	n.a.	n.a.	0.9
13-tigloyloxylupanine	n.a.	n.a.	n.a.	> 10 mM	20	n.a.
Terpene alkaloids						
Aconitine	n.d.	n.a.	n.a.	n.a.	n.a.	n.a.
Protoveratrine B	0	n.a.	n.a.	n.a.	30	n.a.
Solanine	3*	n.a.(0.5 mM)	n.a.(0.5 mM)	< 0.1 mM	50 (0.025 mM)	100 (0.2 mM)
Tropane alkaloids						
Hyoscyamine	n.d.	n.a.	n.a.	n.a.	n.a.	n.a.
Scopolamine	0.3	n.a.	n.a.	n.a.	n.a.	n.a.
Scopine	n.d.	n.a.	n.a.	n.a.	n.a.	n.a.
Tropine	n.d.	n.a.	n.a.	n.a.	15	n.a.
Tropolone alkaloids						
Colchicine	n.d.	n.a.	n.a.	< 10 mM	n.a.	0.7 (1 mM)

[1]70 µM alkaloid solution and 70 µM *Sinapis* DNA; *alkaloid and DNA coelute when separated by gel chromatography on PD10 columns; [2]alkaloids were tested up to 5 mM (n.a. = release < 25% at 5 mM); [3]alkaloids were preincubated with DNA or RNA (n.a. = inhibition < 25% at 10 mM); [4]alkaloids were tested at 1 mM (n.a. = inhibition < 10%). If strong activities were discovered then lower concentrations were also assayed; [5]alkaloids were tested at 5 mM (n.a. = inhibition < 0.6%). If strong activities were discovered then lower concentrations were also assayed.

Abbreviations: DNA, deoxyribonucleic acid; Pol 1, polymerase 1; RNA, ribonucleic acid; RT, reverse transcriptase; IC_{50}, concentration required to cause 50% maximum inhibition; n.d., not determined; n.a., not active.

close contact with many grasses (family Poaceae), such as the cereal, *Hordeum vulgare*. These alkaloids can modulate several receptors of neurotransmitters, such as dopamine, serotonin and noradrenaline. As a consequence, the pharmacological action of ergot alkaloids is rather broad, ranging from vasoconstriction and uterine contraction to hallucinations. These activities can be explained through structural simi-larities between the alkaloid and the different neurotransmitters (Fig. 2.5). It has been suggested that the interactions between *Claviceps* and its host plant are of a symbiotic nature, i.e. infected plants exploit the chemistry of the fungus for their own protection against herbivores (otherwise it would be difficult to explain why a fungal metabolite should interfere with targets that are only present in animals); grasses with endophytes had better protection against herbivores than fungus-free grasses.

Quinolizidine alkaloids (QAs), such as lupanine, sparteine or cytisine, are produced by many members of the Leguminosae. QAs are bitter for many animals (and plants producing them are, therefore, avoided as food). QAs, like many other alkaloids, occur as complex mixtures in plants. We have recently shown (Schmeller *et al.*, 1994), that some QAs preferentially bind to the nicotinic acetylcholine receptor (nAChR), whereas others tend to bind to the muscarinic AChR (Table 2.15). Some QAs exhibit a prominent cross-reactivity. Furthermore, QAs such as lupanine and sparteine inhibit Na^+-channels, thus blocking the signal transduction in nerve cells at a second critical point (Körper *et al.*, 1998). In addition, QAs slightly interfere with protein biosynthesis. A few particular QAs, such as anagyrine, cytisine and the bipiperidine alkaloid, ammodendrine (which co-occurs with QAs in many plants), are mutagenic and lead to malformations. These results suggest that QAs are indeed defence chemicals with a broad range of targets, which might be affected simultaneously (review in Wink, 1992, 1993a,b,c).

In other instances, such apparent hybrid activities could be missing or less prominent, as shown by the example of epibatidine, a powerful alkaloid from the skin of several arrow poison frogs. Synthetic (+)- and natural (−)-epibatidine have potent agonistic activity at ganglionic-type nicotinic receptors. The epibatidines have little or no activity at a variety of other central receptors, including opioid, muscarinic, adrenergic, dopamine, serotonin and GABA receptors (Badio and Daly, 1994).

Ajmaline, berbamine, berberine, boldine, cinchonine, cinchonidine, ergometrine, harmaline, harmine, lobeline, norharman, papaverine, quinidine, quinine, sanguinarine and solanine affect more than one of the basic molecular targets. In addition, the same compounds can bind to one or several neuroreceptors that are only present in animals (Tables 2.15 and 2.16). It is likely that these interactions are responsible (at least in part) for the allelopathic, antibacterial effects and animal toxicity, which are exhibited by these alkaloids (Wink and Twardowski, 1992; Wink and Latz-Brüning, 1995). According to Tables 2.15 and 2.16, many alkaloids are compounds with a very broad activity spectrum. Because of their wide activity, we can consider them 'multipurpose' defence substances. Since plants cannot predict or choose their competitors, infesting microorganisms, insects and other herbivores, such preformed 'multipurpose' compounds are certainly a means of being prepared for most situations.

But, why have neurotransmitter mimics that also intercalate DNA? Plant-herbivore interactions are mutual processes. If the plant produces a nAChR inhibitor, it is likely that some insects develop a resistance, in such a way that they modify the binding site of their nAChR, so that the inhibitor can no longer bind. It was recently shown that *Danaus plexippus*, which sequesters cardiac glycosides from its host plant, has a modified binding domain for its Na^{+},K^{+}-ATPase, which no longer interacts with cardiac glycosides, thus providing an insensitivity at this target (Holzinger *et al.*, 1992; Holzinger and Wink, 1996). However, if the same compound affected other molecular targets at the same time, such as DNA, an adaptation through target site modification would be much more difficult and unlikely, since it would require several concomitant mutations. According to this hypothesis, the evolution of allelochemicals affecting more than one target would be a strategy, firstly, to counteract adaptations by specialists; and secondly, to help in fighting off different groups of enemies. In conclusion, it can be said that Nature has obviously tried 'to catch as many flies with one clap as possible' in the selection of allelochemicals during evolution.

2.5 Accumulation of allelochemicals in plants

2.5.1 The importance of high concentrations

In order to deter the feeding of herbivores, allelochemicals need to be accumulated in sufficient quantities. Storage of water soluble compounds in large concentrations (up to 200–500 mM) in the vacuole is an important requisite for chemical defence in most plants; therefore, we have termed these vacuoles 'defence and signal compartment' (Wink, 1993a). Many vacuolar allelochemicals are positioned at a favourable site for defence because they are stored in substantial amounts in epidermal and subepidermal cells (Wink *et al.*, 1984; Wink, 1992, 1997). The sequestration of lipophilic metabolites in ducts, laticifers, trichomes, glandular hairs or in the cuticle (reviews in Wink, 1993a,b,c 1997) is also important as a defence measure. If a small herbivore or microbe attacks such a plant, it will encounter a high concentration immediately at the periphery upon wounding or entering the tissue, which might deter further feeding (Wink, 1999).

A further question is whether the inhibitory concentrations determined in *in vitro* experiments (Tables 2.15 and 2.16) relate in any way to the *in vivo* situation. A simple calculation may help to assess this problem: alkaloids are usually stored in high concentrations at sites that are important for growth and reproduction and can reach 1–2% of the dry weight (Wink, 1993a). Assuming an alkaloid concentration of 100 mg per 100 g fresh weight (which is approximately equivalent to 1% dry weight) and a small herbivore with a body weight of 1000 g; if this animal ingested 100 g of an alkaloid-producing plant it would take up 100 mg of alkaloids. Supposing that the alkaloids are completely resorbed and equally distributed in the body, a concentration of 100 mg alkaloids/kg body weight would be obtained. Taking a mean molecular weight of 200 KDa, the alkaloid concentration in this herbivore would be 500 μM, which would be high enough to partially or completely block the binding of acetylcholine, serotonin or noradrenalin at their receptors (compare the values for the concentration required to cause 50% maximum inhibition (IC_{50}) in Table 2.15) or to interfere with DNA and related processes. In reality, incomplete resorption and degradation will prevent such high internal toxin concentrations but even a 5–10 times lower internal concentration can produce severe intoxication and disturbance in most instances. Under these conditions, an interaction with multiple targets appears likely. This might explain why alkaloid-producing plants are usually avoided by most herbivores, with the exception of a few adapted organisms that have evolved tolerance or insensitivity towards these toxins.

2.5.2 *Variability and timing of accumulation of defence chemicals in vulnerable tissues*

In most plants, synthesis and accumulation of secondary metabolites is regulated in space and time. Some plants make and store a particular compound in one organ only; in other instances, a compound is made in the roots or leaves but exported to other plant parts via xylem, phloem or apoplastically. In general, vulnerable tissues and organs are defended more than old senescing ones; many seeds, seedlings, buds and young tissues either sequester or synthesize large amounts of defence chemicals. Plant parts that are important for survival and multiplication, such as flowers, fruits and seeds, are nearly always heavily defended. It is difficult to summarize this topic, as the relationships differ from system to system and need to be elucidated for each group of plants. This variation is an important strategy against herbivores, since if all plants followed the same strategy it would be easy for herbivores to adapt.

The profile of SMs in plants usually consists of a series of related compounds; generally a few major metabolites and several minor components are present that differ in the position of their substituents or their stereochemistry. The differences may appear small from a chemical perspective but they can be important for pharmacological activity; the addition of a simple substituent may open another molecular target (see Tables 2.15–2.16) and can thus contribute to an improved fitness. The profile usually varies between plant organs, within developmental stages and sometimes even diurnally (e.g. in lupin alkaloids) (Wink, 1992). Moreover, marked differences can usually be seen between individual plants of a single population, and even more so between members of different populations. In some plants, the population differences appear to be genetically fixed and can be distinguished as particular 'chemotypes'; examples are monoterpene profiles in *Thymus vulgaris*, which lead to differential herbivory. This variation, which is part of the apparent evolutionary 'arms race' between plants and herbivores, makes adaptation by herbivores more difficult.

In most examples discussed so far, variation in patterns and concentrations was found for most secondary compounds but, nevertheless, we consider this defence constitutive; prefabricated metabolites are included that are only activated in case of immediate emergency (e.g. cyanogenic glycosides, coumaroylglycosides, bidesmosidic saponins, ranunculin, alliin, glucosinolates). If a plant is wounded by a herbivore or infected by a pathogen, the defence chemistry often becomes activated (Harborne, 1993; Baldwin, 1994). Two strategies can be distinguished. Either the concentration of the compounds already present is increased by a *de novo* synthesis, as has been observed for several defence chemicals, such as nicotine in *Nicotiana tabacum* (Baldwin, 1994), lupin alkaloids in *Lupinus polyphyllus* (Wink, 1992) and several others. Alternatively, the *de novo* synthesis of compounds (usually termed 'phytoalexins') that were hitherto absent is induced. These compounds include several isoflavones, pterocarpans, furocoumarins, chalcones, stilbenes and others. Many of these metabolites have antifungal properties, so that they are sometimes considered to be part of a specific antimicrobial defence system in plants (see Chapter 5 by J. Reichling in the present volume). However, since most of these compounds also affect herbivores, the plant defence induced appears to be a more general phenomenon. The biochemical mechanisms, starting with binding of elicitors to membrane receptors, via intercellular signalling, to gene activation and protein biosynthesis, have been studied in these induced systems, so that a considerable body of information is already available on the underlying mechanisms. During the last decade, it has been shown that a first important step in the signal pathway is the hydrolysis of unsaturated fatty acids from the cell membrane lipids by phospholipase. The fatty acids are converted into jasmonic acid, which appears as a major signal compound in wounded and infected plants. It will be a challenging task to elucidate all the steps in these induced signal pathways (see Chapter 4 in the present volume).

In conclusion, variation in chemical structures, in organ-specific profiles of allelochemicals within and between different groups of plant and in defence strategies can be interpreted as a means of avoiding the adaptation of specialized herbivores and microorganisms.

2.5.3 Costs of chemical defence

The pathways leading to most groups of secondary metabolites have been elucidated (reviewed in Gershenzon and Kreis, 1999; Petersen *et al.*, 1999; Roberts and Strack, 1999; Selmar, 1999; Wink, 1999; Wink and Waterman, 1999) and several of the specific enzymes involved have been purified and studied in detail (Luckner, 1990). Because the enzymes of biosynthesis also show a regular turnover, mRNA must be transcribed and translated into proteins. Both transcription and translation require substantial amounts of ATP, as does the biosynthesis itself: a prerequisite for defence or signal compounds to be present in relatively high amounts at the right place and time. Storage in the vacuole usually requires energy as well; energy for the uphill transport and often for trapping the metabolite in the vacuole is provided by a H^+-ATPase. As a consequence, both biosynthesis and sequestration (and the corresponding transcription and translation) must be costly processes for the plants producing secondary metabolites. Sequestration in resin ducts, oil cells or trichomes demands the formation of new structures, which can figure as costs for a plant expressing them (Wink, 1999).

Is there any evidence that these costs matter? Plants are autotrophic and usually not limited by energy or carbon dioxide; theoretically, it should be rather easy for them to provide the ATP and carbon necessary for the synthesis of defence substances. But circumstantial evidence suggests that chemical defence is, nevertheless, carried out parsimoniously and not luxuriously. Legumes, for example, produce quinolizidine alkaloids as effective defence chemicals. In brooms, we find several species with spikes and thorns which provide mechanical defence (Wink, 1992). In all these species, alkaloid contents are very low, suggesting that chemical protection has been substituted by mechanical defence. A similar feature has been reported for cacti; spiny species usually contain small amounts of alkaloids, whereas soft-bodied and spineless taxa are alkaloid rich (e.g. *Lophophora williamsii* producing mescaline). Species which have colonized volcanic islands often encounter a niche free from herbivores and sometimes they have apparently reduced their alkaloid content. For example, whereas *Echium* species from continental Europe produce pyrrolizidine alkaloids (PAs) in substantial amounts, *Echium* species from the Canary Islands usually sequester only very low amounts of PAs. These few examples, which could be augmented by several population genetics studies, imply, that chemical defence can to be costly even for autotrophic plants.

2.6 Animal responses: detoxification mechanisms and adaptations

If a herbivore feeds on a plant rich in secondary metabolites, the following scenarios can be postulated (Bernays and Chapman, 1994; Rosenthal and Berenbaum, 1991, 1992):

1. A herbivore does not taste the allelochemicals, ingests them and is killed immediately, for example, in the case of an alkaloid-rich plant.
2. A herbivore eats only a small part of the plant and suffers from unpleasant symptoms afterwards. Most vertebrate herbivores can associate both events and will subsequently avoid this plant for feeding.
3. A herbivore does not feed on a single plant species but samples from a wide range of plants. Using this strategy, it will ingest a wide variety of allelochemicals but will avoid an acute intoxication.
4. A herbivore is adapted on a particular plant species (as a food specialist) and can apparently feed on it although it contains active allelochemicals.

In nature, we usually find situations 3 and 4, which provide some protection against herbivores. A plant has a few adapted herbivores, which are often attracted to the noxious chemistry of their host plants, or is afflicted by browsers, which might take a part but not destroy the complete plant. This situation is obviously better than being attacked by a multitude of herbivores, which might devour the whole plant. A number of plants (e.g. with alkaloids), whose constituents interact with important molecular targets, are often avoided even by the casual browser.

On the other hand, a number of adapted herbivores, especially arthropods, are known. This argument is sometimes used to negate the defence function of secondary metabolites. An example from medicine easily explains why this argument does not count. Our immune system apparently works against the majority of bacteria, fungi and viruses and we are usually unaware of its existence. A few viruses, microbes and parasites have overcome the defence barrier by 'clever' biochemical strategies (e.g. by antigenic variation), so that they escape the immune system and make us sick. Nobody would call the immune system and the antibodies useless because of these few adapted specialists. It is one goal of modern medicine to elucidate the underlying intricate adaptations.

A similar argument also applies for plant secondary metabolites. The question is, therefore: What are the mechanisms which enable a herbivore to feed on chemically defended plants? Because there are so many herbivores, generalizations are difficult and dangerous but a few general strategies can be observed (Bernays and Chapman, 1994; Harborne, 1993). Many insects and vertebrates have evolved highly potent degrading enzymes, which can hydrolyse or hydroxylate (among them inducible cytochrome P_{450} hydroxylases) a compound taken up from the gut. In a second step, these compounds are conjugated with glucuronic acid, glutathione or sulfate, so that they become more water soluble. In vertebrates, these conjugates are then excreted via the bile duct or more commonly via the kidneys. A detoxification of defence chemicals by rumen or gut microbes has also been discussed. Experiments with quinolizidine alkaloids suggest that they are not degraded by rumen microorganisms (R. Aguiar and M. Wink, personal communication). A short intestinal transition time or the ingestion of clay, which can adsorb dietary toxins (geophagy), can be additional mechanisms that have been observed in some herbivores.

Quite a number of insect herbivores but only few vertebrates have adapted to the defence chemistry of a given host plant in a particularly close way (for overviews see Harborne, 1993; Bernays and Chapman, 1994; Roberts and Wink, 1998; Blum, 1981; Hartmann and Witte, 1995; Meinwald, 1990; Braekman *et al.*, 1998; Brown and Trigo, 1995). As a first step in adaptation, these organisms must develop an insensitivity or tolerance against the dietary toxin. The underlying mechanisms are understood for only a few species. In the case of *Manduca sexta* and other Sphingidae that can feed on alkaloid rich plants, a very active degradation and excretion system appears to be the main mechanism. In the case of the monarch butterfly (*Danaus plexippus*), whose larvae feed on *Asclepias* plants that are rich in cardenolides, it was discovered that its Na^+,K^+-ATPase (the molecular target of cardenolides) is insensitive to cardiac glycosides. Molecular analysis revealed that an amino acid of the ouabain binding site is exchanged, so that cardiac glycosides can no longer bind (Holzinger and Wink, 1996). This insect can now tolerate cardiac glycosides; however, adaptations are even more advanced, since *Danaus* can actively sequester cardenolides and store them in its integument as acquired defence compounds against predators. A similar strategy has been observed for the sequestration of cardiac glycosides in aphids, bugs and Lepidoptera, for pyrrolizidine alkaloids in beetles and Lepidoptera (for reviews see Meinwald, 1990; Hartmann and Witte, 1995; Brown and Trigo, 1995), and for quinolizidine alkaloids in aphids and moths (Wink, 1992). In the case of some arctiid moth, the acquired PAs serve as a precursor for pheromones, which are dissipated from coremata that are only developed if the larvae feed on PAs (Schneider *et al.*, 1982). While females transport PAs to the eggs, the male can also contribute to the chemical defence of its clutch by transferring PAs with its spermatophore (as a 'nuptial gift').

Insects which defend themselves chemically, often bear warning coloration (aposematism) to advertise their inpalatability to potential predators. A fascinating coevolution has been observed in several groups of insects, which show the same coloration but do not sequester the corresponding toxins. These mimics, nevertheless, gain protection by this strategy. The few examples given show the intricate and complex interactions that can be encountered in insects that have adapted to the defence chemistry of their host plants (Wink, 1999; see also Chapter 3 by P. Proksch in this volume for further discussions).

2.7 Concluding remarks

Although relatively few of the several tens of thousands of known secondary metabolites have been analyzed biochemically or ecologically in any detail, we can nevertheless generalize that many of them are allelochemicals that serve as defence compounds against herbivores or microorganisms.

If studied in detail, it is usually possible to define the mode of action of each allelochemical and, in particular, the molecular target with which it interferes. Knowing the targets, the toxic or pharmacological effects observed can generally be explained. This knowledge will help in understanding the role of these compounds in nature but is also useful for the exploitation of these active components in medicine or agriculture. Thus, chemical ecology and biotechnology share a common evolutionary base. Given the richness of structures and potential targets which might have been selected during evolution, 'bioprospection' in combination with chemical ecology, pharmacology and medicine will remain an interesting topic for science in the future.

References

Abdel-Fattah, A.-F.M., Matsumoto, K., Gammaz, H.A.-K. and Watanabe, H. (1995) Hypothermic effect of harmala alkaloid in rats: involvement of serotonergic mechanism. *Pharmacol. Biochem. Behav.*, **52** 421-26.

Abel, G. (1987) Chromosomenschädigende Wirkung von β-Asaron in menschlichen Lymphocyten. *Planta Med.*, **53** 251-53.

Abel, G. and Schimmer, O. (1983) Induction of structural chromosome aberrations and sister chromatid exchanges in human lymphocytes *in vitro* by aristolochic acid. *Hum. Genet.*, **64** 131-33.

Abel, G. and Schimmer, O. (1986) Chromosome-damaging effects of heraclenin in human lymphocytes. *Mutat. Res.*, **169** 51-54.

Abel, G., Erdelmeier, C., Meier, B. and Sticher, O. (1985) Iso-Pimpinellin, ein Furanocumarin aus *Heracleum sphondylium* mit chromosomenschädigender Aktivität. *Planta Med.*, **51** 250-52.

Abu-Ghalyun, Y., Masalmeh, A. and Al-Khalil, S. (1997) Effects of allocryptopine, an alkaloid isolated from *Glaucium arabicum* on rat isolated ileum and urinary bladder. *Gen. Pharmacol.*, **29** 621-23.

Achenbach, H., Waibel, R. and Zwanzger, M. (1992) 9-Methoxy- and 7,9-dimethoxytariacur-ipyrone, natural nitro-compounds with a new basic skeleton from *Aristolochia brevipes*. *J. Nat. Prod.*, **55** 918-22.

Akiba, S., Kato, E., Sato, T. and Fujii, T. (1992) Biscoclaurine alkaloids inhibit receptor-mediated phospholipase A2 activation probably through uncoupling of a GTP-binding protein from the enzyme in rat peritoneal mast cells. *Biochem. Pharmacol.*, **44** 45-50.

Akiba, S., Nagatomo, R., Ishimoto, T. and Sato, T. (1995) Effect of berbamine on cytosolic phospholipase A2 activation in rabbit platelets. *Eur. J. Pharmacol. Mol. Pharmacol.*, **291** 343-50.

Akuzawa, S., Yamaguchi, H., Masuda, T. and Ueno, Y. (1992) Radical-mediated modification of deoxyguanine and deoxyribose by luteoskyrin and related anthraquinones. *Mutat. Res.*, **266** 63-69.

Alberts, B., Bray, D., Lewis, J., Raff, M., Roberts, K. and Watson, J.D. (1994) *Molecular Biology of the Cell*. 3rd edn. Garland Publishers, New York and London.

Allouche, S., Polastron, J. and Jauzac, P. (1996) The deltaopioid receptor regulates activity of ryanodine receptors in the human neuroblastoma cell line, SK-N-BE. *J. Neurochem.*, **67** 2461-70.

Ameri, A. (1997a) Effects of the alkaloids 6-benzoylheteratisine and heteratisine on neuronal activity in rat hippocampal slices. *Neuropharmacology*, **36** 1039-46.

Ameri, A. (1997b) Inhibition of rat hippocampal excitability by the plant alkaloid 3-acetylaconitine mediated by interaction with voltage-dependent sodium channels. *Naunyn Sch. Arch. Pharmacol.*, **355** 273-80.

Anderson, R.D. and Berger, N.A. (1994) Mutagenicity and carcinogenicity of topoisomerase-interactive agents. *Mutat. Res.*, **309** 109-42.

Ashby, J., Elliot, B.M. and Styles, J.A. (1980) Norharman and ellipticine: a comparison of their abilities to interact with DNA *in vitro*. *Cancer Lett.*, **9** 21-33.

Ashcroft, F.M., Kerr, A.J., Gibson, J.S. and Williams, B.A. (1991) Amantadine and sparteine inhibit ATP-regulated potassium currents in the insulin-secreting beta-cell line, HIT-T15. *Br. J. Pharmacol.*, **104** 579-84.

Ashwood-Smith, M.J., Towers, G.H.N., Abramowski, Z., Poulton, G.A. and Liu, M. (1982) Photobiological studies with dictamnine, a furoquinoline alkaloid. *Mutat. Res.*, **102** 401-12.

Auerbach, C. and Robson, J.M. (1944) Production of mutations by allylisothiocyanate. *Nature*, **154** 81.

Averbeck, D. (1989) Recent advances in psoralen phototoxicity mechanism. *Photochem. Photobiol.*, **50** 859-82.

Badio, B. and Daly, J.W. (1994) Epibatidine, a potent analgetic and nicotinic agonist. *Mol. Pharmacol.*, **45** 563-69.

Badria, F.A., McChesney, J.D., Halim, A.F., Zaghloul, A.M. and El Sayed, K.A. (1995) Time course and inhibition of stavaroside K, veratramine and cevine-induced hemolysis by other pregnane glycosides and Veratrum alkaloids. *Pharmazie*, **50** 421-23.

Baker, M.D. and Ritchie, J.M. (1994) The action of capsaicin on type I delayed rectifier K^+ currents in rabbit Schwann cells. *Proc. R. Soc. Lond., Series B*, **255** 259-65.

Baldwin, I. (1994) Chemical changes rapidly induced by folivory, in *Insect Plant Interactions* (ed. E.A. Bernays) CRC Press, Boca Raton, pp. 1-23.

Behninger, C., Abel, G., Röder, E., Neuberger, V. and Göggelmann, W. (1989) Wirkung eines Alkaloidextraktes von *Symphytum officinale* auf menschliche Lymphocytenkulturen. *Planta Med.*, **55** 518-22.

Bembenek, M.E., Abell, C.W., Chrisey, L.A., Rozwadowska, M.D., Gessner, W. and Brossi, A. (1990) Inhibition of monoamine oxidases A and B by simple isoquinoline alkaloids: racemic and optically active 1,2,3,4-tetrahydro-, 3,4-dihydro- and fully aromatic isoquinolines. *J. Med. Chem.*, **33** 147-52.

Berger, P., Elsworth, J.D., Reith, M.E.A., Tanen, D. and Roth, R.H. (1990) Complex interaction of cocaine with the dopamine uptake carrier. *Eur. J. Pharmcol.*, **176** 251-52.

Bermejo, A., Protais, P., Blazquez, M.A., Rao, K.S., Zafra-Polo, M.C. and Cortes, D. (1995) Dopaminergic isoquinoline alkaloids from roots. *Nat. Prod. Lett.*, **6** 57-62.

Bernays, E.A. and Chapman, R.F. (1994) *Host-Plant Selection by Phytophagous Insects*. Chapman and Hall, New York.

Berridge, M.J. and Bootman, M.D. (1996) Calcium signaling, in *Signal Transduction* (ed. C.-H. Heldin and M. Purton). Chapman and Hall, London.

Beutler, J.A., Karbon, E.W., Brubaker, A.N., Malik, R., Curtis, D.R. and Enna, S.J. (1985) Securinine alkaloids: a new class of GABA receptor antagonist. *Brain Res.*, **330** 135-40.

Bickmeyer, U. and Wiegand, H. (1993) Tetrandrine effects on calcium currents in cultured neurons of fetal mice. *Neuro Report*, **4** 938-40.

Bickmeyer, U., Hare, M.F. and Atchison, W.D. (1996) Tetrandrine blocks voltage-dependent calcium entry and inhibits the bradykinin-induced elevation of intracellular calcium in NG108-15 cells. *Neurotoxicology*, **17** 335-42.

Bidasee, K.R., Besch, H.R. Jr., Gerzon, K. and Humerickhouse, R.A. (1995) Activation and deactivation of sarcoplasmic reticulum calcium release channels: molecular dissection of mechanisms via novel semisynthetic ryanoids. *Mol. Cell. Biochem.*, **149** 145-60.

Blömeke, B., Poginsky, B., Schmutte, C., Marquardt, H. and Westendorf, J. (1992) Formation of genotoxic metabolites from anthraquinone glycosides present in *Rubia tinctorum* (L.). *Mutat. Res.*, **265** 263-72.

Blum, M.S. (1981) *Chemical Defences of Arthropods*. Academic Press, New York.

Bösch, R., Friederich, U., Lutz, W.K., Brocker, E., Bachmann, M., Schlatter, C. (1987) Investigations on DNA binding in rat liver and in Salmonella and on mutagenicity in the Ames test by emodin, a natural anthraquinone. *Mutat. Res.*, **188** 161-68.

Boyland, E., Busby, E.R., Dukes, C.E., Grover, P.L. and Manson, D. (1964) Further experiments on implantation of materials into the urinary bladder of mice. *Br. J. Cancer*, **18** 575-81.

Bracher, F. and Hildebrand, D. (1995) β-Carboline alkaloids. Part 6. Total synthesis of the phosphodiesterase inhibitor, infractine. *Pharmazie*, **50** 182-83.

Braekman, J.C., Daloze, D. and Pasteels, J.M. (1998) Alkaloids in animals, in *Alkaloids: Biochemistry, Ecology and Medicinal Applications* (ed. M. Roberts and M. Wink), Plenum Press, New York, pp. 349-78.

Breton, P., Asseffa, A., Grzegorzewski, K., Akiyama, S.K., White, S.L., Cha, J.K. and Olden, K. (1990) Swainsonine modulation of protein kinase C activity in murine peritoneal macrophages. *Cancer Commun.*, **2** 333-38.

Broschard, T.H., Wiessler, M., von der Lieth, C.W., Schmeiser, H.H. (1994) Translesional synthesis on DNA templates containing site-specifically placed deoxyadenosine and deoxyguanosine adducts formed by the plant carcinogen, aristolochic acid. *Carcinogenesis*, **15** 2331-40.

Brossi, A. (1993) Mammalian alkaloids. II, in *The Alkaloids*, Vol. 43 (ed. G. Cordell), Academic Press, New York, pp. 119-183.

Brown, A.M., Patch, T.L. and Kaumann, A.J. (1992) Ergot alkaloids as 5-HT1C receptor agonists: relevance to headache. *Front. Headache Res.*, **2** 247-51.

Brown, J.P. (1980) A review of the genetic effects of naturally occurring flavonoids, anthraquinones and related compounds. *Mutat. Res.*, **75** 243-77.

Brown, K.S. and Trigo, J.R. (1995) The ecological activity of alkaloids, in *The Alkaloids*, Vol. 47 (ed. G. Cordell), Academic Press, New York, pp. 227-54.

Bruggeman, I.M. and van der Hoeven, J.C.M. (1984) Lack of activity of the bacterial mutagen, emodin, in HGPRT and SCE assay with V79 Chinese hamster cells. *Mutat. Res.*, **138** 219-24.

Buckingham, J. (1996) *Dictionary of Natural Products on CD-ROM*, Version 5:1 edition, Chapman and Hall, London.

Calsou, P., Sage, E., Moustacchi, E. and Salles, B. (1996) Preferential repair incision of cross-links *versus* monoadducts in psoralen-damaged plasmid DNA by human cell-free extracts. *Biochemistry USA*, **35** 14963-69.

Caria, H., Chaveca, T., Laires, A. and Rueff, J. (1995) Genotoxicity of quercetin in the micronucleus assay in mouse bone marrow erythrocytes, human lymphocytes, V79 cell line and identification of kinetochore-containing (CREST staining) micronuclei in human lymphocytes. *Mutat. Res.*, **343** 85-94.

Chang, G.-J., Wu, M.-H., Wu, Y.-C. and Su, M.-J. (1996) Electrophysiological mechanisms for antiarrhythmic efficacy and positive inotropy of liriodenine, a natural aporphine alkaloid from *Fissistigma glaucescens*. *Br. J. Pharmacol.*, **118** 1571-83.

Changeux, J.P. (1993) Chemical signaling in the brain. *Sci. Am.*, **269** 58-62.

Chen, K., Kokate, T.G., Donevan, S.D., Carroll, F.I. and Rogawski, M.A. (1996) Ibogaine block of the NMDA receptor: *in vitro* and *in vivo* studies. *Neuropharmacology*, **35** 423-31.

Chen, P., Lavin, M.F., Teh, B.S., Seow, W.K. and Thong, Y.H. (1992) Induction of apoptosis by tetrandrine: comparison with other immunosuppressive agents. *Int. J. Immunother.*, **8** 85-90.

Chen, S., Liu, G. and Min, Z. (1987) The actions of some tetrahydroisoquinoline alkaloids on dopamine and serotonin receptors in rat brain. *Yaoxue Xuebao*, **22** 341-46.

Cheng, J.-T., Chang, T.K. and Chen, I.-S. (1994) Skimmianine and related furoquinolines function as antagonists of 5-hydroxytryptamine receptors in animals. *J. Auton. Pharmacol.*, **14** 365-74.

Chiou, W.-F., Chou, C.-J., Liao, J.-F., Sham, A.Y.-C. and Chen, C.-F. (1994) The mechanism of the vasodilator effect of rutaecarpine, an alkaloid isolated from *Evodia rutaecarpa*. *Eur. J. Pharmacol.*, **257** 59-66.

Chiou, W.-F., Liao, J.-F., Yau-Chik Shum, A. and Chen, C.-F. (1996) Mechanisms of vasorelaxant effect of dehydroevodiamine: a bioactive isoquinazolinocarboline alkaloid of plant origin. *J. Cardiovasc. Pharmacol.*, **27** 845-53.

Chu, F.S. (1991) Mycotoxins: food contamination, mechanism, carcinogenic potential and preventive measures. *Mutat. Res.*, **259** 291-306.

Chulia, S., Ivorra, M.D., Lugnier, C., Vila, E., Noguera, M.A. and D'Ocon, P. (1994) Mechanism of the cardiovascular activity of laudanosine: comparison with papaverine and other benzylisoquinolines. *Br. J. Pharmacol.*, **113** 1377-85.

Chulia, S., Noguera, M.A., Ivorra, M.D., Cortes, D. and D'Ocon, M.P. (1995) Vasodilator effects of liriodenine and norushinsunine, two aporphine alkaloids isolated from *Annona cherimolia*, in rat aorta. *Pharmacology*, **50** 380-87.

Clark, A.M. (1960) The mutagenic activity of some pyrrolizidine alkaloids in *Drosophila*. *Zeits. Vererbung.*, **91** 74-80.

Clark, A.M. (1982) Endogenous mutagens in green plants, in *Environmental Mutagenesis, Carcinogenesis and Plant Biology* (ed. E.J. Klekowski Jr.), Praeger Publ., New York, pp. 97-132.

Clarke, C.H. and Wade, M.J. (1975) Evidence that caffeine, 8-methoxypsoralen and steroidal diamines are frameshift mutagens for *E. coli* K12. *Mutat. Res.*, **28** 123-25.

Coates, P.A., Blagbrough, I.S., Lewis, T., Potter, B.V.L. and Rowan, M.G. (1995) An HPLC assay for the norditerpenoid alkaloid, methyllycaconitine, a potent nicotinic acetylcholine receptor antagonist. *J. Pharm. Biomed. Anal.*, **13** 1541-44.

Conn (1980) Cyanogenic glycosides, in *Secondary Plant Products* (eds. E.A. Bell and B.V. Charlwood), Springer, Berlin, pp. 461-92.

Cross, D.L., Redmond, L.M. and Strickland, J.R. (1995) Equine fescue toxicosis: signs and solutions. *J. Anim. Sci.*, **73** 899-908.

D'Ocon, P., Amparo Blazquez, M., Bermejo, A. and Anselmi, E. (1992) Tetrandrine and isotetrandrine, two bisbenzyltetrahydroisoquinoline alkaloids from Menispermaceae, with rat uterine smooth muscle relaxant activity. *J. Pharm. Pharmacol.*, **44** 579-82.

da Silva, B.A., de Araujo Filho, A.P., Mukherjee, R. and Chiappeta, A.D.A. (1993) Bis-nordihydrotoxiferine and vellosimine from *Strychnos divaricans* root: spasmolytic properties of bis-nordihydrotoxiferine. *Phytother. Res.*, **7** 419-24.

Daimon, H., Sawada, S., Asakura, S., Sagami, F. (1997) Analysis of cytogenetic effects and DNA adduct formation induced by safrole in Chinese hamster lung cells. *Teratogen. Carcinogen. Mutagen.*, **17** 7-18.

Dalvi, R.R. (1985) Sanguinarine: its potential as a liver toxic alkaloid present in the seeds of *Argemone mexicana. Experientia*, **41** 77-78.

Daly, J.W., Garraffo, H.M. and Spande, T.F. (1993) Amphibian alkaloids, in *The Alkaloids*, Vol. 43 (ed. G.A. Cordell), Academic press, San Diego, pp. 185-288.

Darroch, S.A., Taylor, W.C., Choo, L.K. and Mitchelson, F. (1990) Structure-activity relationships of some Galbulimima alkaloids related to himbacine. *Eur. J. Pharmacol.*, **182** 131-36.

De Freitas, M.R., Cortes, S.D.F. and Filho, J.M.B. (1996) Modification of Ca^{2+} metabolism in the rabbit aorta as a mechanism of spasmolytic action of warifteine, a bisbenzylisoquinoline alkaloid isolated from the leaves of *Cissampelos sympodialis* Eichl. (Menispermaceae). *J. Pharm. Pharmacol.*, **48** 332-36.

De Hondt, H.A., Fahmy, A.M. and Abdelbaset, S.A. (1984) Chromosomal and biochemical studies on the effect of khat extract on laboratory rats. *Environ. Mutagen.*, **6** 851-60.

Decker, M.W., Anderson, D.J., Brioni, J.D., Donnelly-Roberts, D.L., Kang, C.H., O'Neill, A.B., Piattoni-Kaplan, M., Swanson, S. and Sullivan, J.P. (1995) Erysodine, a competitive antagonist at neuronal nicotinic acetylcholine receptors. *Eur. J. Pharmacol.*, **280** 79-89.

DeFarias, F.P., Carvalho, M.F., Lee, S.H., Kaczorowski, G.J. and Suarez-Kurtz, G. (1996) Effects of the K^+-channel blockers paspalitrem-C and paxilline on mammalian smooth muscle. *Eur. J. Pharmacol.*, **314** 123-28.

Di Giovanni, J., Decina, P.C., Prichett, W.P., Cantor, J., Aalfs, K.K., Coombs, M.M. (1985) Mechanism of mouse skin tumor promotion by chrysarobin. *Cancer Res.*, **45** 2584-89.

Dong, Y., Yang, M.M.-P. and Kwan, C.-Y. (1997) *In vitro* inhibition of proliferation of HL-60 cells by tetrandrine and *Coriolus versicolor* peptide derived from Chinese medicinal herbs. *Life Sci.*, **60** PL135-PL140.

Dunnick, J.K., Prejean, J.D., Haseman, J., Thompson, R.B., Giles, H.D. and McConnel, E.E. (1982) Carcinogenesis bioassay of allylisothiocyanate. *Fund. Appl. Toxicol.*, **2** 114-20.

Dworetzky, S.I., Boissard, C.G., Lum-Ragan, J.T., McKay, M.C., Post-Munson, D.J., Trojnacki, J.T., Chang, C.-P. and Gribkoff, V.K. (1996) Phenotypic alteration of a human BK (hSlo) channel by hSlo beta subunit coexpression: changes in blocker sensitivity, activation/relaxation and inactivation kinetics and protein kinase A modulation. *J. Neurosci.*, **16** 4543-50.

Edashige, K., Utsumi, T. and Utsumi, K. (1991) Inhibition of 12-*O*-tetradecanoyl phorbol-13-acetate promoted tumorigenesis by cepharanthine, a biscoclaurine alkaloid, in relation to the inhibitory effect on protein kinase C. *Biochem. Pharmacol.*, **41** 71-78.

Edwards, G. and Weston, A.H. (1996) The pharmacology of potassium channel superfamilies: modulation of KATP and BKCa, in *Mol. Cell. Mech. Cardiovasc. Regul.* (ed. M. Endoh), Sendai *Int. Symp.*, Meeting Date 1995, Springer, Tokyo, Japan, pp. 93-109.

Eglen, R.M., Harris, G.C., Cox, H., Sullivan, A.O., Stefanich, E. and Whiting, R.L. (1992) Characterization of the interaction of the cervane alkaloid, imperialine, at muscarinic receptors *in vitro. Naunyn Sch. Arch. Pharmacol.*, **346** 144-51.

Eiden, L.E., Giraud, P., Affolter, H.-U., Herbert, E. and Hotchkiss, A.J. (1984) Alternative modes of enkephalin biosynthesis regulation by reserpine and cyclic AMP in cultured chromaffin cells. *Proc. Natl. Acad. Sci. USA*, **81** 3949-53.

Elguero, J., Campillo, N. and Paez, J.A. (1996) Non-conventional analgesics: epibatidine, a potent nicotinic analgesic. *An. R. Acad. Pharm.*, **62** 303-21.

Elliger, C.A., Henika, P.R. and MacGregor, J.T. (1984) Mutagenicity of flavones, chromones and acetophenones in *Salmonella typhimurium*: new structure-activity relationships. *Mutat. Res.*, **135** 77-86.

Enomoto, M. (1987) Safrole, in *Naturally Occurring Carcinogens of Plant Origin* (ed. I. Hirono), Elsevier, Amsterdam, pp. 139-59.

Erdo, S.L., Molnar, P., Lakics, V., Bence, J.Z. and Tomoskozi, Z. (1996) Vincamine and vincanol are potent blockers of voltage-gated Na^+-channels. *Eur. J. Pharmacol.*, **314** 69-73.

Ewart, R. and Bradford, M. (1988) Inhibition of adenosine 3′,5′-cyclic monophosphate phosphodiesterase by colchicine: implications for glucagon and corticosteroid secretion. *Life Sci.*, **42** 2587-92.

Faddejeva, M.D., Belyaeva, T.N., Rosanov, Yu, M., Sedova, V.M., Sokolovskaya, E.L. (1984) Studies on the complex formation with DNA and the effect on DNA hydrolysis, RNA synthesis and cellular membrane ATPase systems of some antitumor agents including alkaloids. *Studia Biophysica*, **104** 267-69.

Farnsworth, N.R., Bingal, A.S., Fong, H.H.S., Saleh, A.A., Christensen, G.M. and Saufferer, S.M. (1976) Oncogenic and tumour promoting spermatophytes and pteridophytes and their active principles. *Cancer Treat. Rep.*, **60** 1171-214.

Fernando, R.C., Schmeiser, H.H., Nicklas, W. and Wiessler, M. (1992) Detection and quantitation of dG-AAI and dA-AAI adducts by P-32-postlabeling methods in Urothelium and exfoliated cells in urine of rats treated with aristolochic acid-I. *Carcinogenesis*, **13** 1835-39.

Forsyth, C.S., Speth, R.C., Wecker, L., Galey, F.D. and Frank, A.A. (1996) Comparison of nicotinic receptor binding and biotransformation of coniine in the rat and chick. *Toxicol. Lett.*, **89** 175-83.

Frei, H., Würgler, F.E., Juon, H., Hall, C.B. and Graf, U. (1985) Aristolochic acid is mutagenic and recombinogenic in Drosophila genotoxicity tests. *Arch. Toxicol.*, **56** 158-66.

Friese, J., Gleitz, J., Gutser, U.T., Heubach, J.F., Matthiesen, T., Wilffert, B. and Selve, N. (1997) *Aconitum* spp. alkaloids: the modulation of voltage-dependent Na^+-channels, toxicity and antinociceptive properties. *Eur. J. Pharmacol.*, **337** 165-74.

Fujita, H. and Kakishima, H. (1989) Further evidence for photoinduced genotoxicity of dictamnine as shown by prophage induction. *Chem-Biol. Interact.*, **72** 105-11.

Furuya, T., Asada, Y. and Mori, H. (1987) Pyrrolizidine alkaloids, in *Naturally Occurring Carcinogens of Plant Origin* (ed. I. Hirono), Elsevier, Amsterdam, pp. 25-51.

Gatehouse, D., Stemp, G., Pascoe, S., Wilcox, P., Oliver, J. and Tweats, D.J. (1991) Investigations into the induction of aneuploidy and polyploidy in cultured mammalian cells by the antitussive agent, noscapine. *Mutat. Res.*, **252** 195.

Gawin, F.H. (1991) Cocaine addiction: psychology and neurophysiology. *Science*, **251** 1580-86.

Gershenzon, J. and Kreis, W. (1999) Biochemistry of terpenoids: monoterpenes, sesquiterpenes, diterpenes, sterols, cardiac glycosides and steroid saponins, in *Biochemistry of Plant Secondary Metabolism* (ed. M. Wink), Annual Plant Reviews, Vol. 2, Sheffield Academic Press, Sheffield, pp. 222-299.

Gilani, A.H., Shaheen, F., Christopoulos, A. and Mitchelson, F. (1997) Interaction of ebeinone, an alkaloid from *Fritillaria imperialis*, at two muscarinic acetylcholine receptor subtypes. *Life Sci.*, **60** 535-44.

Göggelmann, W. and Schimmer, O. (1983) Mutagenicity testing of β-asarone and commercial Calamus drugs with *Salmonella typhimurium. Mutat. Res.*, **121** 191-94.

Goss, P.E., Baker, M.A., Carver, J.P. and Dennis, J.W. (1995) Inhibitors of carbohydrate processing: a new class of anticancer agents. *Clin. Cancer Res.*, **1** 935-44.

Götzl, E. and Schimmer, O. (1993) Mutagenicity of aristolochic acids (I, II) and aristolic acid I in new YG strains in *Salmonella typhimurium* highly sensitive to certain mutagenic nitroarenes. *Mutagenesis*, **8** 17-22.

Gribkoff, V.K., Lum-Ragan, J.T., Boissard, C.G., Post-Munson, D.J., Meanwell, N.A., Starrett, J.E. Jr., Kozlowski, E.S., Romine, J.L., Trojnacki, J.T. *et al.* (1996) Effects of channel modulators on cloned large-conductance calcium-activated potassium channels. *Mol. Pharmacol.*, **50** 206-17.

Gröger, D. (1988) Vorkommen und Biochemie der Acridon-Alkaloide: Ein Fortschrittsbericht. *Pharmazie*, **43** 815-26.

Groopman, J.D. and Cain, L.G. (1990) Interactions of fungal and plant toxins with DNA: aflatoxins, sterigmatocystin, safrole, cycasin, and pyrrolizidine alkaloids, in *Chemical Carcinogenesis and Mutagenesis I* (eds. C.S. Cooper and P.L. Grover), Springer Verlag, Berlin, Heidelberg, pp. 373-407.

Haeberlein, H., Tschiersch, K.P., Boonen, G. and Hiller, K.O. (1996) *Chelidonium majus*, components with *in vitro* affinity for the $GABA_A$ receptor: Positive cooperation of alkaloids. *Planta Med.*, **62** 227-31.

Häfele, F. and Schimmer, O. (1988) Mutagenicity of furoquinoline alkaloids in the Salmonella microsome assay: mutagenicity of dictamnine is modified by various enzyme inducers and inhibitors. *Mutagenesis*, **3** 349-53.

Han, B.Y. and Liu, G.Q. (1988) Effect of tetrahydroisoquinoline alkaloids on α-adrenoceptors in rat brain. *Yaoxue Xuebao*, **23** 806-11.

Harborne, J.B. (1993) *Introduction to Ecological Biochemistry*, 4th edn., Academic Press, London.

Hardick, D.J., Cooper, G., Scott-Ward, T., Blagbrough, I.S., Potter, B.V.L. and Wonnacott, S. (1995) Conversion of the sodium channel activator, aconitine, into a potent alpha-7-selective nicotinic ligand. *FEBS Lett.*, **365** 79-82.

Hartley, S.E. and Jones, C.G. (1997) Plant chemistry and herbivory or why the world is green, in *Plant Ecology* (ed. M.J. Crawley), 2nd edn., pp. 284-324.

Hartmann, T. (1991) Alkaloids, in *Herbivores: Their Interaction with Secondary Metabolites:*, Vol. 1 (ed. G.A. Rosenthal and M.R. Berenbaum), Academic Press, San Diego, pp. 79-121.

Hartmann, T. and Witte, L. (1995) Chemistry, biology and chemoecology of the pyrrolizidine alkaloids, in *Alkaloids: Chemical and Biological Perspectives*, Vol. 9 (ed. S.W. Pelletier), Pergamon, Oxford, pp. 155-233.

Hashizume, T., Yamaguchi, H., Sato, T. and Fujii, T. (1991) Suppressive effect of bis-coclaurine alkaloids on agonist-induced activation of phospholipase A2 in rabbit platelets. *Biochem. Pharmacol.*, **41** 419-23.

Hasrat, J.A., Pieters, L., De Backer, J.P., Vauquelin, G. and Vlietinck, A.J. (1997) Screening of medicinal plants from Suriname for 5-HT1A ligands: bioactive isoquinoline alkaloids from the fruit of *Annona muricata. Phytomedicine*, **4** 133-40.

Heidemann, A., Völkner, W. and Mengs, U. (1996) Genotoxicity of aloeemodin *in vitro* and *in vivo. Mutat. Res.*, **367** 123-33.

Herbert, R., Kattah, A.E. and Knagg, E. (1990) The biosynthesis of the phenethylisoquinoline alkaloid, colchicine: early and intermediate stages. *Tetrahedron*, **20** 7119-38.

Hirono, I. (1987a) Cycasin, in *Naturally Occurring Carcinogens of Plant Origin* (ed. I. Hirono), Elsevier, Amsterdam, pp. 3-24.

Hirono, I. (1987b) *Naturally Occurring Carcinogens of Plant Origin: Toxicology, Pathology and Biochemistry.* Vol. 2, Bioactive molecules, Elsevier, Amsterdam.

Hirono, I. and Yamada, K. (1987) Bracken fern, in *Naturally Occurring Carcinogens of Plant Origin* (ed. I. Hirono), Elsevier, Amsterdam, pp. 87-120.

Hoffmann, G.R. and Morgan, R.W. (1984) Putative mutagens and carcinogens in Foods. V. Cycad azoxyglycosides. *Environ. Mutagen.*, **6** 103-16.

Holzinger, F. and Wink, M. (1996) Mediation of cardiac glycoside insensitivity in the monarch butterfly (*Danaus plexippus*): role of an amino acid substitution in the ouabain binding site of Na^+, K^+-ATPase. *J. Chem. Ecol.*, **22** 1921-37.

Holzinger, F., Frick, C. and Wink, M. (1992) Molecular basis for the insensitivity of the monarch (*Danaus plexippus*) to cardiac glycosides. *FEBS Lett.*, **314** 477-80.

Honerjaeger, P., Dugas, M. and Zong, X.G. (1992) Mutually exclusive action of cationic veratridine and cevadine at an intracellular site of the cardiac sodium channel. *J. Gen. Physiol.*, **99** 699-720.

Horie, S., Yano, S., Aimi, N., Sakai, S. and Watanabe, K. (1992) Effects of hirsutine, antihypertensive indole alkaloid from *Uncaria rhynchophylla*, on intracellular calcium in rat thoracic aorta. *Life Sci.*, **50** 491-98.

Horie, S., Yamamoto, L.T., Futagami, Y., Yano, S., Takayama, H., Sakai, S.-I., Aimi, N., Ponglux, D., Shan, J. *et al.* (1995) Analgesic, neuronal Ca^{2+}-channel-blocking and smooth muscle relaxant activities of mitragynine, an indole-alkaloid, from the Thai folk medicine 'kratom'. *Wakan Lyakugaku Zasshi*, **12** 366-67.

Hou, Y.F. and Liu, G.Q. (1988) The effects of tetrandrine, berbamine and some other tetrahydroisoquinolines on [^{3}H]QNB binding to muscarinic cholinergic receptors in rat brain. *Yaoxue Xuebao*, **23** 801-805.

Huang, J. and Johnston, G.A.R. (1990) (+)-Hydrastine, a potent competitive antagonist at mammalian $GABA_A$ receptors. *Br. J. Pharmacol.*, **99** 727-30.

Huang, X., Shi, J., Xie, X., Zhang, W. and Zhu, Y. (1993) Effects of rhynchophylline and isorhynchophylline on the 45Ca-transport in rabbit aorta. *Zhongguo Yaolixue Tongbao*, **9** 428-30.

Hue, B., Le Corronc, H., Kuballa, B. and Anton, R. (1994) Effects of the natural alkaloid, boldine, on cholinergic receptors of the insect central nervous system. *Pharm. Pharmacol. Lett.*, **3** 169-72.

IARC Monographs (1983) Vol. 31, IARC Lyon, pp. 163-70.

IARC Monographs (1986) Vol. 40, IARC Lyon, pp. 291-371.

Inui, M. (1992) Molecular machinery of calcium release from cardiac sarcoplasmic reticulum, in *Myocard. Mol. Biol.* (ed. M. Tada), Jpn. Sci. Soc. Press, Tokyo, Japan, pp. 181-88.

Ishibashi, M., Ohizumi, Y., Sasaki, T., Nakamura, H., Hirata, Y. and Kobayashi, J. (1987) Pseudodistomins A and B, novel antineoplastic piperdine alkaloids with calmodulin antagonistic activity from the Okinawan tunicate, *Pseudodistoma kanoko. J. Org. Chem.*, **52** 450-53.

Ishidate, M. Jr., Harnois, M.C. and Sofuni, T. (1988) A comparative analysis of data on the clastogenicity of 951 chemical substances tested in mammalian cell cultures. *Mutat. Res.*, **195** 151-213.

Ishii, H. and Natori, S. (1990) Mutagenicity of isoquinoline alkaloids, especially of the aporphine type. *Mutat. Res.*, **240** 267-79.

Ito, N., Hagiwara, A., Tamano, S., Kagawa, M., Shibata, M., Kurata, Y. and Fukushima, S. (1989) Lack of carcinogenicity of quercetin in F344/Du Crj rats. *Jap. J. Cancer Res.*, **80** 317-25.

Ivie, G.W., MacGregor, J.T. and Hammock, B.D. (1980) Mutagenicity of psoralen epoxides. *Mutat. Res.*, **79** 73-77.

Ivorra, M.D., Cercos, A., Zafra-Polo, M.C., Perez-Prieto, J., Saez, J., Cortes, D. and D'Ocon, P. (1992a) Selective chiral inhibition of calcium entry promoted by bis-benzyltetrahydroisoquinolines in rat uterus. *Eur. J. Pharmacol.*, **219** 303-309.

Ivorra, M.D., Lugnier, C., Schott, C., Catret, M., Noguera, M.A., Anselmi, E. and D'Ocon, P. (1992b) Multiple actions of glaucine on cyclic nucleotide phosphodiesterases, $alpha_1$-adrenoceptor and benzothiazepine binding site at the calcium-channel. *Br. J. Pharmacol.*, **106**, 387-94.

Ivorra, M.D., Chulia, S., Lugnier, C. and D'Ocon, M.P. (1993a) Selective action of two aporphines at $alpha_1$-adrenoceptors and potential-operated calcium channels. *Eur. J. Pharmacol.*, **231** 165-74.

Ivorra, M.D., Lugnier, C., Catret, M., Anselmi, E., Cortes, D. and D'Ocon, P. (1993b) Investigations of the dual contractile/relaxant properties shown by antioquine in rat aorta. *Br. J. Pharmacol.*, **109** 502-509.

Ivorra, M.D., Martinez, F., Serrano, A. and D'Ocon, P. (1993c) Different mechanism of relaxation induced by aporphine alkaloids in rat uterus. *J. Pharm. Pharmacol.*, **45** 439-43.

Jones, D.H. and Kim, H.L. (1981) Toxicity and mutagenicity of hymenoxon, a sesquiterpene lactone. *Toxicol. Lett.*, **9** 395-401.

Jovel, E., Kroeger, P. and Towers, N. (1996) Hydroquinone: the toxic compound of *Agaricus hondensis*. *Planta Med.*, **62** 185.

Kadota, S., Sun, X.-L., Basnet, P., Namba, T. (1996) Effects of alkaloids from *Corydalis decumbens* on contraction and electrophysiology of cardiac myocytes. *Phytotherapy Res.*, **10** 18-22.

Kamasaki, A., Takahashi, H., Tsumura, N., Niwa, J., Fujita, T. and Urasawa, S. (1982) Genotoxicity of flavoring agents. *Mutat. Res.*, **105** 387-92.

Kanamori, H., Sakamoto, I., Mizuta, M., Hashimoto, K. and Tanaka, O. (1984) Studies on the mutagenicity of *Swertiae herba. Chem. Pharm. Bull. Tokyo*, **32** 2290-95.

Kardos, J., Blasko, G., Kerekes, P., Kovacs, I. and Simonyi, M. (1984) Inhibition of [^{3}H] GABA binding to rat brain synaptic membranes by bicuculline-related alkaloids. *Biochem. Pharmacol.*, **33** 3537-45.

Kazic, T., Djordjevic, N. and Radulovic, S. (1989) Impairment of calcium permeability as a possible mode of action of ergot alkaloids: dihydroergosine, ergosinine and dihydroergotamine on the terminal ileum of the guinea-pig. *Period. Biol.*, **91** 281-87.

Kebabian, J.W. and Neumeyer, J.L. (1994) *The RBI Handbook of Receptor Classification.* RBI, Natick.

Kettenes, J.J., Van den, B., Salemink, C.A. and Kahn, I. (1981) Biological activity of the alkaloids of *Papaver bracteatum Lindl. J. Ethnopharmacology*, **3** 21-38.

Kilty, J.E., Lorang, D. and Amara, S.G. (1991) Cloning and expression of a cocaine-sensitive rat dopamine transporter. *Science*, **254** 578-79.

Kleibeuker, J.H., Cats, A., Zwart, N., Mulder, N.H., Hardonk, M.J. and deVries, E.G.E. (1995) Excessively high cell proliferation in sigmoid colon after an oral purge with anthraquinone glycosides. *J. Nat. Cancer Inst.*, **87** 452-53.

Klier, B., Schimmer, O. and Eilert, U. (1990) Untersuchungen zur metabolischen Aktivierung von Dictamnin *in vitro*. *Arch. Pharmazie*, **323** 681.

Knasmüller, S., Bresgen, N., Kassie, F., Mersch-Sundermann, V., Gelderblom, W., Zohrer, E. and Eckl, P.M. (1997) Genotoxic effects of three Fusarium mycotoxins, fumonisin B-1, moniliformin and vomitoxin, in bacteria and primary cultures of rat hepatocytes. *Mutat. Res.*, **391** 39-48.

Knaus, H.-G., McManus, O.B., Lee, S.H., Schmalhofer, W.A., Garcia-Calvo, M., Helms, L.M.H., Sanchez, M., Giangiacomo, K., Reuben, J.P. *et al.* (1994) Tremorgenic indole alkaloids potently inhibit smooth muscle high-conductance calcium-activated potassium channels. *Biochemistry*, **33** 5819-28.

Koerper, S., Wink, M. and Fink, H.A. (1998) Differential effects of alkaloids on sodium currents of isolated single skeletal muscle fibres. *FEBS Lett.*, **436** 251-55.

Kozikowski, A.P., Fauq, A.H., Miller, J.H. and McKinney, M. (1992) Alzheimer's therapy: an approach to novel muscarinic ligands based upon the naturally occurring alkaloid, himbacine. *Bioorg. Med. Chem. Lett.*, **2** 797-802.

Kozikowski, A.P., Campiani, G., Sun, L.-Q., Wang, S., Saxena, A. and Doctor, B.P. (1996) Identification of a more potent analog of the naturally occurring alkaloid, huperzine A: predictive molecular modeling of its interaction with AChE. *J. Am. Chem. Soc.*, **118** 11357-62.

Kraus, C., Abel, G. and Schimmer, O. (1985) Untersuchung einiger Pyrrolizidinalkaloide auf chromosomenschädigende Wirkung in menschlichen Lymphozyten *in vitro*. *Planta Med.*, **51** 89-91.

Krey, A.K. and Hahn, F.E. (1969) Berberine: complex with DNA. *Science*, **166** 755-57.

Krumbiegel, G., Hallensleben, J., Mennicke, W.H., Rittmann, N. and Roth, H.J. (1987) Studies on the metabolism of aristolochic acids I and II. *Xenobiotica*, **17** 981-91.

Lai, D.Y. and Woo, Y.-T. (1987) Naturally occurring carcinogens: an overview. *Environ. Carcinogen Rev.* (Part C of *J. Environ. Sci. Health*), **5** 121-73.

Larson, B.T., Samford, M.D., Camden, J.M., Piper, E.L., Kerley, M.S., Paterson, J.A. and Turner, J.T. (1995) Ergovaline-binding and activation of D2 dopamine receptors in GH4ZR7 cells. *J. Anim. Sci.*, **73** 1396-400.

Lazarovici, P., Rasouly, D., Friedman, L., Tabekman, R., Ovadia, H. and Matsuda, Y. (1996) K252a and staurosporine microbial alkaloid toxins as prototype of neurotropic drugs. *Adv. Exp. Med. Biol.*, **391** 367-77.

Lebanidze, M.G. and Gedevanishvili, M.D. (1985) Alkaloid akuammine as a stimulant for smooth muscle cells. *Soobshch. Akad. Nauk. Gruz. SSR*, **119** 541-44.

Leewanich, P., Tohda, M., Matsumoto, K., Subhadhirasakul, S., Takayama, H., Aimi, N. and Watanabe, H. (1997) Inhibitory effects of corymine, an alkaloidal component from the leaves of *Hunteria zeylanica*, on glycine receptors expressed in Xenopus oocytes. *Eur. J. Pharmacol.*, **332** 321-26.

Lever, J.R., Carroll, F.I., Patel, A., Abraham, P., Boja, J., Lewin, A. and Lew, R. (1993) Radiosynthesis of a photoaffinity probe for the cocaine receptor of the dopamine transporter: 3β-(*p*-chlorophenyl)tropan-2β-carboxylic acid m-([125I]-iodo)-*p*-azidophenethyl ester ([125I]-RTI-82). *J. Labelled Compd. Radiopharm.*, **33** 1131-37.

Levin, D.A. (1976) The chemical defenses of plants to pathogens and herbivores. *Ann. Rev. Ecol. Syst.*, **7** 121-59.

Lewin, G., Le Menez, P., Rolland, Y., Renouard, A. and Giesen-Crouse, E. (1992) Akuammine and dihydroakuammine, two indolomonoterpene alkaloids displaying affinity for opioid receptors. *J. Nat. Prod.*, **55** 380-84.

Liu, G., Han, B. and Wang, E. (1989a) Blocking actions of l-stephanine, xylopine and seven other tetrahydroisoquinoline alkaloids on α-adrenoceptors. *Zhongguo Yaoli Xuebao*, **10** 302-306.

Liu, G., Hou, Y., Pan, L. and Lu, Y. (1989b) Effects of some aporphines and their oxygenated, dehydrogenated derivatives on M-cholinergic receptors. *Zhongguo Yaoke Daxue Xuebao*, **20** 114-16.

Liu, Y., Shenouda, D., Bilfinger, T.V., Stefano, M.L., Magazine, H.I. and Stefano, G.B. (1996) Morphine stimulates nitric oxide release from invertebrate microglia. *Brain Res.*, **722** 125-31.

Lodish, H., Baltiomore, D., Berk, A., Zipursky, S.L., Matsudaira, P. and Darnell, J. (1995) *Molecular Cell Biology*, W.H. Freeman, New York.

Low, A.M., Berdik, M., Sormaz, L., Gataiance, S., Buchanan, M.R., Kwan, C.Y. and Daniel, E.E. (1996) Plant alkaloids, tetrandrine and hernandezine, inhibit calcium-depletion stimulated calcium entry in human and bovine endothelial cells. *Life Sci.*, **58** 2327-35.

Loza, I., Orallo, F., Verde, I., Gil-Longo, J., Cadavid, I. and Calleja, J.M. (1993) A study of glaucine-induced relaxation of rat aorta. *Planta Med.*, **59** 229-31.

Lu, H.R. and De Clerck, F. (1993) R 56 865, a sodium/calcium-overload inhibitor, protects against aconitine-induced cardiac arrhythmias *in vivo*. *J. Cardiovasc. Pharmacol.*, **22** 120-25.

Luckner, M. (1990) *Secondary Metabolism in Microorganisms, Plants and Animals*. Springer, Berlin, Heidelberg.

Madrero, Y., Elorriaga, M., Martinez, S., Noguera, M.A., Cassels, B.K., D'Ocon, P. and Ivorra, M.D. (1996) A possible structural determinant of selectivity of boldine and derivatives for the α-1_A-adrenoceptor subtype. *Br. J. Pharmacol.*, **119** 1563-68.

Maelicke, A., Schrattenholz, A., Storch, A., Schroder, B., Gutbrod, O., Methfessel, C., Weber, K.-H., Pereira, E.E.F., Albuquerque, M.A. *et al.* (1995) Noncompetitive agonism at nicotinic acetylcholine receptors: functional significance for CNS signal transduction. *J. Recept. Signal Transduc. Res.*, **15** 333-53.

Maier, P., Schawalder, H.P., Weibel, B. and Zbinden, G. (1985) Aristolochic acid induces 6-thioguanine-resistant mutants in an extrahepatic tissue in rats after oral application. *Mutat. Res.*, **143** 143-48.

Maiti, M. and Chaudhuri, K. (1981) Interaction of berberine chloride with naturally occurring deoxyribonucleic acids. *Ind. J. Biochem. Biophys.*, **18** 245-50.

Maiti, M., Nandi, R. and Chaudhuri, K. (1982) Sanguinarine: a monofunctional intercalating alkaloid. *FEBS Lett.*, **142** 280-84.

Malchow, R.P., Qian, H. and Ripps, H. (1994) A novel action of quinine and quinidine on the membrane conductance of neurons from the vertebrate retina. *J. Gen. Physiol.*, **104** 1039-55.

Manolache, M., Gebauer, J. and Röhrborn, G. (1985) Mutagenic activity of aristolochic acid in the V79/HGPRT point mutation assay. *Mutat. Res.*, **147** 133.

Markstein, R., Seiler, M.P., Jaton, A. and Briner, U. (1992) Structure-activity relationship and therapeutic uses of dopaminergic ergots. *Neurochem. Int.*, **20** 211S-214S.

Marrazzini, A., Betti, C., Barale, R., Bernacchi, F., Loprieno, N. (1991) Cytogenetic effects of possible aneuploidizing agents. *Mutat. Res.*, **252** 195-96.

Marta, M. and Pomponi, M. (1987) A new hypothesis on physostigmine anticholinesterase mechanism. *Acta Med. Rom.*, **25** 433-37.

Martindale: Reynolds, J.E. (1993) *Martindale: The Extra Pharmacopoeia*, 3rd edn., Pharmaceutical Press, London.

Masuda, T. and Ueno, Y. (1984) Microsomal transformation of emodin into a direct mutagen. *Mutat. Res.*, **125** 135-44.

Mathur, A.C., Sharma, A.K. and Verma, V. (1980) Cytopathological effects of aristolochic acid on male house flies causing sterility. *Experientia*, **36** 245-46.

Matsuda, M., Kobayashi, T., Nagao, S., Ohta, T. and Nozoe, S. (1996) Laccarin, a new alkaloid from the mushroom, *Laccaria vinaceoavellanea*. *Heterocycles*, **43** 685-90.

Matsumoto, K., Mizowaki, M., Suchitra, T., Murakami, Y., Takayama, H., Sakai, S.-i., Aimi, N. and Watanabe, H. (1996) Central antinociceptive effects of mitragynine in mice: contribution of descending noradrenergic and serotonergic systems. *Eur. J. Pharmacol.*, **317** 75-81.

Matsumoto, S. and Shimizu, T. (1995) Flecainide blocks the stimulatory effect of veratridine on slowly-adapting pulmonary stretch receptors in anesthetized rabbits without changing lung mechanics. *Acta Physiol. Scand.*, **155** 297-302.

Matsuoka, A., Hirosawa, A., Natori, S., Iwasaki, S., Sofuni, T. and Ishidate M. Jr. (1989) Mutagenicity of ptaquiloside, the carcinogen in bracken, and its related illudane-type sesquiterpenes. II. Chromosomal aberration tests with cultured mammalian cells. *Mutat. Res.*, **215** 179-85.

Matsushima, T., Araki, A., Yagame, O., Maramatsu, M., Koyama, K., Ohsawa, K., Natori, S. and Tomimori, H. (1985) Mutagenicities of xanthone derivatives in *Salmonella typhimurium* TA100, TA98, TA97 and TA2637. *Mutat. Res.*, **150** 141-46.

Mattocks, A.R. (1986) *Chemistry and Toxicology of Pyrrolizidine Alkaloids*. Academic Press, London.

McConnell, O.J., Saucy, G. and Jacobs, R. (1993) Use for topsentin compounds and pharmaceutical compositions containing same. Regents of the University of California, Harbor Branch Oceanographic Institute Inc., USA. Patent US 5290777A.

McDanell, R., McLean, A.E.M., Hanley, A.B., Heaney, R.K. and Fenwick, G.R. (1988) Chemical and biological properties of indole glucosinolates (Glucobrassicins): a review. *Food Chem. Toxicol.*, **26** 59-70.

McKenna, D.J., Towers, G.H.N. and Abbott, F.S. (1984) Monoamine oxidase inhibitors in South American plants. *J. Ethnopharmacol.*, **12** 179-211.

Meinwald, J. (1990) Alkaloids and isoprenoids as defensive and signalling agents among insects. *Pure Appl. Chem.*, **62** 1325-28.

Melchior, C. and Collins, M. (1982) The route and significance of endogenous synthesis of alkaloids in animals. *CRC Crit. Rev. Toxicol.*, pp. 313-55.

Mengs, U. (1988) Tumour induction in mice following exposure to aristolochic acid. *Arch. Toxicol.*, **61** 504-505.

Mengs, U. and Klein, M. (1988) Genotoxic effects of aristolochic acid in the mouse micronucleus test. *Planta Med.*, **54** 502-503.

Mengs, U., Lang, W. and Poch, J.A. (1982) The carcinogenic action of aristolochic acid in rats. *Arch. Toxicol.*, **51** 107-19.

Mitsui, N., Noro, T., Kuroyanagi, M., Miyase, T., Umehara, K. and Ueno, A. (1989) Studies of enzyme inhibitors. Part VI. Monoamine oxidase inhibitors from Cinchonae cortex. *Chem. Pharm. Bull.*, **37** 363-66.

Mix, D.B., Guinaudeau, H. and Shamma, M. (1982) The aristolochic acids and aristolactams. *J. Nat. Prod.*, **45** 657-66.

Mizuta, M. and Kanamori, H. (1985) Mutagenic activities of dictamnine and gamma-fagarine from *Dictamni radicis cortex* (Rutaceae). *Mutat. Res.*, **144** 221-25.

Moore, M.M., Brock, K.H., Doerr, C.R. and DeMarini, D.M. (1987) Mutagenesis of L5178Y/ $TK^{+/-}$-3.7.2C mouse lymphoma cells by the clastogen ellipticine. *Environ. Mutagen.*, **9** 161-70.

Moreno, J.J. (1993) Effect of aristolochic acid on arachidonic acid cascade and *in vivo* models of inflammation. *Immunopharmacology*, **26** 1-9.

Moreno, P.R.H., Vargas, V.M.F., Andrade, H.H.R., Henriques, A.T. and Henriques, J.A.P. (1991) Genotoxicity of the boldine aporphine alkaloid in prokaryotic and eukaryotic organisms. *Mutat. Res.*, **260** 145-52.

Mori, H., Sugie, S., Niwa, K., Takahashi, M. and Kawai, K. (1985) Induction of intestinal tumors in rats by chrysazin. *Br. J. Cancer*, **52** 781-83.

Mori, H., Sugie, S., Niwa, K., Yoshimi, N., Tanaka, T. and Hirono, I. (1986) Carcinogenicity of chrysazin in large intestine and liver of mice. *Jap. J. Cancer Res.*, **77** 871-76.

Mori, H., Yoshimi, N., Iwata, H., Mori, Y., Hara, A., Tanaka, T. and Kawai, K. (1990) Carcinogenicity of naturally occurring 1-hydroxyanthraquinone in rats: induction of large bowel, liver and stomach neoplasms. *Carcinogenesis*, **11** 799-802.

Mori, Y., Yoshimi, N., Iwata, H., Tanaka, T. and Mori, H. (1991) The synergistic effect of 1-hydroxyanthraquinone on methylazoxymethanol acetate-induced carcinogenesis in rats. *Carcinogenesis*, **12** 335-38.

Morimoto, I., Nozaka, T., Watanabe, F., Ishino, M., Hirose, Y. and Okitsu, T. (1983) Mutagenic activities of gentisin and isogentisin from *Gentianae radix* (Gentianaceae). *Mutat. Res.*, **116** 103-17.

Moriyasu, M., Ichimaru, M. and Kato, A. (1990) A semicontinuous assay of the inhibition of cyclic-AMP phosphodiesterase by benzo[c]phenanthridine alkaloids. *J. Liq. Chromatogr.*, **13** 543-55.

Morse, M.A., Wang, C.-X., Amin, S.G., Hecht, S.S. and Chung, F.-L. (1988) Effects of dietary sinigrin or indole-3-carbinol on O^6-methylguanine-DNA transmethylase activity and 4-(methylnitrosamino)-1-(3-pyridyl)-1-butanone-induced DNA methylation and tumorigenicity in F344 rats. *Carcinogenesis*, **9** 1891-95.

Müller, L., Kasper, P. and Madle, S. (1991) The quality of genotoxicity testing of drugs: experiences of a regulatory agency with new and old compounds. *Mutagenesis*, **6** 143-49.

Müller, L., Kasper, P. and Petr, T. (1992) The clastogenicity *in vitro* of quercetin is independent of external metabolization. *Mutat. Res.*, **271** 178.

Müller, S.O., Eckert, I., Lutz, W.K. and Stopper, H. (1996) Genotoxicity of the laxative drug components, emodin, aloe-emodin and danthron, in mammalian cells: topoisomerase II mediated? *Mutat. Res.*, **371** 165-73.

Musk, S.R.R., Smith, T.K. and Johnson, I.T. (1995) On the cytotoxicity and genotoxicity of allyl and phenethyl isothiocyanates and their parent glucosinolates, sinigrin and gluconasturtiin. *Mutat. Res.*, **348** 19-23.

Mutschler, E. (1996) *Arzneimittelwirkungen*. Wissensch. Verlagsges., Stuttgart.

Nagao, M., Morita, N., Yahagi, T., Shimizu, M., Kuroyanagi, M., Fukuoka, M., Yoshihira, K., Natori, S., Fujino, T. and Sugimura, T. (1981) Mutagenicities of 61 flavonoids and 11 related compounds. *Environ. Mutagen.*, **3** 401-19.

Nagao, T., Saito, K., Hirayama, E., Uchikoshi, K., Koyama, K., Natori, S., Morisaki, N., Iwasaki, S. and Matsushima, T. (1989) Mutagenicity of ptaquiloside, the carcinogen in bracken, and its related illudane-type sesquiterpenes. I. Mutagenicity in *Salmonella typhimurium*. *Mutat. Res.*, **215** 173-78.

Nagase, M. and Hagihara, Y. (1986) Effects of raubasine on peripheral circulation in cats. *Yakurito Chiryo*, **14** 5577-89.

Nagata, R., Izumi, K., Iwata, S., Shimizu, T. and Fukuda, T. (1991) Mechanisms of veratramine-induced 5-HT syndrome in mice. *Jpn. J. Pharmacol.*, **55** 139-46.

Nakazawa, K., Watano, T., Ohara-Imaizumi, M., Inoue, K., Fujimori, K., Ozaki, Y., Harada, M. and Takanaka, A. (1991) Inhibition of ion channels by hirsutine in rat pheochromocytoma cells. *Jpn. J. Pharmacol.*, **57** 507-15.

Nanasi, P.P., Kiss, T., Danko, M. and Lathrop, D.A. (1990) Different actions of aconitine and veratrum alkaloids on frog skeletal muscle. *Gen. Pharmacol.*, **21** 863-68.

Nandi, R. and Maiti, M. (1985) Binding of sanguinarine to deoxyribonucleic acids of differing base composition. *Biochem. Pharmacol.*, **34** 321-24.

Nandi, R., Chakraborty, S. and Maiti, M. (1991) Base-dependent and sequence-dependent binding of aristololactam β-D-glucoside to deoxyribonucleic acid. *Biochemistry USA*, **30** 3715-20.

Nataraj, A.J., Black, H.S. and Ananthaswamy, H.N. (1996) Signature p53 mutation at DNA cross-linking sites in 8-methoxypsoralen and ultraviolet A (PUVA)-induced murine skin cancers. *Proc. Natl. Acad. Sci. USA*, **93** 7961-65.

Natori, S. (1987) Mushroom hydrazines, in *Naturally Occurring Carcinogens of Plant Origin* (ed. I. Hirono), Elsevier, Amsterdam, pp. 127-37.

Natori, S. and Ueno, I. (1987) Flavonoids, in *Naturally Occurring Carcinogens of Plant Origin* (ed. I. Hirono), Elsevier, Amsterdam, pp. 53-85.

Neher, E. and Sakman, B. (1992) The patch clamp technique. *Sci. Am.*, **266** 28-35.

Nesic, O. and Pasic, M. (1992) Characteristics of outward current induced by application of dopamine on a small neuron. *Comp. Biochem. Physiol. C.*, **103** 597-606.

Neudecker, T. and Henschler, D. (1985) Allylisothiocyanate is mutagenic in *Salmonella typhimurium*. *Mutat. Res.*, **156** 33-37.

Nimit, Y., Schulze, I., Cashaw, J.L., Ruchirawat, S. and Davis, V.E. (1983) Interaction of catecholamine-derived alkaloids with central neurotransmitter receptors. *J. Neurosci. Res.*, **10** 175-89.

Nozaka, T., Morimoto, I., Ishino, M., Okitsu, T., Kondoh, H., Kyogoku, K., Sugawara, Y. and Iwasaki, H. (1987) Mutagenic principles in *Sinomeni caulis et rhizoma*. I. The structure of a mutagenic alkaloid, *N*-demethyl-*N*-formyldehydronuciferine, in the neutral fraction of the methanol extract. *Chem. Pharm. Bull. Tokyo*, **35** 2844-48.

Nozaka, T., Watanabe, F., Tadaki, S., Ishino, M., Morimoto, I., Kunitomo Jun-ichi, Ishii, H. and Natori, S. (1990) Mutagenicity of isoquinoline alkaloids, especially of the aporphine type. *Mutat. Res.*, **240** 267-79.

Ogawa, S., Hirayama, T., Tokuda, M., Hirai, K. and Fukui, S. (1986) The effect of quercetin, a mutagenicity-enhancing agent, on the metabolism of 2-acetylaminofluorene with mammalian metabolic activation systems. *Mutat. Res.*, **162** 179-86.

Ohashi, Y. and Matsuoka, M. (1985) Localization of pathogenesis-related proteins in the epidermis and intercellular spaces of tobacco leaves after their induction by potassium salicylate or tobacco mosaic virus infection. *Proc. Natl. Acad. Sci.*, **82** 1852-54.

Oliver, J.W., Abney, L.K., Strickland, J.R. and Linnabary, R.D. (1993) Vasoconstriction in bovine vasculature induced by the tall fescue alkaloid, lysergamide. *J. Anim. Sci.*, **71** 2708-13.

Orallo, F., Fernandez Alzueta, A., Loza, M.I., Vivas, N., Badia, A., Campos, M., Honrubia, M.A. and Cadavid, M.I. (1993) Study of the mechanism of the relaxant action of (+)-glaucine in rat vas deferens. *Br. J. Pharmacol.*, **110** 943-48.

Orallo, F., Alzueta, A.F., Campos-Toimil, M. and Calleja, J.M. (1995) Study of the *in vivo* and *in vitro* cardiovascular effects of (+)-glaucine and *N*-carbethoxysecoglaucine in rats. *Br. J. Pharmacol.*, **114** 1419-27.

Ori, K., Mimaki, Y., Sashida, Y., Nikaido, T., Ohmoto, T. and Masuko, A. (1992) Persicanidine A, a novel cerveratrum alkaloid from the bulbs of *Fritillaria persica*. *Chem. Lett.*, 163-66.

Pan, J., Yin, F., Shen, C., Lu, C. and Han, G. (1989) Active constituents of the root of *Berberis poiretti*. *Tianran Chanwu Yanjiu Yu Kaifa*, **1** 23-26.

Papke, R.L. and Heinemann, S.F. (1994) Partial agonist properties of cytisine on neuronal nicotinic receptors containing the beta$_2$-subunit. *Mol. Pharmacol.*, **45** 142-49.

Paulini, H. and Schimmer, O. (1989) Mutagenicity testing of rutacridone epoxide and rutacridone alkaloids in *Ruta graveolens* (L.), using the Salmonella/microsome assay. *Mutagenesis*, **4** 45-50.

Paulini, H., Eilert, U. and Schimmer, O. (1987) Mutagenic compounds in an extract from *Rutae herba* (*Ruta graveolens* L.): mutagenicity is partially caused by furoquinoline alkaloids. *Mutagenesis*, **2** 271-73.

Paulini, H., Waibel, R. and Schimmer, O. (1989) Mutagenicity and structure-mutagenicity relationships of furoquinolines, naturally occurring alkaloids of Rutaceae. *Mutat. Res.*, **227** 179-86.

Paulini, H., Schimmer, O., Ratka, O. and Röder, E. (1991) Isogravacridonchlorine: a potent and direct acting frameshift mutagen from the roots of *Ruta graveolens* (L.) *Planta Med.*, **57** 59-61.

Pelassy, C. and Aussel, C. (1993) Effect of Cinchona bark alkaloids and chloroquine on phospholipid synthesis. *Pharmacology*, **47** 28-35.

Perez Leon, J.A. and Salceda Sacanelles, R. (1996) Postsynaptic glycine receptor. *Ciencia*, **47** 177-89.

Petersen, M., Strack, D. and Matern, U. (1999) Biosynthesis of phenylpropanoids and related compounds, in *Biochemistry of Plant Secondary Metabolism* (ed. M. Wink), Annual Plant Reviews, Vol. 2, Sheffield Academic Press, Sheffield, pp. 151-221.

Pezzuto, J.M., Swanson, S.M., Mar, W., Che, C.-T., Cordell, G.A. and Fong, H.H.S. (1988) Evaluation of the mutagenic and cytostatic potential of aristolochic acid (3,4-methylenedioxy-8-methoxy-10-nitrophenanthrene-1-carboxylic acid) and several of its derivatives. *Mutat. Res.*, **206** 447-54.

Pfau, W., Schmeiser, H.H. and Wiessler, M. (1990a) Aristolochic acid binds covalently to the exocyclic amino group of purine nucleotides in DNA. *Carcinogenesis*, **11** 313-19.

Pfau, W., Schmeiser, H.H. and Wiessler, M. (1990b) ^{32}P-postlabelling analysis of the DNA adducts formed by aristolochic acid I and II. *Carcinogenesis*, **11** 1627-33.

Pfau, W., Pool-Zobel, B.L., von der Lieth, C.W. and Wiessler, M. (1990c) The structural basis for the mutagenicity of aristolochic acid. *Cancer Lett.*, **55** 7-11.

Pfau, W., Schmeiser, H.H. and Wiessler, M. (1991) N_6-adenyl arylation of DNA by aristolochic acid II and a synthetic model for the putative proximate carcinogen. *Chem. Res. Toxicol.*, **4** 581-86.

Pfyffer, G.E. and Towers, G.H.N. (1982a) Photochemical interaction of dictamnine, a furoquinoline alkaloid, with fungal DNA *in vitro* and *in vivo*. *Can. J. Microbiol.*, **28** 468-73.

Pfyffer, G.E., Pfyffer, B.U. and Towers, G.H.N. (1982b) Monoaddition of dictamnine to synthetic double-stranded polydeoxyribonucleotides in UVA and the effect of photomodified DNA on template activity. *Photochem. Photobiol.*, **35** 793-97.

Pinto, M., Guerineau, M. and Paoletti, C. (1982) Mitochondrial and nuclear mutagenicity of ellipticine and derivatives in the yeast *Saccharomyces cerevisiae*. *Biochem. Pharmacol.*, **31** 2161-67.

Pistelli, L., Nieri, E., Bilia, A.R., Marsili, A. and Scarpato, R. (1993) Chemical constituents of *Aristolochia rigida* and mutagenic activity of aristolochic acid IV. *J. Nat. Prod.*, **56** 1605-608.

Poginsky, B., Westendorf, J., Blömeke, B., Marquardt, H., Hewer, A., Grower, P.L. and Phillips, D.H. (1991) Evaluation of DNA-binding activity of hydroxyanthraquinones occurring in *Rubia tinctorum* (L.). *Carcinogenesis*, **12** 1265-71.

Popp, R. and Schimmer, O. (1991) Induction of sister-chromatid exchanges (SCE), polyploidy and micronuclei by plant flavonoids in human lymphocyte cultures: a comparative study. *Mutat. Res.*, **246** 205-13.

Protais, P., Arbaoui, J., Bakkali, E.-H., Bermejo, A. and Cortes, D. (1995) Effects of various isoquinoline alkaloids on *in vitro* 3H-dopamine uptake by rat striatal synaptosomes. *J. Nat. Prod.*, **58** 1475-84.

Puri, E.C. and Müller, D. (1985) Mutagenic properties and carcinogenicity of aristolochic acid. *Mutat. Res.*, **147** 133-34.

Qureshi, S., Tariq, M., El-Feraly, F.S. and Al-Meshal, I.A. (1988) Genetic effects of chronic treatment with cathinone in mice. *Mutagenesis*, **3** 481-83.

Rasolonjanahary, R., Sevenet, T., Gueritte Voegelein, F. and Kordon, C. (1995) Psycholeine, a natural alkaloid extracted from *Psychotria oleoides*, acts as a weak antagonist of somatostatin. *Eur. J. Pharmacol.*, **285** 19-23.

Rasouly, D., Lazarovici, P. and Matsuda, Y. (1995) Biochemical and pharmacological properties of K252 microbial alkaloids, in *Toxic Action of Marine and Terrestrial Alkaloids* (ed. M.S. Blum), Alaken, Fort Collins, pp. 161-190.

Rauwald, H.W., Kober, M., Mutschler, E. and Lambrecht, G. (1992) *Cryptolepis sanguinolenta*: antimuscarinic properties of cryptolepine and the alkaloid fraction at M1, M2 and M3 receptors. *Planta Med.*, **58** 486-88.

Renouard, A., Widdowson, P.S. and Millan, M.J. (1994) Multiple alpha$_2$ adrenergic receptor subtypes. I. Comparison of [^{3}H]RX821002-labeled rat Rα-$_{2A}$ adrenergic receptors in cerebral cortex to human Hα_{2A} adrenergic receptor and other populations of α-$_2$ adrenergic subtypes. *J. Pharmacol. Exp. Ther.*, **270** 946-57.

Roberts, M.F. and Wink, M. (1998) *Alkaloids: Biochemistry, Ecology and Medicinal Applications*. Plenum Press, New York.

Roberts, M.F. and Strack, D. (1999) Biochemistry and physiology of alkaloids and betalains, in *Biochemistry of Plant Secondary Metabolism* (ed. M. Wink), Annual Plant Reviews, Vol. 2, Sheffield Academic Press, Sheffield, pp. 17-78.

Robisch, G., Schimmer, O. and Göggelmann, W. (1983) Aristolochic acid is a direct mutagen in *Salmonella typhimurium*. *Mutat. Res.*, **113** 346-47.

Rodighiero, P., Guiotto, A., Chilin, A., Bordin, F., Baccichetti, F., Carlassare, F., Vedaldi, D., Caffieri, S., Pozzan, A. and Dall'Acqua, F. (1996) Angular furoquinolinones, psoralen analogs: novel antiproliferative agents for skin diseases: synthesis, biological activity, mechanism of action, and computer-aided studies. *J. Med. Chem.*, **39** 1293-302.

Röder, E. (1995) Medicinal plants in Europe containing pyrrolizidine alkaloids. *Pharmazie*, **50** 83-98.

Rommelspacher, H., Nanz, C., Borbe, H., Fehske, K. *et al.* (1980) 1-Methyl-a-carboline (Harmane), apotent endogenous inhibitor of benzodiazepine receptor binding. *Nauyn Sch. Arch. Pharmacol.*, **314** 97-100.

Roquebert, J. and Grenie, B. (1986) α_2-Adrenergic agonist and α_1-adrenergic antagonist activity of ergotamine and dihydroergotamine in rats. *Arch. Int. Pharmacodyn. Ther.*, **284** 30-37.

Rosenfeld, M., Makman, M., Ahn, H. and Thal, L. (1980) Selective influence of ergot alkaloids on cortical and striatal dopaminergic and serotonergic receptors. *Adv. Biochem. Psychopharmacol.*, **23** 83-93.

Rosenkranz, H.S. and Klopman, G. (1990) Novel structural concepts in elucidating the potential genotoxicity and carcinogenicity of tetrandrine, a traditional herbal drug. *Mutat. Res.*, **244** 265-71.

Rosenthal, G.A. (1982) *Plant Nonprotein Amino and Imino Acid.* Academic Press, New York.

Rosenthal, G.A. and Janzen, D.H. (1979) *Herbivores: Their Interaction with Secondary Plant Metabolites*. Academic Press, New York.

Rosenthal, G.A. and Berenbaum, M.R. (1991) *Herbivores: Their Interactions with Secondary Plant Metabolites*, 2nd edn., Academic Press, San Diego.

Rosenthal, G.A. and Berenbaum, M.R. (1992) *Herbivores: Their Interactions with Secondary Plant Metabolites*, 2nd edn., Academic Press, San Diego.

Rossiello, M., Laconi, E., Rao, P.M., Rajalakshmi, S., Sarma, D.S.R. (1993) Induction of hepatic nodules in the rat by aristolochic acid. *Cancer Lett.*, **71** 1-3.

Rubiolo, P., Pieters, L., Calomme, M., Bicchi, C., Vlietinck, A. and van den Berghe, D. (1992) Mutagenicity of pyrrolizidine alkaloids in the *Salmonella typhimurium*/mammalian microsome system. *Mutat. Res.*, **281** 143-47.

Rueff, J., Laires, A., Borba, H., Chaveca, T., Gomez, M.I. and Halpern, M. (1986) Genetic toxicology of flavonoids: the role of metabolic conditions in the induction of reverse mutation, SOS functions and sister-chromatid exchanges. *Mutagenesis*, **1** 179-83.

Saito, K., Nagao, T., Takatsuki, S., Koyama, K., Natori, S. (1990) The sesquiterpenoid carcinogen of bracken fern and some analogues from the Pteridaceae. *Phytochemistry*, **29** 1475-79.

Sakamoto-Hojo, E.T., Takahashi, C.S., Ferrari, I., Motidome, M. (1988) Clastogenic effect of the plant alkaloid, ellipticine, on bone marrow cells of Wistar rats and on human peripheral blood lymphocytes. *Mutat. Res.*, **199** 11-19.

Schiestl, R.H., Shian Chan, W., Gietz, R.D., Mehta, R.D. and Hastings, P.J. (1989) Safrole, eugenol and methyleugenol induce intrachromosomal recombination in yeast. *Mutat. Res.*, **224** 427-36.

Schimmer, O. and Leimeister, U. (1989) The SCE-inducing potency of the furoquinoline alkaloid, γ-fagarine, and a γ-fagarine-containing tincture from *Rutae herba*, in cultured human lymphocytes. *Mutagenesis*, **4** 467-70.

Schimmer, O. and Kühne, I. (1991) Furoquinoline alkaloids as photosensitizers in *Chlamydomonas reinhardtii*. *Mutat. Res.*, **249** 105-10.

Schimmer, O. and Drewello, U. (1994) 9-Methoxytariacuripyrone, a naturally occurring nitroaromatic compound with strong mutagenicity in *Salmonella typhimurium*. *Mutagenesis*, **9** 547-51.
Schimmer, O., Kiefer, J. and Paulini, H. (1991) Inhibitory effects of furocoumarins in *Salmonella typhimurium* TA98 on the mutagenicity of dictamnine and rutacridone, promutagens from *Ruta graveolens* (L.). *Mutagenesis*, **6** 501-506.
Schmeiser, H.H., Pool, L.B. and Wiessler, M. (1984) Mutagenicity of the two main components of commercially available carcinogenic aristolochic acid in *Salmonella typhimurium*. *Cancer Lett.*, **23** 97-101.
Schmeiser, H.H., Pool, L.B. and Wiessler, M. (1986) Identification and mutagenicity of metabolites of aristolochic acid formed by rat liver. *Carcinogenesis*, **7** 59-63.
Schmeiser, H.H., Schoepe, K.-B. and Wiessler, M. (1988) DNA adduct formation of aristolochic acid I and II *in vitro* and *in vivo*. *Carcinogenesis*, **9** 297-303.
Schmeiser, H.H., Jansen, J.W.G., Lyons, J., Scherf, H.R., Pfau, W., Buchmann, A., Bartram, C.R. and Wiessler, M. (1990) Aristolochic acid activates *ras* genes in rat tumors at deoxyadenosine residues. *Cancer Res.*, **50** 5464-69.
Schmeiser, H.H., Scherf, H.R. and Wiessler, M. (1991) Activating mutations at codon-61 of the *c-Ha-ras* gene in thin-tissue sections of tumors induced by aristolochic acid in rats and mice. *Cancer Lett.*, **59** 139-43.
Schmeiser, H.H., Bieler, C.A., Wiessler, M., van Ypersele de Strihou, C. and Cosyns, J.-P. (1996) Detection of DNA adducts formed by aristolochic acid in renal tissue from patients with Chinese herbs nephropathy. *Cancer Res.*, **56** 2025-28.
Schmeller, T. and Wink, M. (1997) Utilization of alkaloids in modern medicine, in *Alkaloids: Biochemistry, Ecology and Medicinal Applications* (ed. M.F. Roberts and M. Wink), Plenum, New York, pp. 435-58.
Schmeller, T., Sauerwein, M., Sporer, F., Wink, M. (1994) Binding of quinolizidine alkaloids to nicotinic and muscarinic acetylcholine receptors. *J. Nat. Prod.*, **57** 1316-19.
Schmeller, T., Sporer, F., Sauerwein, M. and Wink, M. (1995) Binding of tropane alkaloids to nicotinic and muscarinic acetylcholine receptors. *Pharmazie*, **50** 493-95.
Schmeller, T., El-Shazly, A. and Wink, M. (1997a) Allelochemical activities of pyrrolizidine alkaloids: interactions with neuroreceptors and acetylcholine-related enzymes. *J. Chem. Ecol.*, **23** 399-416.
Schmeller, T., Latz-Brüning, B. and Wink, M. (1997b) Biochemical activities of berberine, palmatine and sanguinarine mediating chemical defence against microorganisms and herbivores. *Phytochemistry*, **44** 257-66.
Schmitt, M., Turberg, A., Londershausen, M. and Dorn, A. (1996) Binding sites for Ca^{2+}-channel effectors and ryanodine in Periplaneta americana: possible targets for new insecticides. *Pestic. Sci.*, **48** 375-88.
Schneider, D., Boppré, M., Zweig, J., Horsley, S.B., Bell, T.W., Meinld, J., Hansen, K. and Diehl, E.W. (1982) Scent organ development in Creatonotes moths: regulation by pyrrolizidine alkaloids. *Science*, **215** 1264-65.
Schrattenholz, A., Godovac-Zimmermann, J., Schaefer, H.-J., Albuquerque, E.X. and Maelicke, A. (1993) Photoaffinity labeling of Torpedo acetylcholine receptor by physostigmine. *Eur. J. Biochem.*, **216** 671-77.
Scott, B.R., Pathak, M.A. and Mohn, G.R. (1976) Molecular and genetic basis of furocoumarin reactions. *Mutat. Res.*, **39** 29-74.
Selmar, D. (1999) Biosynthesis of cyanogenic glycosides, glucosinolates and nonprotein amino acids, in *Biochemistry of Plant Secondary Metabolism* (ed. M. Wink), Annual Plant Reviews, Vol. 2, Sheffield Academic Press, Sheffield, pp. 79-150.
Sershen, H., Hashim, A. and Lajtha, A. (1996) Effect of ibogaine on cocaine-induced efflux of [^{3}H]-dopamine and [^{3}H]-serotonin from mouse striatum. *Pharmacol. Biochem. Behav.*, **53** 863-69.

Sershen, H., Hashim, A. and Lajtha, A. (1997) Ibogaine and cocaine abuse: pharmacological interactions at dopamine and serotonin receptors. *Brain Res. Bull.*, **42** 161-68.

Sheldon, R.J., Malarchik, M.E., Burks, T.F. and Porreca, F. (1990) Effects of nerve stimulation on ion transport in mouse jejunum: responses to Veratrum alkaloids. *J. Pharmacol. Exp. Ther.*, **252** 636-42.

Shen, H.M. and Ong, C.N. (1996) Mutations of the p53 tumor suppressor gene and *ras* oncogenes in aflatoxin hepatocarcinogenesis. *Mutat. Res.*, **366** 23-44.

Shoji, N., Umeyama, A., Saito, N., Iuchi, A., Takemoto, T., Kajiwara, A. and Ohizumi, Y. (1987) Asimilobine and lirinidine serotonergic receptor antagonists, from *Nelumbo nucifera. J. Nat. Prod.*, **50** 773-74.

Silva, I.D., Rodrigues, A.S., Gaspar, J., Maia, R., Laires, A. and Rueff, J. (1997a) Involvement of rat cytochrome 1A1 in the biotransformation of kaempferol to quercetin: relevance to the genotoxicity of kaempferol. *Mutagenesis*, **12** 383-90.

Silva, I.D., Rodrigues, A.S., Gaspar, J., Laires, A. and Rueff, J. (1997b) Metabolism of galangin by rat cytochromes P_{450}: relevance to the genotoxicity of galangin. *Mutat. Res.*, **393** 247-58.

Simi, S., Morelli, S., Gervasi, P.G. and Rainaldi, G. (1995) Clastogenicity of anthraquinones in V79 and in three derived cell lines expressing P_{450} enzymes. *Mutat. Res.*, **347** 151-56.

Simonyi, M. (1987) Stereoselective interaction of phthalideisoquinoline and related alkaloids with the GABA receptor, in *Proc. F.E.C.S. Int. Conf. Chem. Biotechnol. Biol. Act. Nat. Prod.*, Meeting date 1985, Vol. 3, 234-43. VCH, Weinheim, FRG.

Snyder, S. (1985) The molecular base of communication between cells. *Sci. Am.*, **253** 132-41.

Solimani, R. (1996) Quercetin and DNA in solution: analysis of the dynamics of their interaction with a linear dichroism study. *Int. J. Biol. Macromol.*, **18** 287-95.

Solimani, R. (1997) The flavonols, quercetin, rutin and morin, in DNA solution: UV-Vis dichroic (and mid-infrared) analysis explain the possible association between the biopolymer and a nucleophilic vegetable-dye. *Biochem. Biophys. Acta. Gen. Subj.*, **1336** 281-94.

Son, J.K., Rosazza, J.P.N. and Duffel, M.W. (1990) Vinblastine and vincristine are inhibitors of monoamine oxidase B. *J. Med. Chem.*, **33** 1845-48.

Song, P.-S. and Tapley, K.J. Jr. (1979) Photochemistry and photobiology of psoralens. *Photochem. Photobiol.*, **29** 1177-97.

Staley, J.K., Ouyang, Q., Pablo, J., Hearn, W.L., Flynn, D.D., Rothman, R.B., Rice, K.C. and Mash, D.C. (1996) Pharmacological screen for activities of 12-hydroxyibogamine: a primary metabolite of the indole alkaloid ibogaine. *Psychopharmacology*, **127** 10-18.

Stefano, G.B., Hartman, A., Bilfinger, T.V., Magazine, H.I., Liu, Y., Casares, F. and Goligorsky, M.S. (1995) Presence of the μ3 opiate receptor in endothelial cells: coupling to nitric oxide production and vasodilation. *J. Biol. Chem.*, **270** 30290-93.

Stefano, G.B., Salzet, B., Rialas, C.M., Pope, M., Kustka, A., Neenan, K., Pryor, S. and Salzet, M. (1997) Morphine- and anandamide-stimulated nitric oxide production inhibits presynaptic dopamine release. *Brain Res.*, **763** 63-68.

Sterner, O., Bergman, R., Kihlberg, J. and Wickberg, B. (1985) The sesquiterpenes of *Lactarius vellereus* and their role in a proposed chemical defense system. *J. Nat. Prod.*, **48** 279-88.

Sterner, O., Carter, R.E. and Nilsson, L.M. (1987) Structure-activity relationships for unsaturated dialdehydes. 1. The mutagenic activity of 18 compounds in the Salmonella/ microsome assay. *Mutat. Res.*, **188** 169-74.

Stiborova, M., Fernando, R.C., Schmeiser, H.H., Frei, E., Pfau, W. and Wiessler, M. (1994) Characterization of DNA adducts formed by aristolochic acids in the target organ (forestomach) of rats by P32-postlabelling analysis using different chromatographic procedures. *Carcinogenesis*, **15** 1187-92.

Stich, H.F. (1991) The beneficial and hazardous effects of simple phenolic compounds. *Mutat. Res.*, **259** 307-24.

Storch, A., Schrattenholz, A., Cooper, J.C., Abdel Ghani, E.M., Gutbrod, O., Weber, K.-H., Reinhardt, S., Lobron, C., Hermsen, B. *et al.* (1995) Physostigmine, galanthamine and codeine act as 'noncompetitive nicotinic receptor agonists' on clonal rat pheochromocytoma cells. *Eur. J. Pharmacol. Mol. Pharmacol.*, **290** 207-19.

Strange, P.G. (1992) *Brain Biochemistry and Brain Disorders*. Oxford University Press, New York.

Strickland, J.R., Cross, D.L., Birrenkott, G.P. and Grimes, L.W. (1994) Effect of ergovaline, loline and dopamine antagonists on rat pituitary cell prolactin release *in vitro*. *Am. J. Vet. Res.*, **55** 716-21.

Su, M.J., Nieh, Y.C., Huang, H.W. and Cgen, C.C. (1994) Dicentrine, an alpha-adrenoreceptor antagonist with sodium and potassium channel-blocking activities. *Naunyn Schmeidebergs Arch. Pharmacol.*, **349** 42-49.

Subramanyam, S.S.P., Reddy, V., Reddy, G.P. and Murthy, D.K. (1974) Cytological effects of argemone oil on mitotic cells of *Allium cepa*. *Proc. Ind. Acad. Sci.*, **79** 216-26.

Sun, H.H., White, C.B., Dedinas, J., Cooper, R. and Sedlock, D.M. (1991) Methylpendolmycin, an indolactam from a *Nocardiopsis* sp. *J. Nat. Prod.*, **54** 1440-43.

Swain, T. (1977) Secondary compounds as protective agents. *Ann. Rev. Plant Physiol.*, **28** 479-501.

Sweetnam, P.M., Lancaster, J., Snowman, A., Collins, J.L., Perschke, S., Bauer, C. and Ferkany, J. (1995) Receptor binding profile suggests multiple mechanisms of action are responsible for ibogaine's anti-addictive activity. *Psychopharmacology*, **118** 369-76.

Sylvia, V.L., Joe, C.O., Stipanovic, R.D., Kim, H.L. and Busbee, D.L. (1985) Alkylation of deoxyguanosine by the sesquiterpene lactone, hymenoxon. *Toxicol. Lett.*, **29** 69-76.

Sylvia, V.L., Kim, H.L., Norman, J.O. and Busbee, D.L. (1987) The sesquiterpene lactone, hymenoxon, acts as a bifunctional alkylating agent. *Cell Biol. Toxicol.*, **3** 39-49.

Tadaki, S., Nozaka, T., Ishino, M., Tanaka, A., Morimoto, I. and Kunitomo, J. (1991) *In vitro* clastogenicity of the aporphine-type alkaloids. *Mutat. Res.*, **253** 280-81.

Tanaka, H., Ahn, J.W., Katayama, M., Wada, K., Maruma, S. and Osaka, Y. (1985) Isolation of two ovicidal substances against two-spotted spider mite, *Tetranychus urticae* KOCH from *Skimmia repens* NAKAI. *Agric. Biol. Chem. Tokyo*, **49** 2189-90.

Tanaka, H., Morooka, N., Haraikawa, K. and Ueno, Y. (1987) Metabolic activation of emodin in the reconstituted cytochrome P_{450} system of the hepatic microsomes of rats. *Mutat. Res.*, **176** 165-70.

Tariq, M., Parmar, N.S., Qureshi, S., El-Feraly, F.S. and Al-Meshal, I.A. (1987) Clastogenic evaluation of cathinone and amphetamine in somatic cells of mice. *Mutat. Res.*, **190** 153-57.

Tayama, S. (1996) Cytogenetic effects of piperonyl butoxide and safrole in CHO-K1 cells. *Mutat. Res.*, **368** 249-60.

Tazima, Y. (1982) Mutagenic and carcinogenic mycotoxins, in *Environmental Mutagenesis, Carcinogenesis and Plant Biology*, Vol. 1 (ed. E.J. Klekowski Jr.), Praeger Publ., New York, pp. 67-95.

Teh, B.S., Seow, W.K., Li, S.Y. and Thong, Y.H. (1990) Inhibition of prostaglandin and leukotriene generation by the plant alkaloids, tetrandrine and berbamine. *Int. J. Immunopharmacol.*, **12** 321-26.

Teuscher, E. and Lindequist, U. (1994) *Biogene Gifte. Biologie, Chemie, Pharmakologie.*, G. Fischer, Stuttgart.

Thomsen, T. and Kewitz, H. (1990) Selective inhibition of human acetylcholinesterase by galanthamine *in vitro* and *in vivo*. *Life Sci.*, **46** 1553-58.

Tinker, A., Sutko, J.L., Ruest, L., Deslongchamps, P., Welch, W., Airey, J.A., Gerzon, K., Bidasee, K.R., Besch, H.R., Jr. and Williams, A.J. (1996) Electrophysiological effects of ryanodine derivatives on the sheep cardiac sarcoplasmic reticulum calcium-release channel. *Biophys. J.*, **70** 2110-19.

Towers, G.H.N. and Abramowski, Z. (1983) UV-mediated genotoxicity of furanoquinoline and of certain tryptophan-derived alkaloids. *J. Nat. Prod.*, **46** 576-81.

Toyoda, M., Rausch, W.D., Inoue, K., Ohno, Y., Fujiyama, Y., Takagi, K. and Saito, Y. (1991) Comparison of solanaceous glycoalkaloid-evoked calcium influx in different types of cultured cells. *Toxicol. In Vitro*, **5** 347-51.

Trist, D.G., Humphrey, P.P.A., Leff, P. and Shankley, N.P. (1997) *Receptor Classification: The Integration of Operational, Structural and Transductional Information.* NY Acad. Sciences, New York.

Tsutsui, T., Hayashi, N., Maizumi, H., Huff, J. and Barrett, J.C. (1997) Benzene-, catechol-, hydroquinone- and phenol-induced cell transformation, gene mutations, chromosome aberrations, aneuploidy, sister chromatid exchanges and unscheduled DNA synthesis in Syrian hamster embryo cells. *Mutat. Res.*, **373** 113-23.

Upender, V., Pollart, D.J., Liu, J., Hobbs, P.D., Olsen, C., Chao, W.-R., Bowden, B., Crase, J.L., Thomas, D.W. *et al.* (1996) The synthesis and biological activity of two analogs of the anti-HIV alkaloid michellamine B. *J. Heterocycl. Chem.*, **33** 1371-84.

Vais, H. and Usherwood, P.N.R. (1995) Novel actions of ryanodine and analogs: perturbers of potassium channels. *Biosci. Rep.*, **15** 515-30.

Van Huizen, F., Wilkinson, M., Cynader, M. and Shaw, C. (1988) Sodium-channel toxins, veratrine and veratridine, modify opioid and muscarinic but not β-adrenergic binding sites in brain slices. *Brain Res. Bull.*, **21** 129-32.

Verpoorte, R. Antimicrobially active alkaloids. *Actual. Chim. Ther.*, **13** 195-209.

Verpoorte, R. (1998) Alkaloids: Biochemistry, Ecology and Medicinal Applications (eds. M.R. Roberts and M. Wink), Plenum, New York, pp. 397-433.

Von der Hude, W. and Braun, R. (1983) On the mutagenicity of metabolites derived from the mushroom poison, gyromitrin. *Toxicology*, **26** 155-60.

Von Zastrow, M., Keith, D.E. Jr. and Evans, C.J. (1993) Agonist-induced state of the δ-opioid receptor that discriminates between opioid peptides and opiate alkaloids. *Mol. Pharmacol.*, **44** 166-72.

Waldmeier, P.C., Wicki, P., Froestl, W., Bittiger, H., Feldtrauer, J.-J. and Baumann, P.A. (1995) Effects of the putative P-type calcium-channel blocker, R,R-(-)-daurisoline on neurotransmitter release. *Naunyn Sch. Arch. Pharmacol.*, **352** 670-78.

Waller, G.R. (1987) *Allelochemicals: Role in Agriculture and Forestry.* American Chemical Society, Washington, DC.

Walles, S.A.S. (1990) DNA damage by hydroquinone and duroquinone. *Mutat. Res.*, **234** 409.

Wang, B., Zhang, Y., Yang, M., Miao, P. and Wang, K. (1994) Study of intercalation binding of harmaline and harmine to DNA by microcalorimetry. *Wuli Huaxue Xuebao*, **10** 82-86.

Wang, G. and Lemos, J.R. (1992) Tetrandrine blocks a slow, large-conductance, calcium-activated potassium channel besides inhibiting a non-inactivating Ca^{2+} current in isolated nerve terminals of the rat neurohypophysis. *Pfluegers Arch.*, **421** 558-65.

Wang, G., Jiang, M., Coyne, M.D. and Lemos, J.R. (1993a) Comparison of effects of tetrandrine on ionic channels of isolated rat neurohypophysial terminals and Y1 mouse adrenocortical tumor cells. *Zhongguo Yaoli Xuebao*, **14** 101-6.

Wang, J.L., Nong, Y., Xia, G.J., Yao, W.X. and Jiang, M.X. (1993b) Effects of liensinine on slow action potential in myocardium and slow inward current in canine cardiac Purkinje fibers. *Yaoxue Xuebao*, **28** 812-16.

Watanabe, K., Miyakado, M., Iwai, T., Izumi, K. and Yanagi, K. (1988) Isolation of aristolochic acid and aristolic acid from *Cocculus triolobus* DC as potent seed germination inhibitors. *Agric. Biol. Chem. Tokyo*, **52** 1079-82.

Watanabe, K., Yano, S., Horie, S. and Yamamoto, L.T. (1997) Inhibitory effect of mitragynine, an alkaloid with analgesic effect from Thai medicinal plant *Mitragyna speciosa*, on electrically-stimulated contraction of isolated guinea-pig ileum through the opioid receptor. *Life Sci.*, **60** 933-42.

Watano, T., Nakazawa, K., Obama, T., Mori, M., Inoue, K., Fujimori, K. and Takanaka, A. (1993) Non-competitive antagonism by hirsuteine of nicotinic receptor-mediated dopamine release from rat pheochromocytoma cells. *Jpn. J. Pharmacol.*, **61** 351-56.

Wess, J., Gdula, D. and Brann, M.R. (1992) Structural basis of the subtype selectivity of muscarinic antagonists: a study with chimeric m2/m5 muscarinic receptors. *Mol. Pharmacol.*, **41** 369-74.

Westendorf, J. (1993) Anthranoid derivatives: general discussion, in *Adverse Effects of Herbal Drugs*, Vol. 2 (eds. P.A.G.M. De Smet, K. Keller, R. Hänsel and R.F. Chandler), Springer Verlag, Berlin, Heidelberg, pp. 105-18.

WHO, Geneva (1988) Pyrrolizidine Alkaloids.

Whong, W.-Z., Lu, C.-H., Stewart, J.D., Jiang, H.-X. and Ong, T. (1989) Genotoxicity and genotoxic enhancing effect of tetrandrine in *Salmonella typhimurium*. *Mutat. Res.*, **222** 237-44.

Williams, M. and Robinson, J.L. (1984) Binding of the nicotinic cholinergic antagonist, dihydro-β-erythroidine, to rat brain tissue. *J. Neurosci.*, **4** 2906-11.

Wink, M. (1987a) Chemical ecology of quinolizidine alkaloids, in *Allelochemicals* (ed. G.R. Waller) American Chemical Society Symposium Series **330** pp. 524-33.

Wink, M. (1988) Plant breeding: importance of plant secondary metabolites for protection against pathogens and herbivores. *Theor. Appl. Genet.*, **75** 225-33.

Wink, M. (1992) The role of quinolizidine alkaloids in plant-insect interactions, in *Insect-Plant Interactions*, Vol. 4 (ed. E.A. Bernays), CRC Press, Boca Raton, pp. 131-66.

Wink, M. (1993a) Allelochemical properties or the raison d'être of alkaloids, in *The Alkaloids*, Vol. 43 (ed. G.A. Cordell), Academic Press, San Diego, pp. 1-118.

Wink, M. (1993b) Production and application of phytochemicals from an agricultural perspective, in *Phytochemistry and Agriculture* (ed. T.A. van Beek and H. Breteler), Clarendon Press, Oxford, pp. 171-213.

Wink, M. (1993c) Quinolizidine alkaloids, in *Methods in Plant Biochemistry*, Vol. 8 (ed. P.G. Waterman), Academic Press, London, pp. 197-239.

Wink, M. (1998) Modes of action of alkaloids, in *Alkaloids: Biochemistry, Ecology and Medicinal Applications*. (ed. M.F. Roberts and M. Wink), Plenum, New York, pp. 301-26.

Wink, M. (1999a) Introduction: biochemistry, role and biotechnology of secondary metabolites, in *Biochemistry of Plant Secondary Metabolism* (ed. M. Wink), Annual Plant Reviews, Vol. 1, Sheffield Academic Press, Sheffield, pp. 1-16.

Wink, M. (1999b) Interference of alkaloids with neuroreceptors and ion channels, in *Bioactive Natural Products* (ed. Atta-Ur-Rahman), Elsevier, Amsterdam pp. 1-127.

Wink, M. and Twardowski, T. (1992) Allelochemical properties of alkaloids: effects on plants, bacteria and protein biosynthesis, in *Allelopathy: Basic and Applied Aspects* (eds. S.J.H. Rizvi and V. Rizvi), Chapmann and Hall, London, pp. 129-50.

Wink, M. and Latz-Brüning, B. (1995) Allelopathic properties of alkaloids and other natural products, in *Allelopathy: Organisms, Processes and Applications*, Vol. 582. ACS Symposium Series (ed. Inderjit, K.M.M. Dakshini and F.A. Einhellig), American Chem. Society, Washington, DC, pp. 117-26.

Wink, M. and Waterman, P.G. (1999) Chemotaxonomy in relation to molecular phylogeny of plants, in *Biochemistry of Plant Secondary Metabolism* (ed. M. Wink), Annual Plant Reviews, Vol. 2, Sheffield Academic Press, Sheffield, pp. 300-341.

Wink, M., Heinen, H.J., Vogt, H. and Schiebel, H.M. (1984) Cellular localization of quinolizidine alkaloids by laser desorption mass spectrometry. *Plant Cell Rep.*, **3**, 230-33.

Wink, M., Latz-Brüning, B. and Schmeller, T. (1998a) Biochemical effects of allelopathic alkaloids, in *Principles and Practices in Chemical Ecology* (ed. Inderjit, K.M.M. Dakshini and C.L. Foy), CRC Press, Boca Raton.

Wink, M., Schmeller, T. and Latz-Brüning, B. (1998b) Modes of action of allelochemical alkaloids: interaction with neuroreceptors, DNA and other molecular targets. *J. Chem. Ecol.*, **24** 1881-937.

Witherup, K.M., Ransom, R.W., Graham, A.C., Bernard, A.M., Salvatore, M.J., Lumma, W.C., Anderson, P.S., Pitzenberger, S.M. and Varga, S.L. (1995) Martinelline and martinellic acid: novel G-protein linked receptor antagonists from the tropical plant, *Martinella iquitosensis* (Bignoniaceae). *J. Am. Chem. Soc.*, **117** 6682-85.

Wölfle, D., Schmutte, C., Westendorf, J. and Marquardt, H. (1991) Hydroxyanthraquinones as tumor promoters: enhancement of malignant transformation of C3H mouse fibroblasts and growth stimulation of primary rat hepatocytes. *Cancer Res.*, **50** 6540-44.

Woo, W.S., Lee, E.B., Kang, S.S., Shin, K.H. and Chi, H.J. (1987) Antifertility principle of *Dictamnus albus* root bark. *Planta Med.*, **53** 399-401.

Wu, W.N., Beal, J.L. and Doskotch, R.W. (1977) Alkaloids of *Thalictrum*: isolation of alkaloids with hypotensive and antimicrobial activity from *Thalictrum revolution. Lloydia* **40** 508-14.

Wu, S.-N., Hwang, T.-L., Jan, C.-R. and Tseng, C.-J. (1997) Ionic mechanisms of tetrandrine in cultured rat aortic smooth muscle cells. *Eur. J. Pharmacol.*, **327** 233-38.

Xing, S.-G., Wu, Z.-L., Whong, W.-Z. and Ong, T. (1989) Enhancing effect of tetrandrine on sister-chromatid exchanges induced by mitomycin C and cigarette smoke condensate in mammalian cells. *Mutat. Res.*, **226** 99-102.

Xu, H. and Tang, X. (1987) Cholinesterase inhibition by huperzine B. *Zhongguo Yaoli Xuebao*, **8** 18-22.

Yamaguchi, T. (1980) Mutagenicity of isothiocyanates, isocyanates and thioureas on *Salmonella typhimurium. Agric. Biol. Chem. Tokyo*, **44** 3017-18.

Yamamura, H.I. and Snyder, S.H. (1974) Muscarinic cholinergic binding in rat brain. *Proc. Natl. Acad. Sci. USA*, **71** 1725-29.

Yano, S., Horiuchi, H., Horie, S., Aimi, N., Sakai, S. and Watanabe, K. (1991) Ca^{2+}-channel blocking effects of hirsutine, an indole alkaloid from *Uncaria* genus, in the isolated rat aorta. *Planta Med.*, **57** 403-405.

Yu, Q.S., Atack, J.R., Rapoport, S.I. and Brossi, A. (1988) Synthesis and anticholinesterase activity of (-)-N1- norphysostigmine, (-)-eseramine, and other N1-substituted analogs of (-)-physostigmine. *J. Med. Chem.*, **31** 2297-300.

Yun-Choi, H.S. and Kim, M.H. (1994) Higenamine-reduced mortalities in the mouse models of thrombosis and endotoxic shock. *Yakhak Hoechi*, **38** 191-6.

Zhu, M., Phillipson, J.D., Yu, H., Greengrass, P.M. and Norman, N.G. (1997) Application of radioligand-receptor binding assays in the search for the active principles of the traditional Chinese medicine, 'Gouteng'. *Phytother. Res.*, **11** 231-32.

3 Chemical defence in marine ecosystems

Peter Proksch

3.1 Introduction

Nearly 10 000 different natural products have been isolated from marine organisms within the last 30 yrs (MarinLit, 1998). This field of research is still a young branch of science when compared to the long tradition of research into terrestrial natural products. The isolation of prostaglandin derivatives from the Caribbean gorgonian, *Plexaura homomalla*, by Weinheimer and Spraggins in 1969 is usually considered to be the starting point of marine chemistry, although first reports on the occurrence of unusual secondary constituents from marine organisms date back almost 50 yrs. In 1951, Bergmann and Feeney reported on the occurrence of unusual nucleosides in the marine sponge, *Tethya crypta*. These compounds have eventually served as leads for the development of modern nucleoside drugs for antiviral chemotherapy. In retrospect, it seems no coincidence that the first biologically-active marine natural products were isolated from invertebrates rather than from algae. We now know that marine invertebrates, such as sponges, tunicates or bryozoans, constitute by far the richest chemical source in the oceans with regard to the structural diversity of compounds as well as to the numbers of metabolites isolated (Faulkner, 1997).

Looking for 'drugs from the sea' – meaning pharmacologically active natural products that might serve as lead structures for new pharmaceuticals–has been, and continues to be, the strongest impetus for research into marine natural products. The numerous scientists engaged in this applied aspect of research, however, maintain also a keen interest in the ecological functions of the metabolites they are studying. Therefore, early reports on marine chemical ecology appeared soon after major interest in this field arose at the beginning of the 1970s (e.g. Bakus and Green, 1974).

Since then, a rapid development has been seen in marine chemical ecology, offering many fascinating insights into the interactions of organisms that are modulated by natural products. As we know today, many if not most natural products from marine organisms play vital as well as diverse roles in ensuring the fitness and survival of their producers. The more important functions ascribed to these compounds may include: intraspecific signalling (pheromones); deterrency of herbivores and predators; suppression of competing neighbours; inhibition of bacterial and fungal invasion; or even protection against UV radiation.

Furthermore, if one considers that not all ecologically relevant natural products that are isolated from a given source (e.g. sponge) are necessarily synthesized by this organism but may result from symbiosis with associated microorganisms (also called 'endosymbionts'), and that these compounds may perhaps also have a signal or regulatory function within the symbiosis, the manifold roles of natural products in the marine ecosystem become tangible.

Given the complexity of ecological functions that marine natural products may serve between organisms as well as within a composite organism, such as a sponge (see above), any treatise on marine chemical ecology is bound to be selective, since a comprehensive treatment is almost impossible. This is also true for the present review, which focuses on three important aspects where natural products play a key role: allelopathy; defence against fouling; and defence against consumers. Readers more deeply interested in these and/or additional aspects of marine chemical ecology are referred to other recent reviews (e.g. Bakus *et al.*, 1986; Paul, 1992a; Pawlik, 1993; Proksch, 1994; Hay, 1996; Proksch and Ebel, 1998).

3.2 Marine natural products in allelopathic interactions

Competition for space is generally intense on marine hard bottom substrates but appears to be most pronounced on tropical coral reefs, which are characterized by an exceedingly high species diversity and remarkable population density, unmatched in any other marine ecosystem (Jackson and Buss, 1975; Jackson, 1977; Branch, 1984; Porter and Targett, 1988; Sale, 1991). Given the high incidence of toxic natural products that have been isolated, especially from marine invertebrates, such as sponges (e.g. Sarma *et al.*, 1993), as well as the bare zones observed around some sponges in their natural habitat (Porter and Targett, 1988; Turon *et al.*, 1996), allelopathic effects through biosynthesis and exudation of toxic secondary metabolites appear, indeed, to be involved in structuring communities of benthic marine invertebrates.

Most of the experiments conducted in order to prove the significance of allelochemicals in competition for space have employed crude extracts (Porter and Targett, 1988; Turon *et al.*, 1996). The compounds responsible for the claimed allelopathic effects, however, have rarely been isolated and characterized. A notable exception is that of the burrowing sponges from the genus *Siphonodictyon*, which burrow into the heads of living corals. Sullivan and co-workers (1981 and 1983) demonstrated, in elegant experiments, that overgrowth and thus killing of *Siphonodictyon coralliphagum* by corals (e.g. *Acropora formosa*) is prevented by exudation of the toxic secondary metabolite, siphonodictidin (**1**) (Fig 3.1), which suppresses photosynthesis of the coral's zooxanthellae and, thereby, coral growth, even at concentrations as low as 10 ppm. Interestingly, the related sponge, *S. mucosa*, which burrows into dead corals (in contrast to the above-mentioned *S. coralliphagum*), lacks the toxic siphonodictidin, thereby corroborating the proposed ecological significance of the respective secondary metabolite.

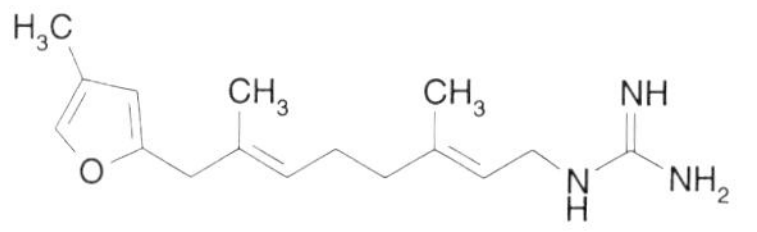

(1)

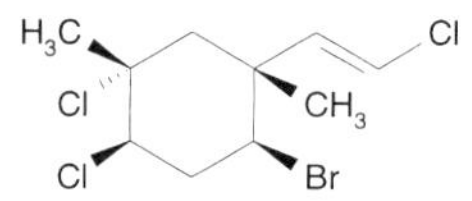

(2)

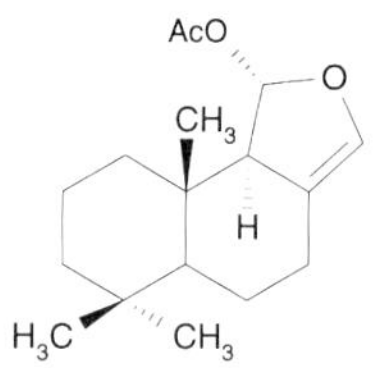

(3)

Figure 3.1 Structures of some marine natural products involved in allelopathic interactions; siphonodictin (**1**), chloromertensene (**2**), 7-deacetoxyolepupuane (**3**).

Allelopathic effects are not restricted to competition between marine invertebrates but are apparently also of importance in interactions between algae and invertebrates. The red alga, *Plocamium hamatum*, present on reefs of North Queensland, was found to cause tissue necrosis of several marine invertebrates when in physical contact with the alga (De Nys *et al.*, 1991). Since allelochemicals, especially the algal secondary metabolite, chloromertensene (**2**) (Fig. 3.1), were suspected of involvement in this interaction, a series of field experiments was conducted that included *P. hamatum* and the soft coral, *Sinularia cruciata*. Healthy algae and soft corals were relocated on mesh grids. In one set of experiments, individuals of both taxa were brought into physical contact; whereas, in the second set of experiments, algae and soft corals were kept in a non-contact situation. Only those soft corals that had physical contact with red algae developed tissue necrosis, indicating that the suspected allelochemicals are not waterborne but act upon physical contact. In further experiments, the algal metabolite, chloromertensene (Fig. 3.1), was coated onto 'artificial algae', which were again brought into physical contact with the soft corals. In all cases where coated 'algae' were in contact with *S. cruciata*, the soft corals exhibited tissue necrosis; whereas, contact with uncoated 'algae' merely caused abrasion but not necrosis (De Nys *et al.*, 1991). In addition to its allelopathic effects, coating with chloromertensene (Fig. 3.1) inhibited fouling of the artificial 'algae' as well as predation, suggesting multiple ecological roles for this natural product.

Another recent example of allelopathic interactions between marine sponges includes two hitherto undescribed sponges of the genera *Dysidea* and *Cacospongia*, that co-occur on reefs of the tropical island, Guam. The *Dysidea* sp. was frequently observed to overgrow adjacent specimens of *Cacospongia* spp. and to cause necrosis of the latter sponge (Thacker *et al.*, 1998). When crude extracts of *Dysidea* sp. or its major secondary metabolite, 7-deacetoxyolepupuane (**3**) (Fig. 3.1), were incorporated into agar strips and placed in contact with *Cacospongia* sp. in the field, typical tissue necrosis of the latter was observed. This suggests that natural products are involved in this allelopathic interaction and are probably the major reason for the success of *Dysidea* sp. over *Cacospongia* sp. The study of *Dysidea* sp. and *Cacospongia* sp., as well as most other studies of marine allelopathy, have not unequivocally established whether the compounds suspected to be involved in allelopathy are present on the surface of the aggressor or are exuded upon contact with other species. However, the tissue-specific localization of any natural product suspected to be involved in allelopathy is of critical importance, since only those compounds that come into contact with other organisms can realistically be expected to be of ecological significance in any scenario involving competition for space.

3.3 Chemical defence against fouling

Surface fouling is a biological phenomenon almost ubiquituously observed in the marine environment. Although the complex interactions involved in biofouling are not yet completely understood, at least three major events can be distinguished: 1) 'conditioning' of a surface by a glycoproteinaceous layer; 2) formation of the 'primary film', which consists of microorganisms, such as bacteria, diatoms and protozoa; and 3) colonization of the conditioned surface by macrophytic algae and/or epibiotic invertebrates (Bakus *et al.*, 1986, 1990; Davis *et al.*, 1989; Melton and Bodnar, 1988). Fouling may be disadvantageous, or even dangerous, especially for filter-feeding invertebrates, since inhalent canals (e.g. in sponges) may be blocked by epibionts, resulting in reduced feeding capacity. Filter-feeding epibionts may also compete with their invertebrate hosts for food.

In spite of the omnipresence of fouling organisms in the marine environment, the surfaces of invertebrates, such as sponges or bryozoans, are often remarkably clean. Given the high incidence of antibiotically active or cytotoxic natural products that have been isolated, especially from sponges or bryozoans (Krebs, 1986; Schmitz, 1994; Munro *et al.*, 1994), a chemical suppression of fouling by involvement of natural products seems highly plausible.

Unequivocal proof of the ecological significance of marine natural products in the suppression of fouling has so far been presented in very few studies, since surface-allocation or continuous exudation of suspected antifouling substances have to be demonstrated in addition to their antibiotic or toxic nature (Paul, 1992c). For example, the presence of antifouling constituents in a sponge extract does not necessarily imply that these compounds are also involved in the suppression of epibiosis in a natural situation, unless it has been demonstrated that potential fouling organisms will indeed come into contact with the respective chemicals.

Given these requirements, very few studies on the ecological significance of antifouling marine constituents appear to be convincing. A notable exception is the study by Walker and co-workers (1985), who demonstrated that the sponge, *Aplysina fistularis*, exudes aerothionin and homoaerothionin (Fig. 3.2) into the surrounding seawater at rates of

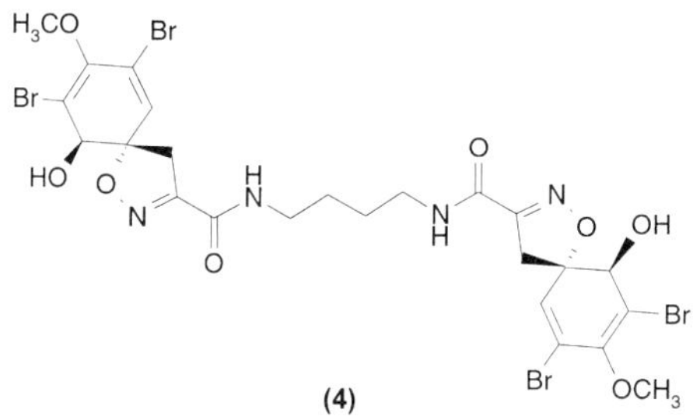

(4)

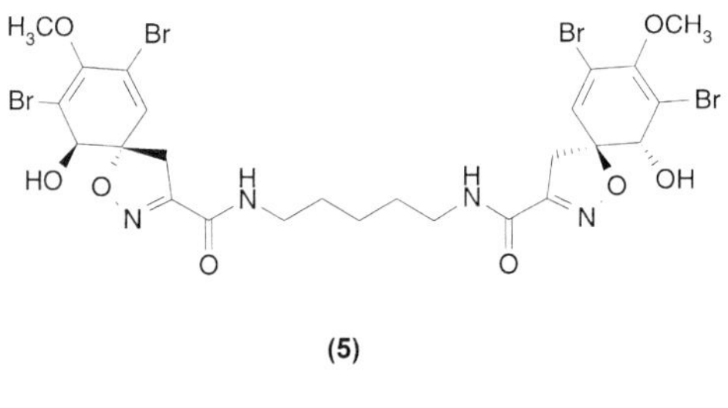

(5)

Figure 3.2 Structure of the chemicals, aerothionin (**4**) and homoaerothionin (**5**), involved in control against fouling.

8.9×10^{-3}–7.7×10^{-4} µg/min/g. Following mechanical injury, the rate of exudation had accelerated by a factor of 10–100, when compared to undisturbed controls. In an earlier investigation, Thompson and co-workers (1985) showed that physiologically relevant concentrations of both aerothionin and homoaerothionin (Fig. 3.2) are able to prevent settlement of fouling organisms on *A. fistularis*.

Recently, antifouling constituents were reported from the brown alga, *Dictyota menstrualis*, which appears to be less frequently covered by fouling organisms than co-occuring algae belonging to other taxa (Schmitt *et al.*, 1995). Laboratory assays indicated that the bryozoan, *Bugula neritina*, which was used as a model for epibionts, did not settle on surfaces of *D. menstrualis*. It was found that rejection of *D. menstrualis* by *B. neritina* occurred only after physical contact with the surface of the alga and was not mediated through waterborne signals. Analysis of a crude lipophilic extract obtained after rubbing the surface of *D. menstrualis* yielded the known diterpenes, pachydictyol A and dictyol E (Fig. 3.3). When exposed to these compounds, larval mortality and abnormal larval development increased significantly, suggesting that pachydictyol A and dictyol E (Fig. 3.3) are the causative agents for the antifouling effects observed.

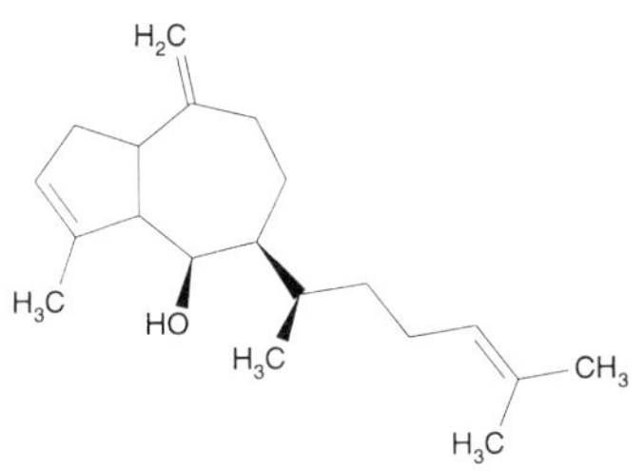

(6)

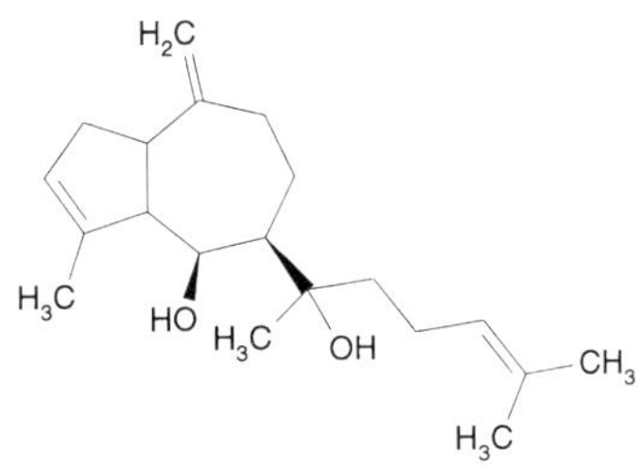

(7)

Figure 3.3 Structures of the diterpenes, pachydictyol A (**6**) and dictyol E (**7**), which are causative agents for antifouling effects.

3.4 Chemical defences of marine invertebrates and algae against consumers

Numerous studies of marine natural products derived from various invertebrates as well as from algae have indicated the significance of these metabolites in chemical defence against predatory or herbivorous fish. Consumer pressure is high in marine ecosystems but especially pronounced on tropical coral reefs, where fish have been estimated to bite the bottom in excess of 150,000 times/m^2/day (Carpenter, 1986). These pronounced stress factors will select for optimized chemical defence of algae as well as soft-bodied marine invertebrates. It is, therefore, no wonder that most reports on antifeedant marine natural compounds concern organisms that occur in the tropics rather than those living in temperate waters.

Early investigators in the field primarily used laboratory assays in order to assess the feeding-deterrent or toxic properties of marine natural products (e.g. Braekman and Daloze, 1986). In some of these studies, freshwater rather than marine fish were employed as test organisms. Whereas these studies yielded interesting insights into the general toxicity of marine natural products and sometimes also into their mode of action (Groweiss *et al.*, 1983), the ecological significance of the data obtained is often questionable. More recent studies on chemical defence of invertebrates or algae against fish focus, therefore, on field rather than laboratory assays and employ naturally occurring assemblages of consumers (e.g. Pawlik *et al.*, 1995; Meyer and Paul, 1995).

In some cases, the origin of fish-deterrent natural products in marine invertebrates can be traced through the food chain. For example, sponges belonging to the genus *Halichondria* were traced as the dietary source of macrolide oxazole alkaloids detected in the nudibranch mollusc, *Hexabranchus sanguineus* (Kernan *et al.*, 1988), also known as 'Spanish dancer' due to its bright coloration. Even though lacking a protective shell, the soft-bodied nudibranchs are rejected by Indo-Pacific reef fish as well as by other potential predators, such as the hermit crab, *Dardanus negistos* (Pawlik *et al.*, 1988). Macrocyclic oxazole alkaloids, such as halichondramide (Fig. 3.4), were traced as the major feeding-deterrent compounds present in the nudibranch, where they are concentrated in the most vulnerable parts, such as the dorsal mantle or the conspicuous egg ribbons (Pawlik *et al.*, 1988). Sponges of the genus *Halichondria* could ultimately be traced as the true sources of the defensive metabolites of *H. sanguineus*. Like other nudibranchs that are morphologically defenceless, the 'Spanish dancer' has become specialized to feed on chemically-protected marine invertebrates, such as sponges, that are largely unpalatable to other generalist consumers. In this trophic interaction, nudibranchs frequently sequester the defence compounds of their prey (Proksch, 1994).

Sequestration of defensive compounds is not restricted to sponge-feeding nudibranchs but also extends to other marine invertebrates. In the mangrove area of the Micronesian island, Truk, the marine flatworm, *Pseudoceros concineus*, was observed feeding on the ascidian, *Eudistoma toealensis*. Despite the lack of morphological defence mechanisms, neither the ascidians nor the flatworms were attacked by fish, suggesting that

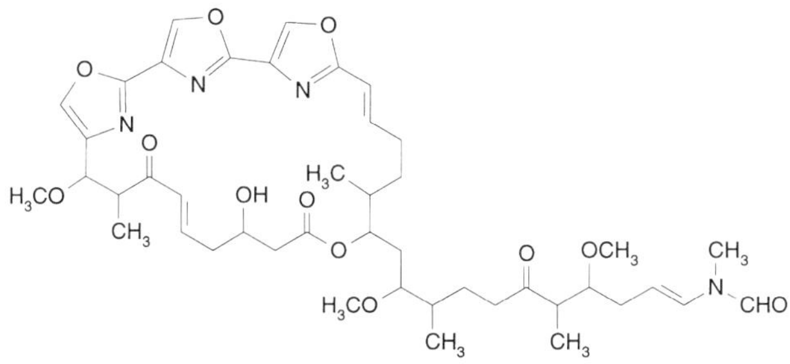

(8)

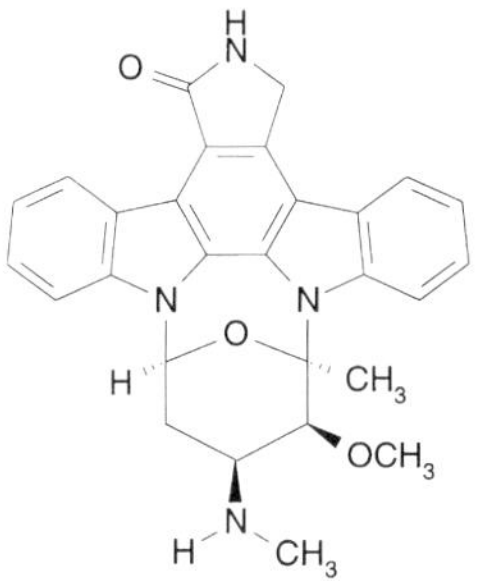

(9)

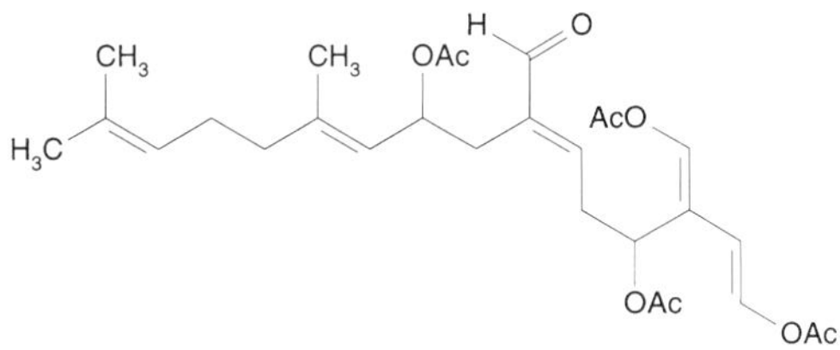

(10)

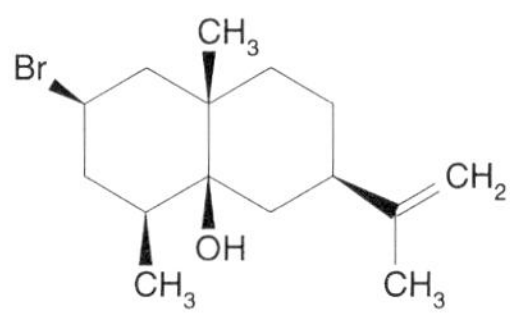

(11)

Figure 3.4 Structures of some chemicals involved in defence of marine invertebrates and algae against consumers: halichondramide (**8**), staurosporine (**9**), halimedatetraacetate (**10**) and a brominated sesquiterpene (**11**).

both organisms were chemically defended. By incorporating crude extracts from the ascidians or from the flatworms into artificial food at physiological concentrations and offering these treated food pellets to a natural fish assemblage, this hypothesis was corroborated. The treated food pellets were largely avoided, whereas control food was readily consumed (Schupp *et al.*, 1998a). Bioassay-guided fractionation of the respective extracts yielded several staurosporine derivatives (Fig. 3.4) as the active constituents responsible for the fish-deterrent effects observed.

Staurosporine is originally known as a bacterial metabolite produced by strains of *Streptomyces* or by other microorganisms. The occurrence of staurosporine and several of its congeners in the ascidian and in the associated flatworms raises questions with regard to the true producer of these defensive compounds, which are also known as potent inhibitors of protein kinase C. It is possible that the staurosporines found in the two marine invertebrates originate from bacteria that are inhaled during filter-feeding or from microorganisms that live in association with *E. toealensis*. In comparison, *de novo* synthesis of these typical microbial metabolites by the ascidian seems less likely. Whereas the true origin of staurosporin derivatives in *E. toealensis* remains obscur, it is clear that the flatworms sequester these compounds from their ascidian prey, since they were found to be lacking in flatworms feeding on other staurosporine-free ascidians (Schupp *et al.*, 1998a).

As in marine invertebrates, natural products play a important role in protecting marine algae from herbivores, although marine plants have apparently been less 'ingenious' in diversifying their chemical armoury than marine invertebrates (Faulkner, 1997, and preceding reviews). The majority of natural products isolated from marine macroalgae are terpenoids, polyketides or aromates (Paul, 1992b). Nitrogenous compounds, which are frequently encountered in marine invertebrates, such as sponges or tunicates, are very rare in macroalgae.

Nevertheless, many of the terpenoids found in green, red or brown algae act as defensive allomones against consumers, such as fish or sea urchins. A vivid example of the ecological significance of terpenoid allomones is provided by the defensive metabolites of green algae from the genera *Halimeda* and *Udotea*. Several species of the latter genera are able to convert biologically weakly-active sesquiterpenoids into highly deterrent defence metabolites following tissue damage caused either by the attack of herbivores or by mechanical injury (Paul, 1992b). *Halimeda* species, for example, have been shown to convert the less active sesquiterpenoid, halimedatetraacetate (Fig. 3.4), into the defensive compound, halimedatrial, following breakdown of cellular compartmentation (Paul and van Alstyne, 1988); whereas, under similar conditions, *Udotea flabellum* converts udoteal to the more active metabolite, petodial (Paul, 1992b). It is possible that the biotransformations described are enzymatically-catalyzed reactions, although experimental proof for this hypothesis is so far lacking.

Most chemical defence mechanisms studied in marine algae seem to be constitutive, as reported, for example, for the tropical green alga, *Neomeris annulata*, which is widely distributed on tropical reefs in the Caribbean and Pacific. Specimens of *N. annulata* growing at Guam accumulate brominated sesquiterpenes as prominent secondary metabolites. Tips of the thalli, which are most vulnerable to attack by herbivores, usually contain higher concentrations of brominated compounds than the tougher middle or basal parts of the algae. When incorporated into artificial diet and tested at natural concentrations such as present in thallus tips, the major brominated sesquiterpene (Fig. 3.4) deterred feeding by three reef herbivores, including the parrot fishes, *Scarus sordidus* and *S. schlegeli*, as well as by the sea urchin, *Diadema savignyi* (Lumbang and Paul, 1996). No synergistic effects were observed when the natural mixture of brominated sesquiterpenes rather that individual compounds was tested for deterrent activity (Lumbang and Paul, 1996).

Defence by means of deterrent natural products is also employed by some benthic marine cyanobacteria, as reported recently for the blue-green alga, *Hormothamnion enteromorphoides* (Pennings *et al.*, 1997). *H. enteromorphoides* periodically dominates shallow reef habitats at Guam, forming erect tufts comparable in size to the thalli of macroalgae. Although at certain times very common, *H. enteromorphoides* is avoided by most herbivores, suggesting that the cyanobacterium is chemically defended. This hypothesis was corroborated by incorporating a crude extract of the cyanobacterium into an artificial diet and offering the spiked diet to a natural assemblage of reef fish, including the parrot fish, *Scarus schlegeli*, as well as to invertebrate grazers, such as the sea urchin, *Diadema savignyi*, or the crab, *Leptodius* sp. (Pennings *et al.*, 1997). Fractionation of the extract suggested that a mixture of cyclic peptides (which occur frequently in cyanobacteria), including the major constituent, laxaphycin A, were responsible for the deterrency of *H. enteromorphoides*.

3.5 Favoured allocation of defensive metabolites in vulnerable and valuable parts of marine invertebrates: examples in support of the 'apparency model' and 'optimal defence theory'

The 'plant apparency model', originally developed for higher terrestrial plants, predicts a correlation between chemical defence and the risk of being discovered by herbivores (Feeny, 1976; Rhoades and Cates, 1976). The 'optimal defence theory', also originally formulated for higher terrestrial plants, suggests that metabolically costly chemical defences should be preferentially invested in the most valuable parts of a plant (McKey, 1979; Rhoades, 1979), such as young developing leaves or seeds. A recent study on the chemical defence of the IndoPacific sponge, *Oceanapia* spp., demonstrated that both defence theories also apply to marine invertebrates.

The conspicuously red-coloured sponge, *Oceanapia* sp., occurs in shallow sandy areas around the Micronesian island, Truk. Part of the sponge, the so-called base, is immersed into the substrate, whereas the fistule and an apical round-shaped structure, the so-called capitum (probably an asexual propagation unit), are exposed in the seawater. Underwater observations have indicated that the exposed parts of the sponge, even though easily accessible to potential predators, are not consumed by the frequently occurring reef fish (Schupp *et al.*, 1998b), suggesting chemical defence of the sponge. This hypothesis was corroborated by incorporating a crude extract derived from fistulae of *Oceanapia* sp. into an artificial diet at the respective natural concentration, and offering treated *versus* non-treated diet cubes to a natural assemblage of reef fish in a field bioassay. Whereas the fish readily consumed the control diet, the treated diet was clearly avoided. The pyridoacridine alkaloids, kuanoniamine C (**12**) and D (**13**) (Fig. 3.5), proved to be responsible for the deterrent properties of the crude extract from *Oceanapia* sp., as shown in a subsequent field feeding experiment. Interestingly, the defensive alkaloids occur at largely different concentrations in the various parts of the sponge analyzed. Whereas total alkaloid concentration amounts to only 0.8% (relative to dry mass) in the base, which is protected in the substrate, alkaloid concentrations equal almost 2% in the exposed fistule and are close to 5% in the asexual propagation unit capitum (Schupp *et al.*, 1998b); thus, providing an example for both the above-mentioned 'apparency model' and for the 'optimal defence theory' in the marine habitat.

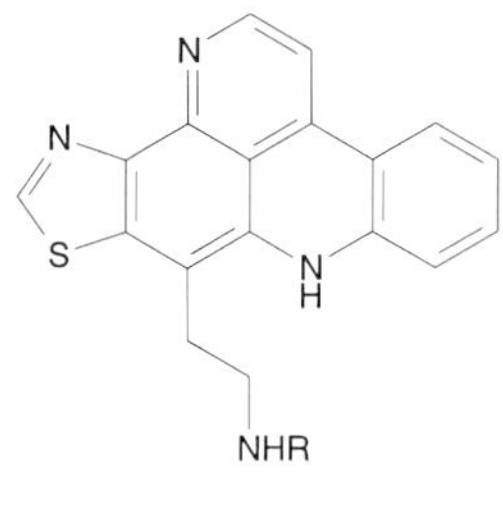

(12): R = COC_2H_5
(13): R = $COCH_3$

Figure 3.5 Structure of the pyridoacridine alkaloids, kuanoniamine C and D, involved in chemical defence against consumers.

Another example from marine invertebrates that fits both the 'apparency model' as well as the 'optimal defence theory' concerns chemical defence of marine invertebrate larvae. Whereas some species of marine invertebrates produce large numbers of small larvae that feed and develop in the plankton and are usually dispersed over large distances ('planktotrophic' larvae), other marine invertebrates produce a smaller number of larger nonfeeding larvae that use yolk for nutrition ('lecitotrophic' larvae) (Lindquist and Hay, 1996). Some marine invertebrates even brood lecitotrophic larvae to an advanced stage of development (Lindquist and Hay, 1996). Lecitotrophic larvae are usually large and conspicuous and should, therefore, be prone to predation by fish or other consumers, especially since the yolk on which these larvae depend is highly nutritious. This latter type of larvae can, therefore, almost certainly be expected to be defended in some way (e.g. by natural products), according to the above-cited 'apparency model' and to the 'optimal defence theory'. An example of the chemical defence of lecitotrophic larvae is provided by the ascidian, *Ecteinascidia turbinata*. *E. turbinata* releases conspicuously coloured larvae that are deterrent to potential fish predators (Young and Bingham, 1987). After mouthing larvae of *E. turbinata*, the pinfish, *Lagodon rhomboides*, was reported to avoid the usually palatable larvae of *Clavelina oblonga*, when the latter were coloured to resemble larvae of *E. turbinata* (Young and Bingham, 1987), thereby suggesting the importance of colour in fish-invertebrate associations.

In a broader survey on the chemical defence of tropical and temperate marine invertebrate larvae, Lindquist and Hay (1996) showed that brooded larvae were significantly more likely to be unpalatable (86% of all species tested) than planktotrophic larvae (33%). Whereas most unpalatable larvae were released during the day (89%), the majority of palatable larvae spawned at night (77%). Furthermore, there was a high incidence of conspicuous coloration in unpalatable larvae (60%), whereas palatable larvae were devoid of aposematic coloration (0%) (Lindquist and Hay, 1996).

3.6 The flexible response: stress-induced accumulation of defence metabolites and activation of protoxins

Most chemical defence mechanisms so far reported for marine invertebrates or algae appear to be static and constitutive and rely on preformed toxic or deterrent compounds that are either continuously exuded (Walker *et al.*, 1985) or are liberated when tissues are injured. However, chemical ecological studies in the terrestrial environment, especially those conducted on the interaction of higher plants with other organisms, have frequently shown that chemical defence mechanisms may be highly dynamic rather than static. Plants are known to respond to an attack by herbivores or pathogens either by a significantly increased accumulation of constitutively present natural products (e.g. Sahm *et al.*, 1995), by *de novo* synthesis of phytoalexins (e.g. Bailey and Mansfield, 1982) or by an enzymatically-catalyzed biotransformation of preformed compounds, such as liberation of hydrocyanic acid (HCN) from cyanogenic glycosides (Jones, 1988), the latter resembling an activation of protoxins. In many cases, the enzymes involved in either an enhanced or a *de novo* synthesis of defence metabolites and even the encoding genes have been isolated and characterized (Hammond-Kosack and Jones, 1996). Marine algae and invertebrates also appear capable of such a 'flexible response' to biotic stress factors, even though very few examples have been described so far.

Grazing by the amphipod, *Ampithoe longimana*, was, for example, reported to induce an increased accumulation of defensive compounds in the marine brown alga, *Dictyota menstrualis* (Cronin and Hay, 1996). *A. longimana* preferentially feeds on thalli of *D. menstrualis*, which contain dictyol-type diterpenes. Field observations as well as controlled feeding experiments have indicated that tissue damage by the amphipod causes an enhanced accumulation of diterpenes, which in turn decrease the palatability of *D. menstrualis* (Cronin and Hay, 1996). Compared to undamaged control plants, thalli damaged by *D. menstrualis* contained 19–34% more dictyol derivatives and were 50% less palatable to the mesograzers. Grazing did not affect protein content or toughness of the thalli, indicating a specific elicitation of diterpenoid accumulation. The nature of the signal(s) responsible for the increase in defensive compounds was not elucidated.

Sponges of the genus *Aplysina* (syn. *Verongia*) provide an example for a wound-induced activation of protoxins. *Aplysinia aerophoba*, which occurs in the Mediterranean as well as in parts of the Atlantic Ocean, is exceptionally rich in brominated isoxazoline alkaloids. The major alkaloids, isofistularin-3 and aerophobin-2 (Fig. 3.6), may account for up to 10% of the dry weight of *A. aerophoba* (Teeyapant *et al.*, 1993). Following disruption of the cellular compartmentation, e.g. by wounding, a rapid, enzymatically-catalyzed bioconversion of isoxazoline alkaloids is observed, leading to the lower molecular weight compound, aeroplysinin-1 (Fig. 3.6) (Teeyapant and Proksch, 1993). In a second enzymatically-catalyzed reaction, the latter is converted into a dienone (Fig. 3.6) (Ebel *et al.*, 1997). These bioconversions, which take place within mere seconds after a sponge has been injured, can also be monitored *in vitro* using crude protein extracts from *A. aerophoba* or other *Aplysina* species as enzyme sources (Ebel *et al.*, 1997). Protein extracts from sponges other than *Aplysina* failed to catalyze the reactions described, indicating that the enzymes involved are of a specific nature.

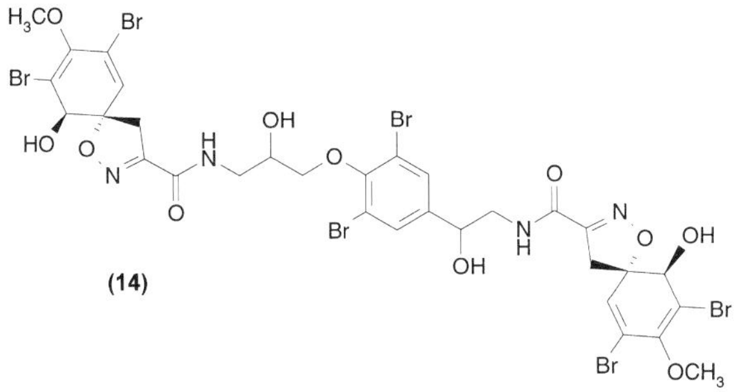

(14)

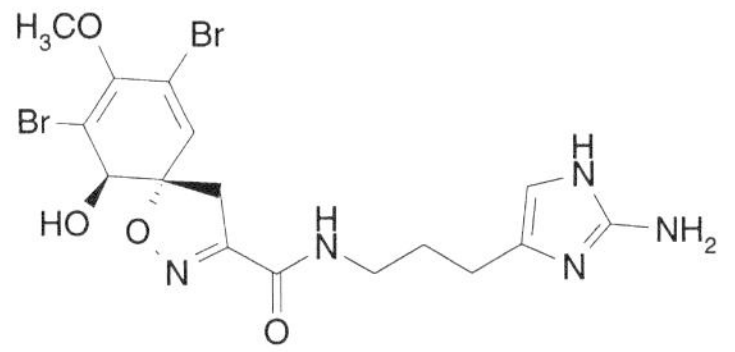

(15)

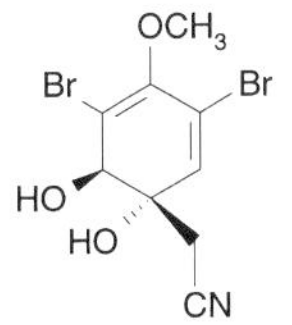

(16)

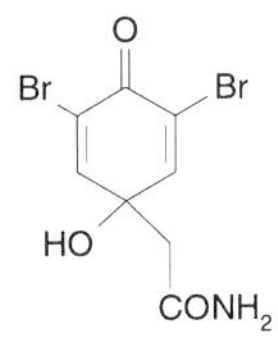

(17)

Figure 3.6 Bioconversion of aerophobin-2 (**15**), or isofistularin-3 (**14**) to aeroplysinin-1 (**16**) and to the dienone (**17**) results in a significant increase of biological activity in chemical defence.

The bioconversions of aerophobin-2 or isofistularin-3 to aeroplysinin-1 and to the dienone (Fig. 3.6) result in a significant increase of the biological activity of the reaction products compared to their substrates. For example, both aeroplysinin-1 and the dienone show pronounced antibacterial activity and may, therefore, protect sponges from invasion of pathogenic microorganisms following wounding of the tissue (Weiss *et al.*, 1996). Both metabolites are also deterrent to fish and marine snails in laboratory assays, and may thus protect the sponge from consumers (Weiss *et al.*, 1996; Ebel *et al.*, 1997). This hypothesis is corroborated by field observations, which indicate that, at least around the Canary Islands, *A. aerophoba* is among the dominating sponge species and shows only rare signs of damage attributable to attack by predators (Teeyapant and Proksch, 1993, unpublished results).

3.7 Geographical patterns of palatability and consumer resistance

Bakus and Green (1974) and Green (1977) were among the first to recognize and describe geographical patterns of marine invertebrate toxicity. Screening experiments aiming at a geographical correlation of sponge toxicity, conducted along a north-south gradient from the state of Washington (USA) to Vera Cruz (Mexico), indicated a clear increase in fish toxicity of sponges from temperate areas to the tropics (Green, 1977). From the sponges collected along the coast of Washington and assayed for fish toxicity, approximately 9% proved to be toxic, whereas this number increased to 75% at Vera Cruz. The significant increase of ichthyotoxic sponges in the tropics is in congruence with the increase in the diversity of fish in the tropics compared to temperate areas. Whereas from the Pacific coast of Canada approximately 227 different species of fish were recorded (Clemens and Wilby, 1946), this number increased to 440 in the Gulf of Mexico (Starck, 1968). At Queensland, Australia, more than 1,000 species of marine fish were recorded (Scott, 1962). This strong increase in the diversity of fish in tropical waters has been hypothesized to be the major driving force for feeding specialization, which will, in turn, select for more effective (chemical) defence strategies in potential prey organisms, such as sponges (Green, 1977).

Since the pioneering work of Bakus and Green, very few investigators have addressed geographical patterns of defensive compounds in the marine environment. A notable exception is the recent study by Bolser and Hay (1996). In this study, tropical and temperate sea urchins were given the choice between a temperate alga from North Carolina and a closely-related, tropical alga from the Bahamas. The algae investigated included members of the genera *Amphiroa* (red alga) and *Udotea* (green alga), as well as *Dictyopteris*, *Dictyota*, *Lobophora*, *Padina* and *Sargassum* (all brown algae). Powders of freeze-dried algae were incorporated into artificial diet and offered to the temperate sea urchin, *Arbacia punctulata*, and to the tropical urchin, *Lytechinus variegatus*. The bioassays usually indicated tropical seaweeds to be less palatable than temperate seaweeds, which were preferred by the sea urchins. The mean amounts of seaweeds from North Carolina consumed during the bioassays were approximately two times higher compared to seaweeds from the Bahamas (Bolser and Hay, 1996). In a similar set of experiments, the authors proved that the deterrency of most of the algae screened was due to lipophilic rather than water-soluble constituents, even though the exact nature of the active compounds was not elucidated.

If it is believed that tropical algae (and invertebrates) generally respond to an increased consumer pressure by optimizing their chemical defence (compared to their temperate congeners), then one should expect that consumers from tropical waters should also have evolved a greater resistance towards these defensive chemicals than those from temperate areas. This hypothesis was tested using diterpenoid metabolites isolated from the tropical brown alga, *Dictyota acutiloba* (Cronin *et al.*, 1997). Known amounts of purified compounds were incorporated into an artificial diet and offered to temperate (pinfish, *Lagodon rhomboides*; sea urchin, *Arbacia punctulata*) and tropical consumers (parrotfishes, *Scarus schlegeli* and *S. sordidus*; surgeonfishes, *Naso literatus* and *N. unicornis*; and sea urchin, *Diadema savignyi*) in laboratory feeding assays. Temperate herbivores were usually deterred by significantly lower diterpenoid concentrations than those from the tropics: whereas feeding of the temperate pinfish, *L. rhomboides*, was already reduced at diterpenoid concentrations equalling approximately one fifth of their natural concentration (as present in thalli of *D. acutiloba*), only one of the compounds analyzed deterred feeding by the tropical parrot fishes (at natural concentration). None of the algal constituents analyzed deterred the tropical surgeonfishes. A similar pattern was found when the temperate sea urchin, *A. punctulata*, was exposed to the diterpenoids in comparison to the tropical urchin, *D. savignyi*. The authors concluded from their study that consistent exposure of consumers to chemically well-defended prey may shape a consumer's response to the respective defensive compounds, thereby gradually leading to an increased resistance (Cronin *et al.*, 1997).

3.8 Conclusions and outlook

Marine chemical ecology, even though still in its infancy when compared to the long tradition of terrestrial chemical ecology (see the pioneering work of Stahl (1904) and others), has provided fascinating insights into the multiple functions natural products may play in interactions between species. However, our present knowledge of marine chemical ecology is still largely descriptive and, furthermore, patchy, focusing on selected aspects and more or less neglecting others. Some important areas of marine chemical ecology that need to be addressed in the future have been summarized below.

The strength of present-day marine chemical ecology is certainly in the area of anticonsumer defence of algae and invertebrates. Whereas, recent years have seen a sharp increase of mainly descriptive studies dealing with marine natural products that deter fish, molluscs or other potential herbivores/predators, very little is known about the actual physiological effects that these compounds have on potential consumers. Some of the key questions that should be addressed in the future for a deeper understanding of the mechanisms of deterency and, thus, of the molecular fundamentals of predator/prey interactions include, for example: How are deterrent compounds perceived; how do they effect the metabolic 'fitness' of consumers; and how are they degraded or excreted?

In contrast to the almost overwhelming chemical diversity of marine natural products that have been reported so far, our knowledge about their biosynthesis, their regulation and their physiology (e.g. transloca-tion, tissue or organ specific accumulation) is at best meagre (Garson, 1994) to virtually nonexistent. However, the site of accumulation of natural products within a given organism, for example, is often of critical importance in order to vigorously assess its possible ecological signi-ficance, as discussed above for compounds having a reputed antifouling activity.

Last but not least, it is becoming increasingly clear that many marine invertebrates, such as sponges, house microbial 'endosymbionts' and, thus, constitute rather complex assemblages of different eukaryotic and prokaryotic life-forms. It is possible that at least some natural products isolated from such an assemblage have their true function in 'house-keeping' (regardless of whether the sponge itself or an associated microorganism is the actual producer), for example, preventing foreign microorganisms from colonizing the sponge. Alternatively, some of these compounds may act as signals in attracting and aggregating the 'right type' of microorganisms. Other activities that these compounds might have, such as deterrency, may be largely side-effects that mask their true 'raison d'être'.

Marine chemical ecology has been, and will continue to be, a largely multidisciplinary effort. In the past, marine natural product chemists have successfully collaborated with marine ecologists and have compiled an extensive set of data in favour of the ecological significance of secondary compounds from marine organisms. In the future, marine chemical ecology should also embrace other disciplines, such as physiology, biochemistry and molecular biology, and move from a descriptive level towards a more causative approach in elucidating the ecological role and mode of action of marine natural products.

Acknowledgements

The author wishes to thank the DFG (SFB 251) and the Fonds der Chemischen Industrie for their continued support of his research on marine chemical ecology.

References

Bailey, J.A. and Mansfield, J.W. (eds.) (1982) *Phytoalexins*. Blackie, Glasgow.

Bakus, G.J. and Green, G. (1974) Toxicity in sponges and holothurians: a geographic pattern. *Science*, **185** 951-53.

Bakus, G.J., Targett, N.M. and Schulte, B. (1986) Chemical ecology of marine organisms: an overview. *J. Chem. Ecol.*, **12** 951-87.

Bakus, G.J., Schulte, B., Wright, M., Green, G. and Gomez, P. (1990) Antibiosis and antifouling in marine sponges: laboratory *versus* field studies, in: *Perspectives in Sponge Biology* (ed. Rützler, K.), Smithsonian Institution Press, Washington, pp. 102-108.

Bergmann, W. and Feeney, R. (1951) Contribution to the study of marine sponges. 32. The nucleosides of sponges. *J. Org. Chem.*, **16** 981-87.

Bolser, R.C. and Hay, M.E. (1996) Are tropical plants better defended? Palatibility and defenses of temperate *vs* tropical seaweeds. *Ecology*, **77** 2269-86.

Braekman, J.C. and Daloze, D. (1986) Chemical defence in sponges. *Pure Appl. Chem.*, **58** 357-64.

Branch, G.M. (1984) Competition between marine organisms: ecological and evolutionary implications. *Oceanogr. Mar. Biol. Ann. Rev.*, **22** 429-593.

Carpenter, R.C. (1986) Partitioning herbivory and its effects on coral algal communities. *Ecol. Monogr.*, **56** 345-65.

Clemens, W.A. and Wilby, G.V. (1946) The fishes of the Pacific coast of Canada. *Bull. Fish. Res. Bd. Can.*, **68** 1-368.

Cronin, G. and Hay, M.E. (1996) Induction of seaweed chemical defenses by amphipod grazing. *Ecology*, **77** 2287-301.

Cronin, G., Paul, V.J., Hay, M.E. and Fenical, W. (1997) Are tropical herbivores more resistant than temperate herbivores to seaweed chemical defenses? Diterpenoid metabolites from *Dictyota acutiloba* as feeding deterrents for tropical *versus* temperate fishes and urchins. *J. Chem. Ecol.*, **23** 289-302.

Davis, A.R., Targett, N.M., McConnell, O.J. and Young, C.M. (1989) Epibiosis of marine algae and benthic invertebrates: natural products chemistry and other mechanisms inhibiting settlement and overgrowth. in *Bioorganic Marine Chemistry* (ed. Scheuer, P.J.), Vol. 3, Springer-Verlag, Berlin, pp. 85-114.

De Nys, R., Coll, J.C. and Price, I.R. (1991) Chemically-mediated interactions between the red alga, *Plocamium hamatum* (Rhodophyta), and the octocoral, *Sinularia cruciata* (Alcyonacea). *Mar. Biol.*, **108** 315-20.

Ebel, R., Brenzinger, M., Kunze, A., Gross, H.J. and Proksch, P. (1997) Wound activation of protoxins in marine sponge, *Aplysina aerophoba*. *J. Chem. Ecol.*, **23** 1451-62.

Faulkner, D.J. (1997) Marine natural products. *Nat. Prod. Rep.*, **14** 259-302.

Feeny, P. (1976) Plant apparency and chemical defense. *Recent Adv. Phytochem.*, **10** 1-40.

Garson, M.J. (1994) The biosynthesis of sponge secondary metabolites: why it is important, in *Sponges in Time and Space* (eds. van Soest, R.W.M., van Kempen, T.M.G. and Braekman, J.-C.), A.A. Balkema, Rotterdam, pp. 427-40.

Green, G. (1977) Ecology of toxicity in marine sponges. *Mar. Biol.*, **40** 207-15.

Groweiss, A., Shmueli, U. and Kashman, Y. (1983) Marine toxins of *Latrunculia magnifica*. *J. Org. Chem.*, **48** 3512-16.

Hammond-Kosack, K.E. and Jones, J.D.G. (1996) Resistance gene-dependent plant defense responses. *Plant Cell*, **8** 1773-91.

Hay, M.R. (1996) Marine chemical ecology: what's known and what's next? *J. Exp. Mar. Biol. Ecol.*, **200** 103-34.

Jackson, J.B.C. (1977) Competition on marine and hard substrata: the adaptive significance of solitary and colonial strategies. *Am. Nat.*, **111** 743-67.

Jackson, J.B.C. and Buss, L. (1975) Allelopathy and spatial competition among coral reef invertebrates. *Proc. Natl. Acad. Sci. USA*, **72** 5160-63.

Jones, D.A. (1988) Cyanogenesis in animal-plant interactions, in: *Cyanide Compounds in Biology* (eds. Evered, D. and Harnett, S.), John Wiley, Chichester, pp. 151-70.

Kernan, M.R., Molinski, T.F. and Faulkner, D.J. (1988) Macrocyclic antifungal metabolites from the Spanish dancer nudibranch, *Hexabranchus sanguineus*, and sponges of the genus *Halichondria*. *J. Org. Chem.*, **53** 5014-20.

Krebs, H.C. (1986) Recent developments in the field of marine natural products with emphasis on biologically-active compounds. *Progr. Chem. Org. Nat. Prod.*, **49** 151-363.

Lindquist, N. and Hay, M.E. (1996) Palatability and chemical defense of marine invertebrate larvae. *Ecol. Monogr.*, **66** 431-50.

Lumbang, W.A. and Paul, V.J. (1996) Chemical defenses of the tropical green seaweed, *Neomeris annulata*, Dickie: effects of multiple compounds on feeding by herbivores. *J. Exp. Mar. Biol. Ecol.*, **201** 185-95.

MarinLit (1998) A marine literature database maintained by the Marine Chemistry Group. University of Canterbury, Christchurch, New Zealand.

McKey, D. (1979) The distribution of secondary compounds within plants, in *Herbivores: Their Interaction with Secondary Plant Metabolites* (eds. Rosenthal, G.A. and Janzen, D.H.), Academic Press, New York, pp. 56-133.

Melton, T. and Bodnar, J.W. (1988) Molecular biology of marine microorganisms: biotechnological approaches to naval problems. *Nav. Res. Rev.*, **40** 24-39.

Meyer, K.D. and Paul, V.J. (1995) Variation in secondary metabolite and aragonite concentrations in the tropical green seaweed, *Neomeris annulata*: effects on herbivory by fishes. *Mar. Biol.*, **122** 537-45.

Munro, M.H.G., Blunt, J.W., Lake, R.J., Litaudon, M., Battershill, C.N. and Page, M.J. (1994) From seabed to sickbed: what are the prospects? in *Sponges in Time and Space* (eds. van Soest, R.W.M., van Kempen, T.M.G. and Braekman, J.-C.), A.A. Balkema, Rotterdam, pp. 473-84.

Paul, V.J. (ed.) (1992a) *Ecological Roles of Marine Natural Products*. Comstock Publishing Associates, Ithaca.

Paul, V.J. (1992b) Seaweed chemical defenses on coral reefs, in *Ecological Roles for Marine Natural Products* (ed. Paul, V.J.), Comstock Publishing Associates, Ithaca, pp. 24-50.

Paul, V.J. (1992c) Chemical defenses of benthic marine invertebrates, in *Ecological Roles of Marine Natural Products* (ed. Paul, V.J.), Comstock Publishing Associates, Ithaca, pp. 164-88.

Paul, V.J. and van Alstyne, K.L. (1988) Antiherbivore defenses in *Halimeda*. *Proc. Sixth. Int. Coral Reef Symp.*, **3** 133-38.

Pawlik, J.R. (1993) Marine invertebrate chemical defenses. *Chem. Rev.*, **93** 1911-22.

Pawlik, J.R., Kernan, M.R., Molinski, T.F., Harper, M.K. and Faulkner, D.J. (1988) Defensive chemicals of the Spanish dancer nudibranch, *Hexabranchus sanguineus*, and its egg ribbons: macrolides derived from a sponge diet. *J. Exp. Mar. Biol. Ecol.*, **119** 99-109.

Pawlik, J.R., Chanas, B., Toonen, R.J. and Fenical, W. (1995) Defenses of Caribbean sponges against predatory reef fish. I. Chemical deterrency. *Mar. Ecol. Prog. Ser.*, **127** 183-94.

Pennings, S.C., Pablo, S.R. and Paul, V.L. (1997) Chemical defenses of the tropical, bentic marine cyanobacterium, *Hormothamnion enteromorphoides*: diverse consumers and synergisms. *Limnol. Oceanogr.*, **42** 911-17.

Porter, J.W. and Targett, N.M. (1988) Allelochemical interactions between sponges and corals. *Biol. Bull.*, **175** 230-39.

Proksch, P. (1994) Defensive roles for secondary metabolites from marine sponges and sponge-feeding nudibranchs. *Toxicon*, **32** 639-55.

Proksch, P. and Ebel, R. (1998) Ecological significance of alkaloids from marine invertebrates, in *Alkaloids: Biochemistry, Ecology and Medicinal Applications* (eds. Roberts, M.F. and Wink, M.), Plenum Press, pp. 379-94.

Rhoades, D.F. (1979) Evolution of plant chemical defense against herbivores, in *Herbivores: Their Interaction with Secondary Plant Metabolites* (eds. Rosenthal, G.A. and Janzen, D.H.), Academic Press, New York, pp. 3-54.

Rhoades, D.F. and Cates, R.G. (1976) Toward a general theory of plant antiherbivore chemistry. *Recent Adv. Phytochem.*, **10** 168-213.

Sahm, A., Pfanz, H., Grünsfelder, M., Czygan, F.-C. and Proksch, P. (1995) Anatomy and phenylpropanoid metabolism in the incompatible interaction of *Lycopersicon esculentum* and *Cuscuta reflexa*. *Bot. Acta*, **108** 358-64.

Sale, P.F. (ed.) (1991) *The Ecology of Fishes on Coral Reefs*. Academic Press, San Diego.

Sarma, A.S., Daum, T. and Müller, W.E.G. (1993) *Secondary Metabolites from Marine Sponges*. Akademie gemeinnütziger Wissenschaften zu Erfurt, Ullstein-Mosby Verlag, Berlin.

Schmitt, T.M., Hay, M.E. and Lindquist, N. (1995) Constraints on chemically-mediated coevolution: multiple functions of seaweed secondary metabolites. *Ecology*, **6** 107-23.

Schmitz, F.J. (1994) Cytotoxic compounds from sponges and associated microfauna, in *Sponges in Time and Space* (eds. van Soest, R.W.M., van Kempen, T.M.G. and Braekman, J.-C.), A.A. Balkema, Rotterdam, pp. 485-94.

Schupp, P., Eder, C., Wray, V., Paul, V.J. and Proksch, P. (1998a) Staurosporine derivatives from the marine tunicate, *Eudistoma toealensis*, and the associated flatworm, *Pseudoceros concineus*. *J. Nat. Prod.*, (submitted).

Schupp, P., Paul, V.J. and Proksch, P. (1998b) Intraspecimen variation of defensive alkaloids in the sponge, *Oceanapia* sp. *Mar. Biol.*, (submitted).

Scott, T.D. (1962) The Marine and Fresh Water Fishes of South Australia. W.L. Hames, Adelaide.

Stahl, E. (1904) Die Schutzmittel der Flechten gegen Tierfraß, in *Festschrift zum 70. Geburtstag von Ernst Haeckel*, G. Fischer Verlag, Jena, pp. 357-74.

Starck, W.A. (1968) A list of fishes of Alligator Reef, Florida with comments on the nature of the Florida fish fauna. *Undersea Biol.*, **1** 1-40.

Sullivan, B., Djura, P., McIntyre, D.E. and Faulkner, D.J. (1981) Antimicrobial constituents of the sponge, *Siphonodictyon coralliphagum*. *Tetrahedron*, **37** 979-82.

Sullivan, B., Faulkner, D.J. and Webb, L. (1983) Siphonodictidine, a metabolite of the burrowing sponge, *Siphonodictyon* sp., that inhibits coral growth. *Science*, **221** 1175-76.

Teeyapant, R. and Proksch, P. (1993) Biotransformation of brominated compounds in the marine sponge, *Verongia aerophoba*: evidence for an induced chemical defense? *Naturwissenschaften*, **80** 369-70.

Teeyapant, R., Kreis, P., Wray, V., Witte, L. and Proksch, P. (1993) Brominated secondary compounds from the marine sponge, *Verongia aerophoba*, and the sponge-feeding gastropod *Tylodina perversa*. *Z. Naturforsch.*, **48c** 640-44.

Thacker, R.W., Becerro, M.A., Lumbang, W.A. and Paul, V.J. (1998) Allelopathic interactions between sponges on a tropical reef. *Ecology*, **79** 1740-50.

Thompson, J.E., Walker, R.P. and Faulkner, D.J. (1985) Screening and bioassays for biologically-active substances from forty marine sponge species from San Diego, California, USA. *Mar. Biol.*, **88** 11-21.

Turon, X., Becerro, M.A., Uriz, M.-J. and Llopis, J. (1996) Small scale association measures in epibenthic communities as a clue for allelochemical interactions. *Oecologia*, **108** 351-60.

Walker, R.P., Thompson, J.E. and Faulkner, D.J. (1985) Exudation of biologically-active metabolites in the sponge, *Aplysina fistularis*. II. Chemical evidence. *Mar. Biol.*, **88** 27-32.

Weinheimer, A.J. and Spraggins, R.L. (1969) The occurrence of two new prostaglandin derivatives (15-epi-PGA_2 and its acetate, methylester) in the gorgonian, *Plexaura homomalla*: chemistry of coelenterates. XV. *Tetrahedron Lett.*, **15** 5185-88.

Weiss, B., Ebel, R., Elbrächter, M., Kirchner, M. and Proksch, P. (1996) Defense metabolites from the marine sponge, *Verongia aerophoba*. *Biochem. Syst. Ecol.*, **24** 1-12.

Young, C.M. and Bingham, B.L. (1987) Chemical defense and aposematic coloration in larvae of the ascidian, *Ecteinascidia turbinata*. *Mar. Biol.*, **96** 539-44.

4 The jasmonate cascade and the complexity of induced defence against herbivore attack

Ian T. Baldwin

4.1 Introduction: a comparison with pathogen-induced defences

Early in the history of experimentation into plant-pathogen interactions, researchers separated along two lines of investigation. On the one hand, geneticists studied gene-for-gene interactions and, on the other, physiologists and biochemists focused on the signal transduction pathways responsible for hypersensitive and systemic responses to pathogen attack. Because these two subdisciplines have very different perspectives on how plants perceive and respond to attacking pathogens (Gabriel and Rolfe, 1990), the development of each is independent of the other. Recently, these subdisciplines have united; and the subsequent advances during the last 5 yrs have been spectacular (Baker *et al.*, 1997).

The gene-for-gene model, first proposed by Flor in a series of studies in the 1940s (e.g. Flor, 1947), is being realized in exquisite molecular detail: three R genes have been sequenced, 20 additional ones have been mapped and the physical interaction of the protein from a bacterial avirulence (*avr*) gene with a plant resistance (*R*) gene product has been described (Tang *et al.*, 1996). Moreover, a detailed biochemical and genetic understanding of the signal cascades mediating local as well as systemic responses is rapidly unfolding. The choreographies of events from pathogen recognition to the production of the defended phenotype (from the pathogen-elicitor-mediated ion flux, oxidative burst, jasmonate accumulation, protein phosphorylation, defence gene activation, messenger ribonucleic acid (mRNA) accumulation, enzyme accumulation and, finally, the accumulation of phytoalexins responsible for the defended phenotype) are being described in model plant-pathogen systems, with the *Arabidopsis thaliana* genome project dramatically accelerating the rate of progress. Many of the discoveries made in *Arabidopsis* appear to be applicable to other plant systems. For example, the *Arabidopsis* R genes, *RPS2* and *RPM1*, appear to have functional equivalents in soybean, bean and pea (see Baker *et al.*, 1997, and citations therein).

In contrast, our understanding of the molecular mechanisms of plant-insect interactions is still in its infancy. So far, only one gene responsible for a specific herbivore-resistance trait has been cloned and sequenced: the $Hs1^{pro\text{-}1}$gene, which codes for a 282 amino acid protein that confers resistance in sugar beets against the sugar beet cyst nematode

(Cai *et al.*, 1997). This exciting discovery indicates that the recognition of a nematode by a resistant host is not fundamentally different from the molecular mechanisms of recognition of microbes (Gebhardt, 1997) and could be the harbinger of rapid advances to come. However, both historical and ecological considerations suggest that progress in the elucidation of the signal cascades responsible for induced resistance against herbivores will proceed more slowly.

History has profoundly influenced the rate of progress. Plant physiologists, biochemists, geneticists and molecular biologists have only recently taken an interest in the phenomena. With the notable exception of C. Ryan's group at Washington State University, whose laboratory is responsible for most of our existing knowledge of herbivore-induced signalling, ecologists have populated the discipline and their efforts have uncovered the taxonomic and functional diversity of the interactions and the evolutionary processes responsible for maintaining the interactions in nature. Because ecologists have been interested in the interaction *per se*, they have avoided model systems. Researchers studying plant-pathogen interactions, on the other hand, have developed well-behaved model systems and these have been instrumental in the rapid progress. The ecological legacy of plant-herbivore interactions will probably serve the field well, particularly in the era of genomics, when gene identification will no longer limit the rate of progress. The challenge will then be to understand the functional consequences of the plethora of genetic changes that are likely to occur in a plant after herbivore attack.

A key experimental advance in the development of model plant-pathogen systems has been the ability to elicit defence responses with purified elicitors (Perez *et al.*, 1997), an advance which has allowed researchers to remove the pathogen from the interaction and focus on the events leading to pathogen recognition. Interestingly, a plant's response to purified elicitors does not simply involve the activation of a small number of defence-related genes; rather, it entails an extensive reprogramming of cellular metabolism (Somssich and Hahlbrock, 1998). The full genetic response to herbivore attack, although still largely unknown, is likely to be equally daunting (Bergey *et al.*, 1996). Organisms whose genomes have been fully sequenced, and for whom the emerging deoxyribonucleic acid (DNA) chip technology is being used to monitor environmentally-induced changes in all transcribed genes ('transcriptome'), highlight the magnitude of the transcriptional changes we should expect after herbivore attack. For example, when DeRisi and co-workers (1997) monitored the changes in expression in 6400 distinct complementary deoxyribonucleic acid (cDNA) sequences of yeast undergoing a metabolic shift from fermentation to respiration, they found 710 and 1030 genes to be up- and downregulated, respectively, by at least a factor of 2. In other words, approximately 27% of the transcriptome of yeast is altered during the diauxic shift. Whilst it is not known whether the proportional change in transcription after herbivore attack will be as extensive, the absolute amount of expression altered by herbivory may be, given the larger transcriptome size of higher plants than that of yeast. Therefore, a major challenge facing this field of research in the coming decades will be to understand the function of the multivariate changes in the transcriptome elicited by herbivore attack. It is in this regard that the ecological legacy of the field will prove invaluable.

A recent book (Karban and Baldwin, 1997) examined the integration of the ecological with the biochemical and physiological subdisciplines studying plant-herbivore interactions. Continuing this examination, this review focuses on the question of specificity of plant responses to herbivore attack. The jasmonate (octadecanoid or Vick-Zimmerman) cascade is demonstrably important in mediating resistance against some herbivores and regulates the expression of traits which are important in resistance. However, since this cascade is also involved in responses to other abiotic stress, such as UV (see Conconi *et al.*, 1996b), it remains an open question whether the jasmonate cascade can specifically tailor a plant's responses to a particular herbivore. Many research groups are scrutinizing the jasmonate cascade and it has recently been the subject of numerous excellent reviews (Creelman and Mullet, 1997a, b; Blechert *et al.*, 1995; Weiler, 1997; Wasternack and Parthier, 1997; Reinbothe *et al.*, 1994; Gross and Parthier, 1994; and Hamberg and Gardner, 1992). This review examines the evidence for: 1) the herbivore-specific responses in plants; 2) the importance of the jasmonate cascade in induced resistance; 3) the resistance traits regulated by the jasmonate cascade; and 4) the potential for the jasmonate cascade to mediate herbivore-specific responses. A preamble on the ecological complexity of resistance to herbivores and a phytocentric definition of the term 'resistance', one that includes traits that allow a plant to experience herbivory without reducing its Darwinian fitness (e.g. 'tolerance-enhancing' traits), highlight the value of the ecological legacy of this field. Since the study of plant-pathogen responses is more advanced in some areas, comparisons with this discipline help predict future developments.

4.2 Induced resistance and tolerance

Most herbivores are physiologically less dependent on their host plants than are most pathogens. This functional autonomy complicates the development of model systems for herbivore-induced responses in two important ways. Firstly, the autonomy influences the spatial scale at which herbivores perceive and attack plants. While pathogens attack plants on a 'fine-grained' (*sensu* Levins, 1968) scale, selecting individual cells to initiate their attack, herbivores frequently operate on a more 'coarse-grained' scale, and the interactions between plants and herbivores can begin long before the herbivore makes physical contact with the plant. Therefore, resistance to herbivores may include plant traits that interfere with host location (Feeny, 1976a; Kennedy and Barbour, 1992), a process which will only be relevant to plant-pathogen interactions when the pathogen is transmitted by a herbivore (Carter, 1973) or pollinator (e.g. Roy, 1993).

Secondly, the functional autonomy of herbivores allows plants to utilize a defensive strategy that is not effective against pathogens: plant traits that facilitate 'top-down' control of herbivore populations by the herbivore's predators, pathogens and parasitoids (Price, 1984; Takabayashi and Dicke, 1996; Turlings *et al.*, 1995). However, these higher-level ecological interactions can wreak havoc with traits that function defensively in the rarefied ecological interaction of a laboratory bioassay. For example, plant chemical defences are frequently sequestered by specialist herbivores for their own defence against their parasitoids and predators (Duffey *et al.*, 1986; Barbosa and Saunders, 1985; and Malcolm and Zalucki, 1996), thereby turning a plant's defences against itself (Fig. 4.1). Many inducible chemical defence factors, such as the well-studied protease inhibitors (PIs) (Jongsma and Bolter, 1997), slow herbivore growth by reducing their digestive efficiency. Because slow-growing herbivores may eat more leaf material to complete development, such traits may not function as defences in nature (Jongsma and Bolter, 1997; Moran and Hamilton, 1980) without the use of the third trophic level (Fig. 4.1). These considerations, in addition to an influential 'devil's advocate' review (Fowler and Lawton, 1985) of the evidence for the defensive function of herbivore-induced responses, have helped to establish rigorous criteria for a trait to be demonstrably 'defensive': i.e. it must be shown to increase plant fitness correlates (Karban and Meyer, 1989) preferably among the complexity of ecological interactions that occur in natural populations. Interestingly, only two recent studies have met these onerous criteria and demonstrated the fitness benefit of an induced response in nature (Agrawal, 1998; Baldwin, 1998).

This 'phytocentric' definition of plant defence, which quantifies the effect of a trait in a currency of plant fitness components rather than in a currency of herbivore performance, must include traits that decrease the fitness consequences of tissue removal and allow plants to 'tolerate' herbivory. Tolerance is operationally defined by quantitative geneticists as the ratio of the fitness of a plant lineage which has been attacked by herbivores to that of plants from the same lineage grown under identical

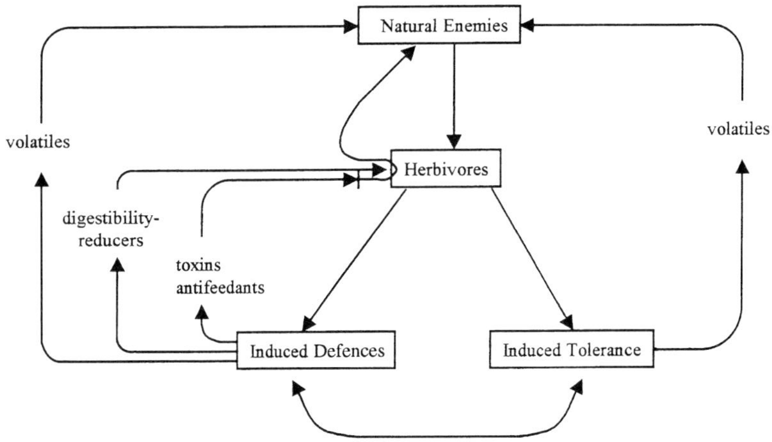

Figure 4.1 Traits that reduce the probability of attack (defences) or reduce its fitness consequences (tolerance) are frequently induced by herbivore attack, and these traits should be evaluated together with a plant's community. The complexity of ecological interactions in a plant's community can: undermine the function of a trait (e.g. when a toxin is sequestered by a plant's herbivores and used to protect the herbivore against its own natural enemies, thereby turning the plant defence against itself); or can be essential for its function (e.g. when a trait slows the growth rate of a herbivore, without which the third trophic level might cause a plant to lose more, rather than less, tissues to the herbivore). Volatile emissions from attacked plants, which are thought to increase the hunting efficiency of the natural enemies by making herbivores more apparent to them, thereby function as an indirect defence.

conditions but without the herbivores (Simms and Rausher, 1987; Rausher, 1992). Whilst this operational definition has done little to identify the specific traits responsible for tolerance, it has established that these traits can be heritable (Mauricio *et al.*, 1997) and, therefore, as likely to evolve as any trait conferring resistance against herbivores.

Physiological studies have implicated a large suite of morphological (the number and dormancy of meristems) and physiological traits (intrinsic rates of growth, storage of reserves, changes in resource acquisition, such as compensatory photosynthesis and increases in mineral nutrient uptake, and alterations in patterns of senescence) as contributing to the buffering of plant reproductive performance against tissue loss to herbivores (reviewed in Trumble *et al.*, 1993; Rosenthal and Kotanen, 1994; Karban and Baldwin, 1997). Since reproductive performance is a whole-plant trait, it is not surprising that some resistance traits operate on a whole-plant scale, as the following two examples illustrate.

Dyer and co-workers (1991) studied carbon-partitioning in two populations of the C_4 Serengeti grass, *Panicum coloratum*, which came from areas with different grazing histories. Plants cloned from the population with intense grazing pressure allocated significantly more of their recently fixed carbon to underground storage in their roots, whilst plants derived from individuals collected from an area with low grazing pressure stored their recently fixed carbon in aboveground stem tissues. Storage in stem tissues may be advantageous to plants growing in habitats with intense competition from other plants, where vertical growth is essential for successful light capture. However, storage of reserves in stems may represent a liability in heavily grazed habitats. In a second study, Larson and Whitham (1997) studied clones of cottonwood trees that differed in their resistance to the leaf-galling aphid, *Pemphigus betae*. These aphids establish feeding sites on newly developing leaves, which compete with other developing leaves as sinks for translocated photoassimilates. Resistant tree genotypes had more natural sinks (i.e. leaf buds) relative to their stem volume than did trees that were more susceptible to aphid attack. The authors suggested that, on trees with a fewer leaf buds, aphids were able to establish stronger and larger sinks and were more successful in establishing leaf galls than they were on clones with a greater number of competing natural sinks.

These two examples demonstrate that herbivores feeding on very different spatial scales, from the large mammalian grazers to gall-forming aphids, have probably exerted selection on the organization of sinks and sources in their host plant, traits which will also influence a plant's competitive ability and, possibly, explain the controversial phenomenon of 'overcompensation' to herbivore attack (Whitham *et al.*, 1991; Paige and Whitham, 1987; Bergelson and Crawley, 1992). For example, apical dominance and the concomitant suppression of lateral meristems may be strongly advantageous in environments with strong competition for aboveground resources (such as light), and, if grazing pressure both removes the apical meristems and alters the competitive dynamics, plants may be able to overcompensate by releasing suppressed lateral reproductive meristems (Aarssen and Irwin, 1991). These examples emphasize the fact that resistance, tolerance and competition are all likely to exert strong selection on similar whole-plant physiological traits but that the direction of the selection on these traits may vary greatly. Tolerance and resistance traits are also integrated at a cellular level and some of this integration may be mediated by the jasmonate cascade.

Herbivore-induced defence responses can exact large resource demands on a plant's metabolism. For example, when *Nicotiana attenuata* responds to herbivore attack by increasing nicotine production, 6% of a plant's whole-plant nitrogen content is in this toxin alone (Baldwin *et al.*, 1998). Similarly, most of the large increases in respiration that occur after wounding can be attributed to the metabolic demands of induced furanocoumarin synthesis in wild parsnip (Zangerl *et al.*, 1997). If such high metabolic demands are commonly incurred when defences are activated, coordinated alterations in metabolism are expected, and evidence continues to accumulate that the jasmonate cascade orchestrates both the upregulation of defence genes and the down regulation of many 'housekeeping' genes (Wasternack and Parthier, 1997; Wasternack *et al.*, 1998). The best studied downregulated response is also, quantitatively, the most important protein synthetic process in green plants, i.e. the synthesis of RuBPCase (Görschen *et al.*, 1997a,b). Since many of the jasmonate-induced defence responses are transcriptionally regulated (Wasternack and Parthier, 1997), the resources required for these new initiatives must be made available within the plant (by reallocation from growth-related processes) or by increasing the rate of resource acquisition. These alterations in primary metabolism may play additional roles in minimizing the fitness consequences of herbivore attack and, therefore, contribute to tolerance-enhancing traits.

In short, there are many ways for a plant to minimize the fitness consequences of herbivore attack, only some of which can be understood by studying plants grown in the ecological isolation of a laboratory environment. The functional autonomy of herbivores and the ecological complexity of plant-herbivore interactions intertwine traits responsible for resistance, tolerance and competitive ability (Herms and Mattson, 1992; Zangerl and Bazzaz, 1992). If we are to understand the function of the complicated changes that occur after herbivory, our functional screens must incorporate some of this ecological complexity.

4.2.1 Herbivore- and wound-induced responses

The prediction that herbivore attack is likely to result in a large shift in gene expression begs the question: What proportion of these induced changes are specific to herbivore attack? It is reasonable to assume that herbivore attack is not recognized by the plant by the introduction of specific recognition gene-products through a type III or IV macromolecule delivery system, as has been demonstrated with some plant pathogens (Baker *et al.*, 1997). Herbivores force their way past a plant's protective barriers with a diverse array of feeding apparatus and, in doing so, cause wounding. Since many biotic and abiotic agents can cause wounding and these wounds can be opportunistically used as points of entry by microbes, overlap between the response to microbes and herbivores is inevitable. Furthermore, many plant-herbivore associations require a microbial partner (Berenbaum, 1988) that herbivores sometimes house symbiotically (to cope with the difficulties of the herbivorous diet); therefore, plant resistance traits may target the microbial fauna as well as the herbivores. Furthermore, some herbivores (e.g. aphids) function as 'Trojan horses', vectoring pathogenic organisms during feeding (Carter, 1973), which should select for defence responses that do not distinguish between microbes and herbivores. However, the ability to distinguish between herbivore and pathogen attack and to respond appropriately should be advantageous, if the responses are costly or if they interfere with each other (Karban and Baldwin, 1997). In this regard, it is interesting to note that salicylic acid (SA) (Fig. 4.2), an important signal in the elicitation of systemic response, can disable both the jasmonate cascade and the wound- and herbivore-induced responses that this cascade initiates (Baldwin *et al.*, 1997; Creelman and Mullet, 1997a; Peña-Cortés *et al.*, 1993; Doares *et al.*, 1995), suggesting that the responses elicited by these two cascades may be incompatible.

The question of specificity in herbivore-induced responses can be separated into two components: specificity both in the responses of plants and also in their effects on herbivores (reviewed by Karban and Baldwin, 1997; see also Stout *et al.*, 1998 for a recent experimental example using tomato). The evidence that responses elicited by herbivores differ from those elicited by the wounding they cause will now be considered. Mechanical simulation of a herbivore's wounding that faithfully mimics the spatial, temporal and mechanical characteristics of herbivory is tedious work. However, the efforts are frequently rewarded with insights into the specificity of the interactions. Three recent comparisons have demonstrated that the responses elicited by herbivores differ from those elicited by the mechanical wounding they cause.

Herbivory and wounding to the leaves of *Nicotiana sylvestris* and *N. attenuata* plants dramatically stimulates nicotine synthesis in the roots (Baldwin *et al.*, 1994a; Baldwin, 1996). The induction is activated by jasmonic acid (Fig. 4.3) production in the wounded leaves and its subsequent transport to the roots (Baldwin *et al.*, 1994b, 1997; Zhang and Baldwin,1997). Strong positive relationships exist between the amount of mechanical wounding, the jasmonate synthesized at the wound site shortly after wounding and the nicotine that accumulates in the entire plant 5 days later (Baldwin *et al.*, 1997; Ohnmeiss *et al.*, 1997). Interestingly, larval feeding dramatically increases the amount of jasmonate produced in the wounded leaf compared to mechanical simulation of larval feeding; however, it suppresses the increase in nicotine production (McCloud and Baldwin, 1997; Baldwin, 1988) and causes the rapid release of mono- and sesquiterpenes that are not released after mechanical wounding (I.T. Baldwin, J. Kahl, R. Aerts, C.A. Preston, unpublished data). All three of these alterations in the response of *Nicotiana* to wounding by herbivore feeding can be mimicked by applying oral secretions from *Manduca sexta* to mechanically-wounded leaves (McCloud and Baldwin, 1997). Similarly, the same insect species and its oral secretions alter the kinetics of the accumulation of transcripts of proteinase inhibitor II (PI II) and 3-hydroxy-3-methylglutaryl-coenzyme A reductase (HMGR) genes in potato (Korth and Dixon, 1997). True herbivory and oral secretions increased the rate but not the amount of transcript accumulation in comparison with mechanically-damaged plants. These studies demonstrate that larvae of *M. sexta* can activate signal transduction pathways, both in *Nicotiana* and potato, that are distinct from those activated by mechanical wounding.

An enormous advance in the understanding of how plants recognize herbivore attack was achieved with the elucidation and synthesis of the first low molecular weight elicitor from an insect herbivore. *N*-(17-hydroxylinolenoyl)-L-glutamine ('volicitin') (Fig. 4.3), isolated from the oral secretions of beet army worm larvae, *Spodoptera exigua*, increases the emission of volatiles when applied to corn (Alborn *et al.*, 1997). These volatiles are thought to function in the 'alarm-call' defences of plants, whereby plant volatiles increase the hunting efficiency of parasitoids or other members of the third trophic level (Turlings *et al.*, 1995). Compounds such as volicitin will be invaluable research tools in unravelling: 1) the mechanisms of herbivore-induced responses, by allowing researchers to use a 'reagent-grade' herbivore and standardize the elicitation of herbivore-induced responses; and 2) the functional consequences of herbivore-induced responses, by allowing researchers to pretreat plants and determine whether the plant or the herbivore benefits from the herbivore-specific alteration in the wound response.

4.2.2 *Jasmonate-induced responses function defensively but are also costly to plant fitness*

Responses elicited by the jasmonate cascade have been shown to be necessary and sufficient for protection against herbivore attack in three genera of plants. The most elegantly established case is that of tomato, *Lycopersicon esculentum*, due to the efforts of C. Ryan and co-workers at Washington State University. Ryan's group identified the first polypeptide hormone, systemin (Fig. 4.3), in plants, which is processed from the 200-amino acid precursor protein, prosystemin (Pearce *et al.*, 1991), and is localized in vascular bundles (Jacinto *et al.*, 1997) throughout the plant. After wounding or herbivore attack, systemin is released into the vascular system and increases the accumulation of transcripts of at least 15 different genes (Bergey *et al.*, 1996), including PI genes. All of the known effects of systemin are mediated by the jasmonate cascade (Pearce *et al.*, 1991). Tomato plants transformed with the prosystemin gene in an antisense orientation had highly attenuated PI responses to wounding

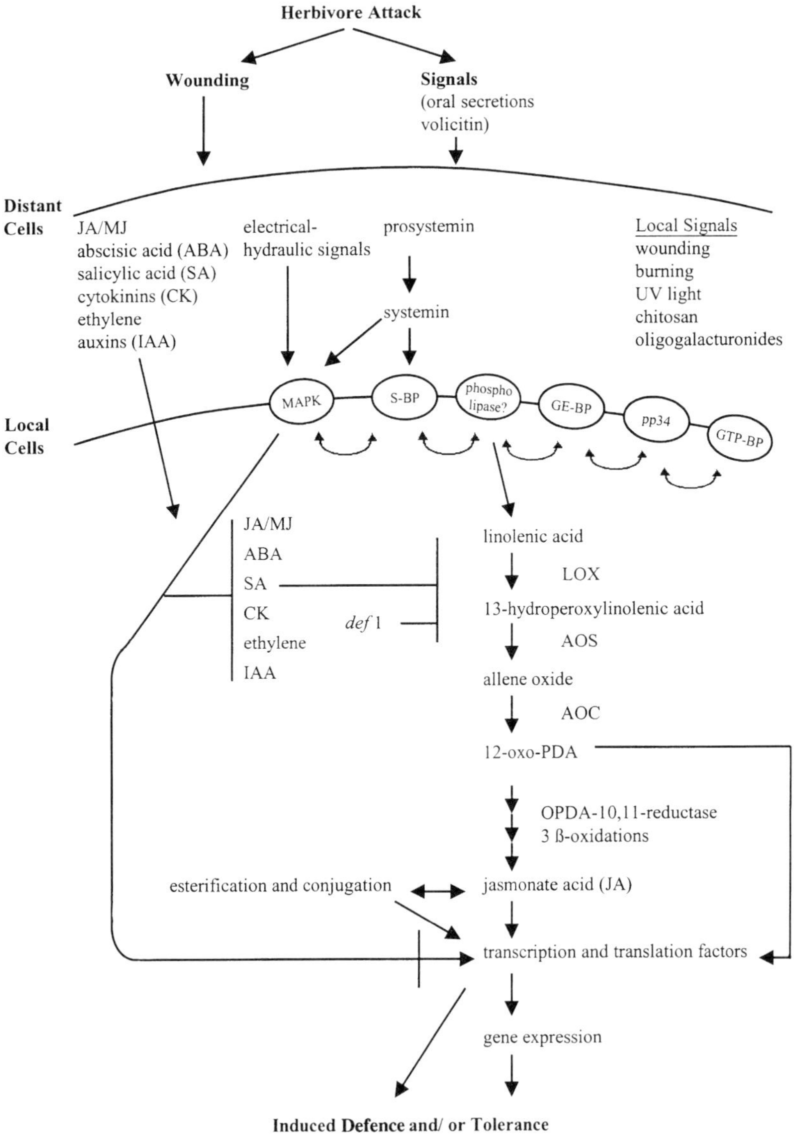

Figure 4.2 The emerging complexity of signalling cascades mediating herbivore-induced response. Responses can be elicited both locally and distally, and the jasmonate signal transduction cascade is important to many responses. The activation of the systemic responses may involve the production of jasmonic acid (JA) and its ester, methyl jasmonate (MJ), systemin, hydraulic and/or electrical signals, or other plant hormones (ethylene, abscisic acid [ABA], cytokinins, salicylic acid [SA]). These products either directly activate the jasmonate cascade (systemin, ABA), interact with its products and modulate its effects (cytokinins, SA, ethylene), or

(McGurl *et al.*, 1992) and herbivory (Orozco-Cardenas *et al.*, 1993). *M. sexta* larvae grew three times faster and consumed more leaf tissue on these transformed plants compared to control plants. More recently, members of the Ryan research group characterized a mutant tomato with a lesion in the jasmonate cascade between the synthesis of 13-hydroperoxylinolinic acid and 12-oxo-phytodienoic acid (PDA) (Fig. 4.2), which was unable to respond to wounding, oligosaccharides or systemin with increases in jasmonic acid (Fig. 4.3) or PIs (Howe *et al.*, 1996). When these mutant tomato plants were attacked by larvae of *M. sexta*, they lost substantially more leaf area and, after 11 days of growth, the larvae weighed 2.8 times as much as they did when fed on control plants with a functioning jasmonate cascade. Moreover, exogenous applications of jasmonic acid and its methyl ester decreased the suitability of tomato foliage to two additional lepidopteran herbivores, *Spodoptera exigua and Helicoverpa zea*, in laboratory feeding trials (Thaler *et al.*, 1996).

Arabidopsis thaliana is also protected by responses activated by the jasmonate cascade. Mutants (*fad3*, *2fad7*, *2fad8*) that are deficient in the jasmonate pathway because they lack the important fatty acid precursor, linolenic acid (Fig. 4.2), contain very low constitutive levels of jasmonate and are unable to respond to wounding with increased levels of jasmonate or transcripts of the jasmonate- and wound-induced vegetative storage protein genes (*vsp*). When these mutants are attacked by the larvae of the fungus gnat, *Bradysia impatiens*, they suffer approximately 80% mortality compared to almost no mortality in wild type plants that are not deficient in the jasmonate cascade (McConn *et al.*, 1997). When these mutants were treated with exogenous methyl jasmonate, the mortality was reduced

---

activate gene expression independently of the jasmonate cascade (hydraulic or electrical signals). In addition, herbivore-specific factors, such as volicitin, may be directly transferred to plants in the herbivore's oral secretions. While the biosynthesis of JA from free linolenic acid is well understood, with many of the key enzymes described, e.g. allene oxide synthetase (AOS), allene oxide cyclase (AOC), the mechanisms responsible for supplying linolenic acid to the lipoxygenases (LOX) which initiate its synthesis are less well known. They may involve: mitogen-activated protein kinases (MAPK), which stimulate phospholipases to release the fatty acids, presumably in chloroplasts; Kex2-like systemin-binding proteins (S-BP), B-glucan-elicitor-binding proteins (GE-BP), 34 kDa phosphoproteins (pp34) and guanosine triphosphate (GTP) binding proteins (GTP-BP). It is not known how JA, its precursors (12-oxo-phytodienoic acid [12-oxo-PDA]), or their conjugates effect changes in gene expression other than that both transcriptional and translational (via ribosomal inactivating proteins) processes are involved. SA, which functions in signal transduction cascades mediating defence responses against viruses, inhibits JA-induced responses, either by inhibiting its biosynthesis or inducing changes in gene transcription. Ethylene, another hormone long implicated in wound responses, appears to synergize some JA-induced responses (PIs) but inhibit others (nicotine). The defenceless (*def 1*) tomato mutant has a lesion in the octadecanoid cascade.

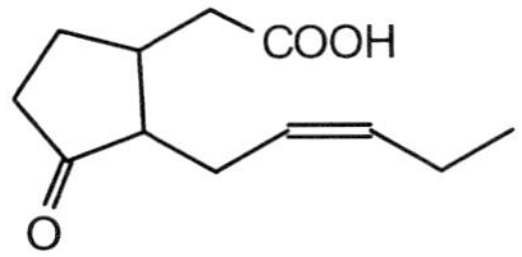

(I) jasmonic acid

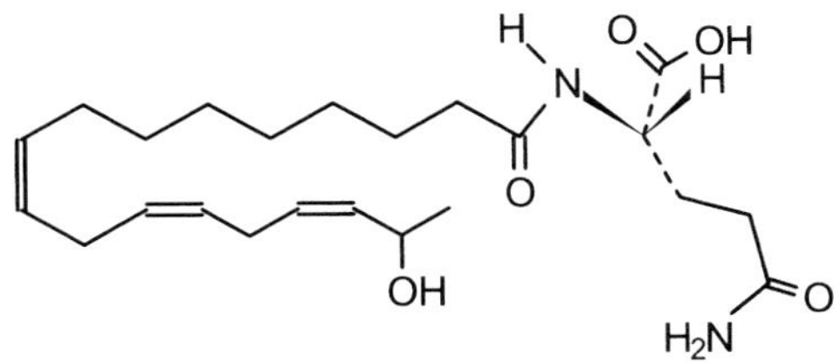

(II) volicitin

AVQSKPPSKRDPPKMQTD

(III) systemin

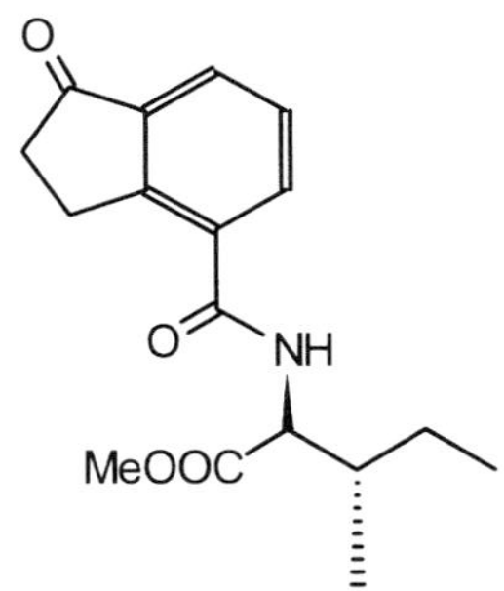

(IV) 1-oxo-indan-4-carboxylic acid ILE-methyl ester

Figure 4.3 Structures of jasmonic acid (**I**), volicitin (**II**), systemin (**III**) and indanone-isoleucine methyl ester (**IV**).

to approximately 12%, demonstrating that the disabled jasmonate cascade is responsible for the plant's susceptibility rather than the responses activated by the cascade. It is not known which traits activated by the jasmonate cascade contribute to the increased resistance in this species; *vsp* gene products, which apparently function as phosphatases or as storage proteins (Staswick, 1994), are unlikely candidates.

Finally, wound- and herbivory-induced jasmonate production is necessary and sufficient for the induction of the root-produced toxin, nicotine, in *N. attenuata* (see section 4.2.1). A field study which utilized the natural history characteristics of this species demonstrated that jasmonate-induced responses function defensively in nature (Baldwin, 1998). This species is an ephemeral member of the annual community in burned sagebrush, blackbrush and piñyon-juniper forests of the Great Basin desert. It synchronizes its growth with the post-fire environment by producing dormant seeds that germinate in response to cellulose combustion product(s) found in wood smoke (Baldwin *et al.*, 1994c; Baldwin and Morse, 1994; Preston and Baldwin, 1999). By synchronizing its germination with fires, this species is able to exploit the ephemeral but nutrient-rich (Lynds and Baldwin, 1998) herbivore- and competition-poor habitats that commonly exist after fires. This window of opportunity is quite short, for as post-fire succession proceeds, herbivores and competitors quickly recolonize the burned habitat. By inducing plants growing in burns of different age, the fitness consequences of jasmonate induction in plants with different probabilities of attack by herbivores can be examined. To accomplish this, the roots of plants were treated with methyl jasmonate to elicit a response in one member of each of 745 matched pairs of plants growing in four burns of different age. Induced plants that were attacked by herbivores realized an 11–18% increase in lifetime viable seed production compared to their noninduced counterparts (Baldwin, 1998).

Interestingly, this same field study also provided insight into why plants have inducibly-expressed rather than constitutively-expressed resistance traits. When plants grew in populations where attack rates were very low because the plants were protected by fencing and insecticides, or when plant pairs simply escaped detection by the herbivore community, jasmonate-treated plants produced 17–26% fewer lifetime viable seeds than their untreated partners (Baldwin, 1998). These results demonstrate that the responses elicited by jasmonate incur a large fitness-based cost to plants. The physiological basis for the reductions in seed production may be, in part, due to the large nitrogen investments in nicotine after induction (Baldwin, 1998). This investment cannot be recouped by metabolism (Baldwin and Ohnmeiss, 1994; and Baldwin *et al.*, 1994a) in the sibling species, *N. sylvestris*, making this nitrogen unavailable for other activities, such as seed production. However, many other physiological and ecological explanations are possible for the decreased seed production observed in jasmonate-induced plants. For example, well-defended plants may have difficulty in attracting pollinators (Euler and Baldwin, 1996).

The results of this field study lend strong support to the cost-benefit model (Givnish, 1986; Zangerl and Bazzaz, 1992; and Harvell, 1990) for the maintenance of inducible defences. Inducible defences are inherently inferior to defences that are constitutively deployed, due to the time lag between the first attack and the activation of the defence. As a result of this delay, a plant could remain vulnerable for hours or even days while waiting for the defence to be activated. The cost-benefit model argues that although defences are beneficial and may increase a plant's fitness when it is under attack, they are costly when they are not needed, which decreases plant fitness. In short, inducibility is thought to be a cost-saving measure, allowing plants to time the production of a defence with the prevailing environmental conditions and forego the payment of defence costs when they are not needed. Other models have been proposed for the evolution of induced defences, most notably the moving-target model (Karban and Baldwin, 1997; Adler and Karban, 1994; and Karban, 1993), which argues that the ability to produce a variable phenotype is, in itself, a defensive trait and a well-defended phenotype need not be costly. These theories, however, remain to be tested empirically.

These studies demonstrate that responses elicted by the jasmonate cascade can protect *Arabidopsis*, *Nicotiana* and *Lycopersicon* plants from herbivore attack but they do not identify which of the many responses that are activated by this cascade are responsible for the resistant phenotype. Induced defences against herbivores can be broadly categorized into two groups: endogenous or direct defences, in which resistance is a direct consequence of some change in phenotype (e.g. production of a thorn, toxin or antinutrient); and indirect defences, in which the change in phenotype is not intrinsically defensive but heightens the vulnerability of the plant's herbivores to their own predators. The jasmonate cascade activates examples of both categories.

4.2.3 Jasmonate-induced direct defences

Researchers who study the mechanisms of induced defences can be broadly categorized as those primarily interested in the mechanisms of alterations within a plant and those interested in the consequences of alterations for herbivores feeding on the plant. Results from both groups of researchers are adding intriguing complexity to an already complicated interaction.

C. Ryan and co-workers have broadened their focus from the elicitation of serine PIs to the characterization of a suite of 13 additional systemic wound-response proteins (SWRPs) that are all induced by systemin and the jasmonate cascade (Bergey *et al.*, 1996). The SWRPs can be categorized as: those directly involved in defence, four protease inhibitors (PIs) and a polyphenyl oxidase (PPO); those directly involved in signal transduction, prosystemin, lipoxygenase (LOX), a nucleotide diphosphate kinase, 48 kDa mitogen-activated protein (MAP) kinase (Stratmann and Ryan, 1997), and an acyl CoA-binding protein; and those associated with proteolysis. From the diversity of functions of these induced proteins, it is clear that systemin plays a central role in reconfiguring protein metabolism after herbivory. The challenge is to understand how the different responses contribute to resistance and tolerance.

PIs clearly function as defence agents; advances in determining the regulation of PI induction have recently been reviewed (Koiwa *et al.*, 1997). The PPO gene family represents another group of SWRPs that are likely to be directly defensive. PPOs oxidize phenolic compounds, in tomato, they are systemically induced by herbivore attack, the jasmonate cascade and systemin (Constabel *et al.*, 1995). The regulation of these oxidases differs dramatically among different plant taxa, from strongly inducible (tomato, tobacco and poplar) to constitutively expressed at high (willow and soybean) or very low (11 other species) levels (Constabel and Ryan, 1998). The change in transcript levels in inducible taxa can be large: tomato plants constitutively expressing the prosystemin gene display a 70-fold increase in PPO transcripts as compared to untransformed plants (Constabel *et al.*, 1995). PPOs in tomato consist of a gene family with a high degree of tissue-specific expression; different PPOs are expressed even among the four different types of trichomes that occur on the leaf surface of tomato (Thipyapong *et al.*, 1997). An elegant study by Thipyapong and Steffens (1997) at Cornell University utilized a β-glucuronidase (GUS) fusion construct with the promotor of *PPO-F*, one of the most wound-inducible of the PPO genes, and demonstrated that signals involved in wound- and pathogen-induced responses (ethylene, jasmonate and salicylic acid) initiated transcript accumulation in different tissues. Jasmonate and wounding treatment induced accumulations in young leaves, ethylene induced accumulations in old leaves, and salicylic acid treatments effected changes in leaves of all ages and in also stems.

The enzyme's defensive role depends on where it is localized. In glandular trichomes, it can function defensively as a 'super glue', trapping small-bodied insects in phenolic exudates that are oxidatively polymerized by PPOs when insects rupture cells and mix the previously separate enzymes and substrates (Kowalski *et al.*, 1992; and Constabel *et al.*,

1996). When expressed in the mesophyll, it can decrease the nutritional values of plant tissues. In its simplest form, this antinutritive function of PPOs involves the oxidation of phenolics to quinones, which, in turn, covalently modify and cross-link dietary proteins during feeding and decrease their digestibility in the herbivore's gut. The case for the defensive function of PPOs comes from studies correlating resistance and induction by different herbivores (Stout *et al.*, 1998), that use PPO inhibitors and labelled phenolic substrates (Felton *et al.*, 1989, 1992) and that alter activity *in planta* by transgenic expression of PPOs (Steffens *et al.*, 1994).

The use of transgenic expression techniques to study the function of an induced response *in planta* is significantly easier for direct gene products, such as PIs, than it is for the secondary metabolites which are frequently produced by multienzyme pathways. However, when the regulation of the pathway is understood, then large changes in secondary metabolite profiles can be produced by altering the expression of only a few genes. For example, transgenic expression in tobacco of the key regulatory enzyme in phenolic metabolism, phenylalanine ammonia lyase (PAL), from bean dramatically altered the concentrations of chlorogenic acid, rutin and other flavonoids. The performance of two Lepidopteran herbivores was examined to study the consequences of these changes (Bi *et al.*, 1997a). Interestingly, in contrast to the results of many studies examining the manipulations of these same phenolics in artificial diets, no effect on larval growth was observed. If the principal defensive role of chlorogenic acid is as a substrate in the formation of quinones by PPOs and their subsequent interaction with dietary proteins (Appel, 1993; and Bi *et al.*, 1997b), then one might not expect any correlation with chlorogenic acid content and resistance without a commensurate increase in PPO activity. These results highlight one of the difficulties of examining the defensive function of individual components of a large suite of chemicals, which are altered in coordination.

Herbivore attack results in multivariate changes in plant chemistry (Baldwin, 1994) that include alterations in metabolites generally recognized as 'primary' (Berenbaum, 1995; and Felton, 1996), small molecular weight toxins (Hammerschmidt and Schultz, 1996) and other small and large molecular weight compounds, which interact to influence the nutritional quality of a plant. Since 'resistance' is an emergent property of this suite of interacting chemical changes, assessing the defensive value of individual components with univariate approaches may not be appropriate in many circumstances (Duffey and Stout, 1996). Moreover, we should expect that many induced changes will play multiple roles in the response. For example, induced isoforms of LOX, which produce 13-hydroperoxy linolenic acid for the biosynthesis of octadecanoid wound-signals (Heitz *et al.*, 1997), may also contribute directly to defence because these hydroperoxides may be toxic to the herbivore or may lower plant nutritional quality by interacting with other plant components (Felton *et al.*, 1994).

The complexity of the chemical interactions that occur within the plant parallels a comparable complexity among the postingestive processes. The best characterized defensive proteins in plants, the serine and cystine PIs, function defensively by targeting the major proteolytic digestive enzymes of insects. After ingestion of leaf material high in PIs, some insect herbivores appear to alter the proteolytic enzymes they secrete into their digestive cavities to maintain digestive function. In larvae of the beet army worms, *Spodoptera exigua*, fed on tobacco plants overexpressing the PI II gene, only 18% of the gut proteinase activity was sensitive to inhibition by PI II. In contrast, 78% of the tryptic activity was sensitive to inhibition in larvae fed on untransformed plants with low PI II levels (Jongsma *et al.*, 1995). Similar compensatory responses were found for the Colorado potato beetle, *Leptinotarsa decemlineata*, feeding on potato plants that were induced with methyl jasmonate to increase the production of their papain inhibitors (Bolter and Jongsma, 1995). Such compensatory responses appear to be common among herbivores fed on diets containing high quantities of PIs (Jongsma and Bolter, 1997; Giri *et al.*, 1998). A second example comes from the larvae of *M. sexta*, in which dietary intake of nicotine from its host plant induces the production of P_{450}-mediated detoxification of the nicotine (Snyder and Glendinning, 1996).

In short, we should expect the induction of counter-responses in herbivores that have evolved to feed on plants with induced defences. While this expectation further complicates matters, it may help to interpret the changes that occur in plants. For example, wild parsnip induces a suite of three furanocoumarins and one phenylpropanoid compound, myristicin (Zangerl *et al.*, 1997), after wounding and herbivore attack. Myristicin, a methylenedioxyphenyl inhibitor of the P_{450}-mediated detoxification of furanocoumarins functions as a synergist (Berenbaum and Neal, 1985; and Berenbaum and Zangerl, 1993). Therefore, this particular combination of induced metabolites makes adaptive sense when one understands the mechanisms by which herbivores cope with plant defences.

4.2.4 Jasmonate-induced indirect defences

In response to attack by herbivores, plants can release volatile compounds that help guide the searching behaviour of the herbivores' predators and parasites. This approach has recently been reviewed

(Takabayashi and Dicke, 1996; Turlings *et al.*, 1995; and Karban and Baldwin, 1997), and the discovery of volicitin (Fig. 4.3) (Alborn *et al.*, 1997) represents only one of many advances in understanding the mechanisms responsible. Tumlinson and co-workers have described the spatial (Paré and Tumlinson, 1998) and temporal (Loughrin *et al.*, 1994) characteristics of this induced response in cotton plants and, with $^{13}CO_2$ pulse-chase experiments, have determined that many of the volatiles released are synthesized *de novo* after elicitation (Paré and Tumlinson, 1997). Elicitation of the release of volatiles after caterpillar feeding can be mimicked by applying oral secretions from the caterpillar (Turlings *et al.*, 1995), volicitin (Alborn *et al.*, 1997), and jasmonic acid (Boland *et al.*, 1995). Since caterpillar feeding and oral secretions dramatically increase the free endogenous jasmonate acid levels above those caused by wounding alone (McCloud and Baldwin, 1997), it is possible that volicitin, like systemin, functions by activating the jasmonate cascade. Moreover, structural analogues of jasmonates and other octadecanoid metabolites, specifically coronatin and the amino acid conjugates of indanones (Fig. 4.3), elicit release of the same volatiles that are released following herbivore attack in lima bean (Boland *et al.*, 1995, 1998). Moreover, in the lima bean, a mixture of fungal cellulases, called cellulysin, elicits the volatile release, apparently by activating the jasmonate cascade (Boland *et al.*, 1998). Other lytic enzymes in oral secretions of herbivores may function in a similar manner. Interestingly, ethylene, which is known to function as a synergist to jasmonate in the induction of PIs in tomato (O'Donnell *et al.*, 1996), also appears to play a role in the emission of volatiles (Boland *et al.*, 1998). In short, the jasmonate cascade and pathways that interact with it to activate direct defences also appear to be involved in indirect defences.

The ecological significance of these volatile emissions for plants in nature is still largely unexplored. The volatile release attracts parasitoids in laboratory bioassays and evidence is accumulating that these emissions act similarly under field conditions (e.g. Shimoda *et al.*, 1997; Khan *et al.*, 1997). Moreover, components of the volatile emissions, particularly the 'green leaf' volatiles may function directly in defence (Avdiushko *et al.*, 1997). Whether these emissions result in greater plant fitness and can, therefore, be considered an established defence response remains to be determined. A plant has little control over the response of its ecological community to an increase in emission and, under some circumstances, a release of volatiles may increase a plant's apparency (*sensu* (Feeny, 1976b)) and attract its herbivores (Takabayashi and Dicke, 1996). For example, herbivore-induced volatile emissions from crabapple appear to serve as host location cues for the Japanese beetle and contribute to the formation of large feeding aggregations of this pest species (Loughrin *et al.*, 1996). Once the genetic basis of the volatile release of these herbivore-induced responses is understood, particularly the components that are released only after herbivore attack and not after mechanical damage, such as the two acyclic C_{11} and C_{16} homoterpenes (Boland *et al.*, 1992), then the potential to examine the functional consequences of these emissions rigorously under field conditions will become feasible.

4.2.5 Can the jasmonate cascade mediate herbivore-specific responses?

The jasmonate cascade is demonstrably important in mediating resistance to many herbivores and regulates the expression of traits that are important in resistance and, perhaps, even tolerance. The experiments that demonstrate these points (reviewed in section 4.2.2) produce large alterations in the cascade and it is not clear that such evidence addresses the role of the cascade in effecting specific plant-herbivore responses. Some researchers (Hammerschmidt and Schultz, 1996) have argued that, since the jasmonate cascade is activated by many different stimuli, e.g. herbivore and pathogen attack, wounding, UV exposure, burning and drought stress (Wasternack and Parthier, 1997), the possibility of specifically-tailored responses is unlikely. Specificity in responses could result from many different mechanisms, including: tissue-specific activation; modification of responses by other signal cascades; and the use of different biosynthetic components within the jasmonate cascade as signals.

Firstly, the jasmonate cascade can be activated in individual cell types within a plant by different 'upstream' events. The best described 'upstream' signalling event is that activated by systemin (Fig. 4.3). Exposure to systemin results in the rapid increase in free linolenic acid in a cell, probably by the activation of an as yet unidentified phospholipase (Fig. 4.2). The increase is more than sufficient to supply the biosynthetic requirements for the observed increases in jasmonate biosynthesis (Conconi *et al.*, 1996a). In addition, a wound- and systemin-inducible LOX isoform has been identified in tomato (Heitz *et al.*, 1997). The enzymology of the jasmonate cascade has recently been reviewed (Mueller, 1997; Creelman and Mullet, 1997a), and a better understanding of how the fatty acids released from the cell membranes are channelled into jasmonate biosynthesis is emerging. Systemin and wounding result in increases in linoleic acid (18:2) as well as linolenic acid (18:3) (Conconi *et al.*, 1996a), and the 18:2 fatty acid could be converted to dihydrojasmonate (Blechert *et al.*, 1995), which sometimes activates defence-related responses (Blechert *et al.*, 1995, 1997). While LOX and allene oxide synthase (AOS) utilize substrates derived from both fatty acids (Heitz *et al.*, 1997; Laudert *et al.*, 1996), the recently characterized allene oxide cyclase (AOC) utilizes only the 18:3 derived product, apparently requiring the extra double bond (Ziegler *et al.*, 1997) in its substrate. These results suggest that AOC, rather than AOS, may be more important in regulating the synthesis of jasmonic acid and may be a more specific target for the molecular manipulation of inducible jasmonate production *in planta* by transgenic techniques. Hence, the details of systemin's activation of the jasmonate cascade is remarkably consistent with the model originally proposed by Ryan and co-workers (Farmer and Ryan, 1992). How systemin accomplishes this activation is less clear (Fig. 4.2).

Recently, MAP kinases activating the jasmonate cascade, which respond to systemin (Stratmann and Ryan, 1997) and wounding (Seo *et al.*, 1995) have been identified. This may represent one of the first steps in transducing the systemically-produced signal into a localized response. Interestingly, one of the MAP kinases (Stratmann and Ryan, 1997) is not activated by products of the jasmonate cascade (jasmonic acid [JA] or 12-oxo-PDA) and, therefore, would help to limit the induction of the jasmonate cascade to those cells elicited by the peptide inducer. Systemin appears to be processed by a protein with Kex2-like properties (Schaller and Ryan, 1994), which cleaves four amino acids from the COOH terminus of the peptide to produce the smallest fragment (MQTD) (Fig. 4.3) that still has PI-inducing activity (Pearce *et al.*, 1993).

Systemin-like molecules have been difficult to find in taxa not closely related to tomato. With a reverse transcriptase-polymerase chain reaction (RT-PCR) strategy for the prosystemin gene and antibodies raised against systemin, systemin analogues are reported from potato, black nightshade and bell pepper (Constabel and Ryan, 1998), but not in more distally-related taxa (including tobacco or *Arabidopsis*). The prosystemin gene in these species has 73–88% sequence similarity with that from tomato, and the deduced systemins from these different species vary in their ability to elicit PI production in tomato (Constabel and Ryan, 1998). These results underscore the potential for structural diversity among peptides activating the octodecanoid cascade. Systemin and a small polypeptide responsible for regulating auxin action in tobacco, which also has a highly conserved sequence of four amino acids in its COOH terminus (van de Sande *et al.*, 1996), remain the only two small signal peptides in plants that function as 'master switches' (Bergey *et al.*, 1996; Ryan, 1996), although the ubiquity of large coordinated changes in plant metabolism after wounding suggest that more exist. In conclusion, while systemin apparently does not allow a plant to distinguish among different types of wounding agents, the potential for specificity clearly exists within this upstream component.

Numerous other components appear to act upstream of the jasmonate cascade and include guanosine triphosphate (GTP)-binding proteins

(Sano and Ohashi, 1995), B-glucan-elicitor-binding protein (GEBP), a 34 kDa phosphoprotein pp34 (Kiowa *et al.*, 1997), electrical signals (Wildon *et al.*, 1992; Herde *et al.*, 1996), and signals which rapidly transverse steam-girdled stems and are, therefore, likely to be hydraulic pressure signals (Stratmann and Ryan, 1997). Many parts of the octadecanoid cascade, as well as their upstream components, interact with the other plant hormones (Fig. 4.2) including auxin, abscisic acid, cytokinins, ethylene and salicylic acid, (Seo *et al.*, 1997; Lee *et al.*, 1996; Sano *et al.*, 1996; Creelman and Mullet, 1994; Baldwin *et al.*, 1997; and O'Donnell *et al.*, 1996) to modify the responses. These components interact to provide mechanisms which tailor the responses activated by the jasmonate cascade both within and between tissue types in a plant.

In addition to generating specificity in jasmonate-induced responses by modifying the manner in which the cascade is activated in particular cell types, specificity can arise from within the cascade itself. It is now evident that the jasmonate pathway produces a family of active metabolites and that different responses can be activated by products derived from the same cascade (Blechert *et al.*, 1997), perhaps even in the same plant system. For example, although jasmonic acid (Fig. 4.3) is the essential active signal for some responses, e.g. volatile emission in lima bean (Boland *et al.*, 1998), its biosynthetic precursor, 12-oxo-PDA acid (Fig. 4.2), is the signal (Parchmann *et al.*, 1997), for mechanotransduction responses (Stelmach *et al.*, 1998). In addition, evidence from the responses elicited by jasmonate mimics derived from indanone (Boland *et al.*, 1995; Zhang *et al.*, 1997; Wasternack *et al.*, 1998) suggest that various putative 'jasmonate receptors' may tailor the responses. For example, only particular components of the volatile bouquet (Boland *et al.*, 1995) and gene expression (Wasternack *et al.*, 1998) induced by jasmonic acid (Fig. 4.3) are elicited by an isoleucine conjugate of indanone (Fig. 4.3). Like the protaglandins of animals, the biosynthetic pathway that produces jasmonic acid (Fig. 4.3) may also produce a family of related and differentially active metabolites (Blechert *et al.*, 1997).

The picture that is emerging, therefore, suggests enormously complex interactions. The cascade is more a reticulated network of interacting cascades than a simple linear sequence of signals. When this complexity is considered in the context of the growing number of jasmonate-independent responses to wounding (Pearce *et al.*, 1998; Ellard-Ivey and Douglas, 1996; Rickauer *et al.*, 1997; and Titarenko *et al.*, 1997), the possibilities for specific responses to herbivore attack are virtually endless. Our relative ignorance of the complexity of these signalling mechanisms should caution against generalizations about the potential for adaptive plant responses. In this regard, the results from an elegant pair of studies of pathogen-induced response are instructive: since many different R genes elicit the induction of the hypersensitive response (HR) in the same plant, researchers assumed that these different R genes activate a common signalling pathway. However, by cloning both *avr* genes from the bacterial pathogen, *Pseudomonas syringae*, and their corresponding R genes in *A. thaliana*, two research groups (Ritter and Dangl, 1996; Reuber and Ausubel, 1996) have demonstrated, with truly isogenic lines of bacteria and plants expressing different *avr* and *R* genes, that although two distinct *avr* genes elicited an HR response, the timing of the HR phenotype and its correlated patterns of gene expression differed dramatically. These results demonstrate that plants launch HRs that are specific to the different elicitors from the same pathogen. Whether herbivore-induced responses will have the same degree of specificity remains to be determined but such studies demonstrate that plants are clearly capable of exquisitely tailoring their induced responses.

4.3 Conclusions and future directions

In summary, the evidence demonstrates that: 1) herbivores can induce responses that are different from those induced by the wounding they cause; 2) the jasmonate cascade is necessary and sufficient for eliciting induced resistance and the traits involved both in direct and indirect defences; and 3) the complexity of the interactions among signalling pathways effected by herbivore attack provide the mechanisms necessary for highly specific responses. Whether plants have been selected to respond specifically to different herbivores remains to be seen but they appear to have the potential to do so.

A large proportion of a plant's genome appears to be devoted to information processing; 23% of the 209 genes on 1.9 Mb of contiguous sequence of chromosome 4 of *A. thaliana* that could be assigned a function were found to be transcription factors or involved in signal transduction (Bevan *et al.*, 1998). This suggests that a large proportion of a plant's genome is devoted to sensing the environment and orchestrating phenotypic responses. Perhaps particular genes will orchestrate the suites of transcriptional changes responsible for adaptive responses to herbivore attack. If these genes are to be located, model systems and ecologically relevant screens must be developed that will make it possible to identify genes that function in ecological interactions, much like the yeast and *E. coli* expression systems that molecular biologists use to understand biochemical function. These 'ecological expression systems' should contain many of the ecological interactions that influence plant performance in environments with herbivores. The fitness costs and benefits of the different induced traits must be examined to determine whether the plasticity is, in fact, adaptive: whether it increases the 'fit' between the environment and the phenotype. Natural populations are, of course, the ultimate test system for ecological function, and the molecular techniques used should be compatible with tests of function in nature.

Induced direct resistance has been demonstrated in over 100 plant-herbivore associations (Karban and Baldwin, 1997), and induced indirect defences in more than 15 plant species and 10 herbivore-carnivore combinations (Takabayashi and Dicke, 1996). The high frequency of induced expression suggests that this mode of deployment has important selective advantages. If these induced responses are examples of adaptive phenotypic plasticity that allow a plant to gather information about its environment and alter its phenotype in an adaptive matter to the ecological contingencies that are so demonstrably important in determining the fitness consequences of herbivory, then an intimate understanding of the natural history of model plant-herbivore systems must be developed.

A. thaliana is clearly an ideal model system for pathogen-induced defences; however, it is not clear that it is ideally suited for herbivore-induced defences. The very traits which make this species an ideal model system for the genome-sequencing project, may limit its usefulness in identifying ecologically important genes. Its rapid life cycle and small stature have allowed it to become an ecological 'escape artist', avoiding interactions that plants which attain larger statures must endure and evolve responses to. Although a number of insects will feed on *A. thaliana* in nature (Mauricio *et al.*, 1997), few can complete their life cycle on an individual plant. Many of these interactions are probably opportunistic associations of herbivores that have specialized on one of the relatives of *Arabidopsis* with larger statures (e.g. *Brassica*, *Arabis*, *Cardaminopsis*). Such interactions will provide insight into the plant traits that these insects use to recognize their hosts but may not be useful in uncovering specific plant responses to these herbivores.

The enormous taxonomic diversity of herbivores (among insects alone, there are more taxa of organisms that eat plants than there are taxa of plants) (Strong *et al.*, 1984) has resulted in an equally bewildering array of feeding modes and host-plant specialization among herbivores. The diversity of associations that herbivores have with plants, associations that range from the highly specific endophytic feeders (leaf miners and gall-formers that have become adept at manipulating plant source-sink relationships) to the more generalist grazers that feed on plants from many taxa during their life, suggests that model systems will need to be developed for each feeding mode. In this regard, the $Hs1^{pro\text{-}1}$ resistance gene (Cai *et al.*, 1997) appears to function by interrupting the ability of the nematode to form a feeding cyst, a structure which may have similarities with some insect-induced galls and provide insights into this feeding mode. It is too early to make generalizations about the similarity of defensive responses across species and systems. On the one hand, the $Hs1^{pro-1}$ resistance gene from sugar beet contains structural motifs that are common to other pathogen resistance genes, namely, the leucine-rich repeat (Gebhardt, 1997), and suggest common modes of action but, on the other hand, the prosystemin gene, which functions as a 'master switch', activating responses to wounding (and not specifically to the wounding caused by herbivores), has only been found in taxa closely related to tomato (Constabel and Ryan, 1998). The common solutions that plants have evolved in response to herbivore attack will become apparent to researchers given the vantage point of hindsight.

Acknowledgements

Supported by the National Science (DEB-9505950) and the Max-Planck-Gesellschaft. Many of the ideas developed in this chapter have benefited from discussion with Dr. D. Hermsmeier. I thank E. Wheeler and C. Preston for insightful editorial suggestions and E. Claußen for typing the bibliography and drawing the figures.

Note added in proof

Important advances since this review was written continue at an impressive rate. In particular readers should note the discovery of the hexadecanoid pathway [Farmer, E.E. *et al.*, (1998) *Planta*, **206** 167-74] and the characterization of the COI 1 mutant [Xic, D.X. *et al.*, (1998) *Science*, **280** 1091-94].

References

Aarssen, L.W. and Irwin, D.L. (1991) What selection: herbivory or competition? *OIKOS*, **60** 261-62.

Adler, F.R. and Karban, R. (1994) Defended fortresses or moving targets? Another model of inducible defenses inspired by military metaphors. *Am. Nat.*, **144** 813-32.

Agrawal, A.A. (1998) Induced responses to herbivory and increased plant performance. *Science*, **279** 1201-202.

Alborn, H.T., Turlings, T.C.J., Jones, T.H., Stenhagen, G., Loughrin, J.H. and Tumlinson, J.H. (1997) An elicitor of plant volatiles from beet armyworm oral secretion. *Science*, **276** 945-49.

Appel, H.M. (1993) Phenolics in ecological interactions: the importance of oxidation. *J. Chem. Ecol.*, **19** 1521-53.

Avdiushko, S.A., Brown, G.C., Dahlman, D.L. and Hildebrand, D.F. (1997) Methyl jasmonate exposure induces insect resistance in cabbage and tobacco. *Environ. Entomol.*, **26** 642-54.

Baker, B., Zambryski, P., Staskawicz, B. and Dinesh-Kumar, S.P. (1997) Signaling in plant-microbe interactions. *Science*, **276** 726-33.

Baldwin, I.T. (1988) The alkaloidal responses of wild tobacco to real and simulated herbivory. *Oecologia*, **77** 378-81.

Baldwin, I.T. (1994) Chemical changes rapidly induced by folivory, in *Insect-Plant Interactions* (ed. E.A. Bernays), CRC Press Inc., Boca Raton, FL, pp. 1-23.

Baldwin, I.T. (1996) Methyl jasmonate-induced nicotine production in *Nicotiana attenuata*: inducing defenses in the field without wounding. *Entomol. Exp. Appl.*, **80** 213-20.

Baldwin, I.T. (1998) Jasmonate-induced responses are costly but benefit plants under attack in native populations. *Proc. Natl. Acad. Sci. USA*, **95** 8113-118.

Baldwin, I.T. and Morse, L. (1994) Up in smoke. II. Germination of *Nicotiana attenuata* in response to smoke-derived cues and nutrients in burned and unburned soils. *J. Chem. Ecol.*, **20** 2373-91.

Baldwin, I.T. and Ohnmeiss, T.E. (1994) Swords into plowshares? *Nicotiana sylvestris* does not use nicotine as a nitrogen source under nitrogen-limited growth. *Oecologia*, **98** 385-92.

Baldwin, I.T., Karb, M.J. and Ohnmeiss, T.E. (1994a) Allocation of ^{15}N from nitrate to nicotine: production and turnover of a damage-induced mobile defense. *Ecology*, **75** 1703-13.

Baldwin, I.T., Schmelz, E.A. and Ohnmeiss, T.E. (1994b) Wound-induced changes in root and shoot jasmonic acid pools correlate with induced nicotine synthesis in *Nicotiana sylvestris* Spegazzini and Comes. *J. Chem. Ecol.*, **20** 2139-57.

Baldwin, I.T., Staszak-Kozinski, L. and Davidson, R. (1994c) Up in smoke. I. Smoke-derived germination cues for the post-fire annual, *Nicotiana attenuata* Torr. Ex. Watson. *J. Chem. Ecol.*, **20** 2345-71.

Baldwin, I.T., Zhang, Z.-P., Diab, N., Ohnmeiss, T.E., McCloud, E.S., Lynds, G.Y. and Schmelz, E.A. (1997) Quantification, correlations and manipulations of wound-induced changes in jasmonic acid and nicotine in *Nicotiana sylvestris*. *Planta*, **201** 397-404.

Baldwin, I.T., Gorham, D., Schmelz, E.A., Lewandowski, C. and Lynds, G.Y. (1998) Allocation of nitrogen to an inducible defense and seed production in *Nicotiana attenuata*. *Oecologia*, **115** 541-52.

Barbosa, P. and Saunders, J.A. (1985) Plant allelochemicals: linkages between herbivores and their natural enemies, in *Chemically-Mediated Interactions Between Plants and other Organisms* (eds. G.A. Cooper-Driver and T. Swain), Plenum, New York, pp. 107-37.

Berenbaum, M.R. (1988) Allelochemicals in insect-microbe-plant interactions; agents provocateurs in the coevolutionary arms race, in *Novel Aspects of Insect-Plant Interactions* (ed. P. Barbosa), John Wiley & Sons Inc., New York, pp. 97-123.

Berenbaum, M.R. (1995) Turnabout is fair play: secondary roles for primary compounds. *J. Chem. Ecol.*, **21** 925-40.

Berenbaum, M.R. and Neal, J.J. (1985) Synergism between myristicin and xanthotoxin, a naturally co-occurring plant toxicant. *J. Chem. Ecol.*, **11** 1349-58.

Berenbaum, M.R. and Zangerl, A.R. (1993) Furanocoumarin metabolism in *Papilio polyxenes*: biochemistry, genetic variability and ecological significance. *Oecologia*, **95** 370-75.

Bergelson, J. and Crawley, M.J. (1992) Herbivory and *Ipomopsis aggregata*: the disadvantages of being eaten. *Am. Nat.*, **139** 870-82.

Bergey, D.R., Howe, G.A. and Ryan, C.A. (1996) Polypeptide signaling for plants defensive genes exhibits analogies to defense signaling in animals. *Proc. Natl. Acad. Sci. USA*, **93** 12053-58.

Bevan, M., Bancroft, I., Bent, E., Love, E., Goodman, H., Dean, C., Bergkamp, R., Dirkse, W., Van Staveren, M., Stiekema, W., Drost, L., Ridley, P., Hudson, S.-A., Patel, K., Murphy, G., Piffanelli, P., Wedler, H., Wedler, E., Wambutt, R., Weitzenegger, T., Pohl, T.M., Terryn, N., Gielen, J., Villarroel, R., DeClerck, R., Van Montagu, M., Lecharny, A., Augborg, S., Gy, I., Kreis, M., Lao, N., Kavanagh, T., Hempel, S., Kotter, P., Entain, K.-D.,

Rieger, M., Schaeffer, M., Funk, B., Mueller-Auer, S., Silvey, M., James, R., Montfort, A., Pons, A., Puigdomenech, P., Douka, A., Voukelatou, E., Milioni, D., Hatzopoulos, P., Piravandi, E., Obermaier, B., Hilbert, H., Düstefhöft, A.T.M., Jones, J.D.G., Eneva, T., Palme, K., Benes, V., Rechman, S., Ansorge, W., Cooke, R., Berger, C., Delseny, M., Voet, M., Volckaert, G., Mewes, H.-W., Klosterman, S., Schueller, C. and Chalwatzis, N. (1998) Analysis of 1.9 Mb of contiguous sequence from chromosome 4 of *Arabidopsis thaliana*. *Nature*, **391** 485-88.

Bi, J.L., Felton, G.W., Murphy, J.B., Howles, P.A., Dixon, R.A. and Lamb, C.J. (1997a) Do plant phenolics confer resistance to specialist and generalist insect herbivores? *J. Agric. Food Chem.*, **45** 4500-504.

Bi, J.L., Murphy, J.B. and Felton, G.W. (1997b) Antinutritive and oxidative components as mechanisms of induced resistance in cotton to *Helicoverpa zea*. *J. Chem. Ecol.*, **23** 97-117.

Blechert, S., Brodschelm, W., Hölder, S., Kammerer, L., Kutchan, T.M., Müller, J.M., Xia, Z.Q. and Zenk, M.H. (1995) The octadecanoid pathway: signal molecules for the regulation of secondary pathways. *Proc. Natl. Acad. Sci. USA*, **92** 4099-105.

Blechert, S., Bockelmann, C., Brümmer, O., Füßlein, M., Gundlach, H., Haider, G., Hölder, S., Kutchan, T.M., Weiler, E.W. and Zenk, M.H. (1997) Structural separation of biological activities of jasmonates and related compounds. *J. Chem. Soc.*, **23** 3549-59.

Boland, W., Feng, Z., Donath, J. and Gäbler, A. (1992) Are acyclic C_{11} and C_{16} homoterpenes plant volatiles indicating herbivory? *Naturwissenschaften*, **79** 368-71.

Boland, W., Hopke, J., Donath, J., Nüske, J. and Bublitz, F. (1995) Jasmonic acid and coronatin induce odor production in plants. *Angew. Chemie* (International Edition in English) **34** 1600-602.

Boland, W., Hopke, J. and Piel, J. (1998) Induction of plant volatile biosynthesis by jasmonates, in *Natural Product Analysis* (ed. P. Schreier), Viehweg Verlag, Wiesbaden, pp. 255-70.

Bolter, C.J. and Jongsma, M.A. (1995) Colorado potato beetles (*Leptinotarsa decemlineata*) adapt to proteinase inhibitors induced in potato leaves by methyl jasmonate. *J. Insect Physiol.*, **41** 1071-78.

Cai, D., Kleine, M., Kifle, S., Harloff, H.-J., Sandal, N.N., Marcker, K.A., Klein-Lankhorst, R.M., Salentijn, E.M.J., Lange, W., Stiekema, W.J., Wyss, U., Grundler, F.M.W. and Jung, C. (1997) Positional cloning of a gene for nematode resistance in sugar beet. *Science*, **275** 832-34.

Carter, W. (1973) *Insects in Relation to Plant Disease*, Wiley Interscience, New York.

Conconi, A., Miquel, M., Browse, J.A. and Ryan, C.A. (1996a) Intracellular levels of free linolenic and linoleic acids increase in tomato leaves in response to wounding. *Plant Physiol.*, **111** 797-803.

Conconi, A., Smerdon, M.J., Howe, G.A. and Ryan, C.A. (1996b) The octadecanoid signalling pathway in plants mediates a response to ultraviolet radiation. *Nature*, **383** 826-29.

Constabel, C.P. and Ryan, C.A. (1998) A survey of wound- and methyl jasmonate-induced leaf polyphenol oxidase in crop plants. *Phytochemistry*, **47** 507-11.

Constabel, C.P., Bergey, D.R. and Ryan, C.A. (1995) Systemin activates synthesis of wound-inducible tomato leaf polyphenol oxidase via the octadecanoid defense signaling pathway. *Proc. Natl. Acad. Sci. USA*, **92** 407-11.

Constabel, C.P., Bergey, D.R. and Ryan, C.A. (1996) Polyphenol oxidase as a component of the inducible defense response in tomato against herbivores. *Rec. Adv. Phytochem.*, **30** 231-52.

Creelman, R.A. and Mullet, J.E. (1994) Jasmonic acid distribution and action in plants: regulation during development and response to biotic and abiotic stress. Colloquium paper by the colloquium entitled 'Self-Defense by Plants: Induction and Signalling Pathways', **92** 4114-19.

Creelman, R.A. and Mullet, J.E. (1997a) Biosynthesis and action of jasmonates in plants. *Annu. Rev. Plant Physiol. Plant Mol. Biol.*, **48** 355-81.

Creelman, R.A. and Mullet, J.E. (1997b) Oligosaccharides, brassinolides and jasmonates: nontraditional regulators of plant growth, development and gene expression. *Plant Cell*, **9** 1211-23.

DeRisi, J.L., Iyer, V.R. and Brown, P.O. (1997) Exploring the metabolic and genetic control of gene expression on a genomic scale. *Science*, **278** 680-86.

Doares, S.H., Narváez-Vásquez, J., Conconi, A. and Ryan, C.A. (1995) Salicylic acid inhibits synthesis of proteinase inhibitors in tomato leaves induced by systemin and jasmonic acid. *Plant Physiol.*, **108** 1741-46.

Duffey, S.S. and Stout, M.J. (1996) Antinutritive and toxic components of plant defense against insects. *Arch. Insect Biochem. Physiol.*, **32** 3-37.

Duffey, S.S., Bloem, K.A. and Campbell, B.C. (1986) Consequences of sequestration of plant natural products in plant-insect-parasitoid interactions, in *Interactions of Plant Resistance and Parasitoids and Predators of Insects* (eds. D.J. Boethel and R.D. Eikenbarry), Ellis Horwood, Chicester, UK, pp. 31-60.

Dyer, M.I., Acra, M.A., Wang, G.M., Coleman, D.C., Freckman, D.W., McNaughton, S.J. and Strain, B.R. (1991) Source-sink carbon relations in two panicum coloratum ecotypes in responce to herbivory. *Ecology*, **72** 1472-83.

Ellard-Ivey, M. and Douglas, C.J. (1996) Role of jasmonates in the elicitor- and wound-inducible expression of defense genes in parsley and transgenic tobacco. *Plant Physiol.*, **112** 183-92.

Euler, M. and Baldwin, I.T. (1996) The chemistry of defense and apparency in the corollas of *Nicotiana attenuata*. *Oecologia*, **107** 102-12.

Farmer, E. and Ryan, C.A. (1992) Octadecanoid precursors of jasmonic acid activate the synthesis of wound-inducible proteinase inhibitors. *Plant Cell*, **4** 129-34.

Feeny, P. (1976) Plant apparency and chemical defense. *Rec. Adv. Phytochem.*, **10** 1-40.

Felton, G.W. (1996) Nutritive quality of plant protein: sources of variation and insect herbivore responses. *Arch. Insect Biochem. Physiol.*, **32** 107-30.

Felton, G.W., Donato, K., Del Vecchio, R.J. and Duffey, S.S. (1989) Activation of plant foliar oxidases by insect feeding reduces nutritive quality of foliage for noctuid herbivores. *J. Chem. Ecol.*, **15** 2667-93.

Felton, G.W., Donato, K.K., Broadway, R.M. and Duffey, S.S. (1992) Impact of oxidized plant phenolics on the nutritional quality of dietary protein to a noctuid herbivore, *Spodoptera exigua*. *J. Insect Physiol.*, **38** 277-85.

Felton, G.W., Bi, J.L., Summers, C.B., Mueller, A.J. and Duffey, S.S. (1994) Potential role of lipoxygenases in defense against insect herbivory. *J. Chem. Ecol.*, **20** 651-67.

Fowler, S.V. and Lawton, J.H. (1985) Rapidly induced defences and talking trees: the Devil's advocate position. *Am. Nat.*, **126** 181-95.

Flor, H.H. (1947) Inheritance of reaction to rust in flax. *J. Agric. Res.*, **74** 214-62.

Gabriel, D.W. and Rolfe, B.G. (1990) Working models of specific recognition in plant-microbe interactions. *Annu. Rev. Phytopathol.*, **28** 365-91.

Gebhardt, C. (1997) Plant genes for pathogen resistance: variation on a theme. *Trends Plant Sci.*, **2** 243-44.

Giri, A.P., Harsulkar, A.M., Deshpande, V.V., Sainani, M.N., Gupta, V.S. and Ranjekar, P.K. (1998) Chickpea defensive proteinase inhibitors can be inactivated by podborer gut proteinases. *Plant Physiol.*, **116** 393-401.

Givnish, T.J. (1986) Economics of biotic interactions, in *On the Economy of Plant Form and Function* (ed. T.J. Givnish), Cambridge University Press, Cambridge, Mass., pp. 667-79.

Görschen, E., Dunaeva, M., Hause, B., Reeh, I., Wasternack, C. and Parthier, B. (1997a) Expression of the ribosome-inactivating protein, JIP60, from barley in transgenic tobacco leads to an abnormal phenotype and alterations on the level of translation. *Planta*, **202** 470-73.

Görschen, E., Dunaeva, M., Reeh, I. and Wasternack, C. (1997b) Overexpression of the jasmonate-inducible 23 kDa protein (JIP23) from barley in transgenic tobacco leads to the repression of leaf proteins. *FEBS Lett.*, **419** 58-62.

Gross, D. and Parthier, B. (1994) Novel natural substances acting in plant growth regulation. *J. Plant Growth Reg.*, **13** 93-114.

Hamberg, M. and Gardner, H.W. (1992) Oxylipin pathway to jasmonates: biochemistry and biological significance. *Biochim. Biophys. Acta*, **1165** 1-18.

Hammerschmidt, R. and Schultz, J.C. (1996) Multiple defenses and signals in plant defense against pathogens and herbivores. *Rec. Adv. Phytochem.*, **30** 121-54.

Harvell, C.D. (1990) The ecology and evolution of inducible defenses. *Q. Rev. Biol.*, **65** 323-40.

Heitz, T., Bergey, D.R. and Ryan, C.A. (1997) A gene encoding a chloroplast-targeted lipoxygenase in tomato leaves is transiently induced by wounding, systemin and methyl jasmonate. *Plant Physiol.*, **114** 1085-93.

Herde, O., Atzorn, R., Fisahn, J., Wasternack, C., Willmitzer, L. and Peña-Cortés, H. (1996) Localized wounding by heat initiates the accumulation of proteinase inhibitor II in abscisic acid-deficient plants by triggering jasmonic acid biosynthesis. *Plant Physiol.*, **112** 853-60.

Herms, D.A. and Mattson, W.J. (1992) The dilemma of plants: to grow or defend. *Q. Rev. Biol.*, **67** 283-335.

Howe, G.A., Lightner, J., Browse, J. and Ryan, C.A. (1996) An octadecanoid pathway mutant (JL5) of tomato is compromised in signaling for defense against insect attack. *Plant Cell*, **8** 2067-77.

Jacinto, T., McGurl, B., Franceschi, V., Delano-Freier, J. and Ryan, C.A. (1997) Tomato prosystemin promoter confers wound-inducible, vascular bundle-specific expression of the β-glucuronidase gene in transgenic tomato plants. *Planta*, **203** 406-12.

Jongsma, M.A. and Bolter, C. (1997) The adaptation of insects to plant protease inhibitors. *J. Insect Physiol.*, **43** 885-95.

Jongsma, M.A., Bakker, P.L., Peters, J., Bosch, D. and Stiekma, W. (1995) Adaptation of *Spondoptera exigua* larvae to plant proteinase inhibitors by induction of gut proteinase activity insensitive to inhibition. *Proc. Natl. Acad. Sci. USA*, **92** 8041-45.

Karban, R. (1993) Costs and benefits of induced resistance and plant density for a native shrub, *Gossypium thurberi. Ecology*, **74** 9-19.

Karban, R. and Meyer, J.H. (1989) Induced plant responses to herbivory. *Annu. Rev. Ecol. System.*, **20** 331-48.

Karban, R. and Baldwin, I.T. (1997) *Induced Responses to Herbivory*, Chicago University Press, Chicago.

Kennedy, G.G. and Barbour, J.D. (1992) Resistance variation in natural and managed systems, in *Plant Resistance to Herbivores and Pathogens* (eds. R.S. Fritz and E.L. Simms), University of Chicago Press, Chicago, pp. 13-41.

Khan, Z.R., Ampong-Nyarko, K., Chiliswa, P., Hassanali, A., Kimani, S., Lwande, W. and Overholt, W.A. (1997) Intercropping increases parasitism of pests. *Nature*, **388** 631-32.

Koiwa, H., Bressan, R.A. and Hasegawa, P.M. (1997) Regulation of protease inhibitors and plant defense. *Trends Plant Sci.*, **2** 379-83.

Korth, K.L. and Dixon, R.A. (1997) Evidence for chewing insect-specific molecular events distinct from a general wound response in leaves. *Plant Physiol.*, **115** 1299-305.

Kowalski, S.P., Eannetta, N.T., Hirzel, A.T. and Steffens, J.C. (1992) Purification and characterization of polyphenol oxidase from glandular trichomes of *Solanum berthaultii. Plant Physiol.*, **100** 677-84.

Larson, K.C. and Whitham, T.G. (1997) Competition between gall aphids and natural plant sinks: plant architecture affects resistance to galling. *Oecologia*, **109** 575-82.

Laudert, D., Pfannschmidt, U., Lottspeich, F., Holländer-Czytko, H. and Weiler, E.W. (1996) Cloning, molecular and functional characterization of *Arabidopsis thaliana* allene oxide synthase (CYP 74), the first enzyme of the octadecanoid pathway to jasmonates. *Plant Mol. Biol.*, **31** 323-35.

Lee, J., Parthier, B. and Löbler, M. (1996) Jasmonate signalling can be uncoupled from abscisic acid signalling in barley: identification of jasmonate-regulated transcripts which are not induced by abscisic acid. *Planta*, **199** 625-32.

Levins, R. (1968) *Evolution in Changing Environments*, Princeton University Press, Princeton.

Loughrin, J.H., Heath, R.R., Turlings, C.J. and Tumlinson, J.H. (1994) Diurnal cycle of emission of induced volatile terpenoids by herbivore-injured cotton plants. *Proc. Natl. Acad. Sci. USA*, **91** 11836-40.

Loughrin, J.H., Potter, D.A., Hamiltonkemp, T.R. and Byers, M.E. (1996) Role of feeding-induced plant volatiles in aggregative behavior of the Japanese beetle (Coleoptera, Scarabaeidae). *Environ. Entomol.*, **25** 1188-91.

Lynds, G.Y. and Baldwin, I.T. (1998) Fire, nitrogen and defensive plasticity. *Oecologia*, **115** 531-40.

Malcolm, S.B. and Zalucki, M.P. (1996) Milkweed latex and cardenolide induction may resolve the lethal plant defence paradox. *Entomol. Exp. Appl.*, **80** 193-96.

Mauricio, R., Rausher, M.D. and Burdick, D.S. (1997) Variation in the defense strategies of plants: are resistance and tolerance mutually exclusive? *Ecology*, **78** 1301-11.

McCloud, E.S. and Baldwin, I.T. (1997) Herbivory and caterpillar regurgitants amplify the wound-induced increases in jasmonic acid but not nicotine in *Nicotiana sylvestris*. *Planta*, **203** 430-35.

McConn, M., Creelman, R.A., Bell, E., Mullet, J.E. and Browse, J. (1997) Jasmonate is essential for insect defense in *Arabidopsis*. *Proc. Natl. Acad. Sci. USA*, **4** 5473-77.

McGurl, B., Pearce, G., Orozco-Cardenas, M. and Ryan, C.A. (1992) Structure, expression and antisense inihibition of the systemin precursor gene. *Science*, **255** 1570-73.

Moran, N. and Hamilton, W.D. (1980) Low nutritive quality as a defence against herbivores. *J. Theoret. Biol.*, **86** 247-54.

Mueller, M.J. (1997) Enzymes involved in jasmonic acid biosynthesis. *Physiol. Plant.*, **100** 653-63.

O'Donnell, P.J., Calvert, C., Atzorn, R., Wasternack, C., Leyser, H.M.O. and Bowles, D.J. (1996) Ethylene as a signal mediating the wound response of tomato plants. *Science*, **274** 1914-17.

Ohnmeiss, T., McCloud, E.S., Lynds, G.Y. and Baldwin, I.T. (1997) Within-plant relationships among wounding, jasmonic acid and nicotine: implications for defence in *Nicotiana sylvestris*. *New Phytol.*, **137** 441-52.

Orozco-Cardenas, M., McGurl, B. and Ryan, C.A. (1993) Expression of an antisense prosystemin gene in tomato plants reduces resistance toward *Manduca sexta* larvae. *Proc. Natl. Acad. Sci. USA*, **90** 8273-76.

Paige, K.N. and Whitham, T.G. (1987) Overcompensation in response to mammalian herbivory: the advantage of being eaten. *Am. Nat.*, **129** 407-16.

Parchmann, S., Gundlach, H. and Mueller, M.J. (1997) Induction of 12-oxo-phytodienoic acid in wounded plants and elicited plant cell cultures. *Plant Physiol.*, **115** 1057-64.

Paré, P.W. and Tumlinson, J.H. (1997) *De novo* biosynthesis of volatiles induced by insect herbivory in cotton plants. *Plant Physiol.*, **114** 1161-67.

Paré, P.W. and Tumlinson, J.H. (1998) Cotton volatiles synthesized and released distal to the site of insect damage. *Phytochemistry*, **47** 521-26.

Pearce, G., Johnson, S., Strydom, D. and Ryan, C.A. (1991) A polypeptide from tomato leaves induces wound-inducible proteinase inhibitor proteins. *Science*, **253** 895-98.

Pearce, G., Johnson, S. and Ryan, C.A. (1993) Structure-activity of deleted and substituted systemin, an 18-amino acid polypeptide inducer of plant defensive genes. *J. Biol. Chem.*, **268** 212-16.

Pearce, G., Marchand, P.A., Griswold, J., Lewis, N.G. and Ryan, C.A. (1998) Accumulation of feruloyltyramine and *p*-coumaroyltyramine in tomato leaves in response to wounding. *Phytochemistry*, **47** 659-64.

Peña-Cortés, H., Albrecht, T., Prat, S., Weiler, E.W. and Willmitzer, L. (1993) Aspirin prevents wound-induced gene expression in tomato leaves by blocking jasmonic acid biosynthesis. *Planta*, **191** 123-28.

Perez, V., Huet, J.C., Nespoulous, C. and Pernollet, J.-C. (1997) Mapping the elicitor and necrotic sites of *Phytophthora* elicitins with synthetic peptides and reporter genes controlled by tobacco defense gene promoters. *Mol. Plant-Microbe Interact.*, **6** 750-60.

Preston, C.A. and Baldwin, I.T. (1999) Positive and negative signals regulate germination in the post-fire annual, *Nicotiana attenuata*. *Ecology*, (in press).

Price, P.W. (1984) *Insect Ecology*. Second Edition, Wiley, New York.

Rausher, M.D. (1992) Natural selection and the evolution of plant-insect interactions, in *Insect Chemical Ecology and Evolutionary Approach* (eds. B.D. Roitberg and M.B. Isman), Chapman and Hall, New York, pp. 20-88.

Reinbothe, S., Mollenhauer, B. and Reinbothe, C. (1994) JIPs and RIPs: the regulation of plant gene expression by jasmonates in response to environmental cues and pathogens. *Plant Cell*, **6** 1197-209.

Reuber, T.L. and Ausubel, F.M. (1996) Isolation of arabidopsis genes that differentiate between resistance responses mediated by the *RPS2* and *RPM1* disease resistance genes. *Plant Cell*, **8** 241-49.

Rickauer, M., Brodschelm, W., Bottin, A., Véronési, C., Grimal, H. and Esquerré-Tugayé, M.T. (1997) The jasmonate pathway is involved differentially in the regulation of different defence responses in tobacco cells. *Planta*, **202** 155-62.

Ritter, C. and Dangl, J.L. (1996) Interference between two specific pathogen recognition events mediated by distinct plant disease resistance genes. *Plant Cell*, **8** 251-57.

Rosenthal, J.P. and Kotanen, P.M. (1994) Terrestrial plant tolerance to herbivory. *Trends Ecol. Evolut.*, **9** 145-48.

Roy, B.A. (1993) Floral mimicry by a plant pathogen. *Nature*, **362** 56-58.

Ryan, C.A. (1996) A polypeptide gets the nod. *Trends Plant Sci.*, **1** 365-66.

Sano, H. and Ohashi, Y. (1995) Involvement of small GTP-binding proteins in defense signal-transduction pathways of higher plants. *Proc. Natl. Acad. Sci. USA*, **92** 4138-44.

Sano, H., Seo, S., Koizumi, N., Niki, T., Iwamura, H. and Ohashi, Y. (1996) Regulation by cytokinins of endogenous levels of jasmonic and salicylic acids in mechanically-wounded tobacco plants. *Plant Cell Physiol.*, **37** 762-69.

Schaller, A. and Ryan, C.A. (1994) Identification of a 50 kDa systemin-binding protein in tomato plasma membranes having Kex2p-like properties. *Proc. Natl. Acad. Sci. USA*, **91** 11802-806.

Seo, S., Sano, H. and Ohashi, Y. (1997) Jasmonic acid in wound signal transduction pathways. *Physiol. Plant.*, **101** 740-45.

Seo, S., Okamoto, M., Seto, H., Ishizuka, K., Sano, H. and Ohashi, Y. (1995) Tobacco MAP kinase: a possible mediator in wound signal transduction pathways. *Science*, **270** 1988-92.

Shimoda, T., Takabayashi, J., Ashihara, W. and Takafuji, A. (1997) Response of predatory insect, *Scolothrips takahashii*, toward herbivore-induced plant volatiles under laboratory and field conditions. *J. Chem. Ecol.*, **23** 2033-49.

Simms, E.L. and Rausher, M.D. (1987) Costs and benefits of plant resistance to herbivory. *Am. Nat.*, **130** 570-81.

Snyder, M.J. and Glendinning, J.I. (1996) Causal connection between detoxification enzyme activity and consumption of a toxic plant compound. *J. Comp. Physiol. A: Sensory Neural and Behavioral*, 255-61.

Somssich, I.E. and Hahlbrock, K. (1998) Pathogen defence in plants: a paradigm of biological complexity. *Trends Plant Sci.*, **3** 86-90.

Staswick, P.E. (1994) Storage proteins of vegetative plant tissues. *Annu. Rev. Plant Physiol. Plant Mol. Biol.*, **45** 303-22.

Steffens, J.C., Harel, E. and Hunt, M. (1994) Polyphenol oxidase. *Rec. Adv. Phytochem.*, **28** 276-304.

Stelmach, B.A., Müller, A., Hennig, P., Laudert, D., Andert, L. and Weiler, E.W. (1998) Quantitation of the octadecanoid 12-oxo-phytodienoic acid, a signalling compound in plant mechanotransduction. *Phytochemistry*, **47** 539-46.

Stout, M.J., Workman, K.V., Bostock, R.M. and Duffey, S.S. (1998) Specificity of induced resistance in the tomato, *Lycopersicon esculentum. Oecologia*, **113** 74-81.

Stratmann, J.W. and Ryan, C.A. (1997) Myelin basic protein kinase activity in tomato leaves is induced systemically by wounding and increases in response to systemin and oligosaccharide elicitors. *Proc. Natl. Acad. Sci. USA*, **94** 11085-89.

Strong, D.R., Lawton, J.H. and Southwood, Sir R. (1984) *Insects on Plants: Community, Patterns and Mechanisms*, Havard University Press, Cambridge, MA.

Takabayashi, J. and Dicke, M. (1996) Plant-carnivore mutualism through herbivore-induced carnivore attractancts. *Trends Plant Sci.*, **1** 109-13.

Tang, X., Frederick, R.D., Zhou, J., Halterman, D.A., Jia, Y. and Martin, G.B. (1996) Initation of plant disease resistance by physical interaction of ArvPto and Pto kinase. *Science*, **274** 2060-62.

Thaler, J.S., Stout, M.J., Karban, R. and Duffey, S.S. (1996) Exogenous jasmonates simulate insect wounding in tomato plants (*Lycopersicon exculentum*) in the laboratory and field. *J. Chem. Ecol.*, **22** 1767-79.

Thipyapong, P. and Steffens, J.C. (1997) Tomato polyphenol oxidase: differential response of the polyphenol oxidase F promoter to injuries and wound signals. *Plant Physiol.*, **115** 409-18.

Thipyapong, P., Joel, D.M. and Steffens, J.C. (1997) Differential expression and turnover of the tomato polyphenol oxidase gene family during vegetative and reproductive development. *Plant Physiol.*, **113** 707-18.

Titarenko, E., Rojo, E., León, J. and Sánchez-Serrano, J.J. (1997) Jasmonic acid-dependent and -independent signaling pathways control wound-induced gene activation in *Arabidopsis thaliana. Plant Physiol.*, **115** 817-26.

Trumble, J.T., Kolodny-Hirsch, D.M. and Ting, I.P. (1993) Plant compensation for arthropod herbivory. *Annu. Rev. Entomol.*, **38** 93-119.

Turlings, T.C.J., Loughrin, J.H., McCall, P.J., Röse, U.S.R., Lewis, W.J. and Tumlinson, J.H. (1995) How caterpillar-damaged plants protect themselves by attracting parasitic wasps. *Proc. Natl. Acad. Sci. USA*, **92** 4169-74.

van de Sande, K., Pawlowski, K., Czaja, I., Wieneke, U., Schell, J., Schmidt, J., Walden, R., Matvienko, M., Wellink, J., van Kammen, A., Franssen, H. and Bisseling, T. (1996) Modification of phytohormone response by a peptide encoded by ENOD40 of legumes and a nonlegume. *Science*, **273** 370-73.

Wasternack, C. and Parthier, B. (1997) Jasmonate-signalled plant gene expression. *Trends Plant Sci.*, **2** 302-307.

Wasternack, C., Ortel, B., Miersch, O., Kramell, R., Beale, M., Greulich, F., Feussner, I., Hause, B., Krumm, T., Boland, W. and Parthier, B. (1998) Diversity in octadecanoid-induced gene expression of tomato. *J. Plant Physiol.*, **152** 1-8.

Weiler, E.W. (1997) Octadecanoid-mediated signal transduction in higher plants. *Naturwissenschaften*, **84** 340-49.

Whitham, T.G., Maschinski, J., Larson, K.C. and Paige, K.N. (1991) Plant responses to herbivory: the continuum from negative to positive and underlying physiological mechanisms, in *Plant-Animal Interactions: Evolutionary Ecology in Tropical and Temperate Regions* (eds. P.W. Price, T.M. Lewinsohn, G.W. Fernandes and W.W. Benson), John Wiley and Sons Inc., New York, pp. 227-56.

Wildon, D.C., Thain, J.F., Minchin, P.E.H., Gubb, I.R., Reilly, A.J., Skipper, Y.D., Doherty, H.M., O'Donnell, P.J. and Bowles, D.J. (1992) Electrical signalling and systemic proteinase inhibitor induction in the wounded plant. *Nature*, **360** 62-65.

Zangerl, A.R. and Bazzaz, F.A. (1992) Theory and pattern in plant defense allocation, in *Plant Resistance to Herbivores and Pathogens: Ecology, Evolution and Genetics* (eds. R.S. Fritz and E.L. Simms), University of Chicago Press, Chicago, pp. 363-91.

Zangerl, A.Z., Arntz, A.M. and Berenbaum, M.R. (1997) Physiological price of an induced chemical defense: photosynthesis, respiration, biosynthesis and growth. *Oecologia*, **109** 433-41.

Zhang, Z.-P. and Baldwin, I.T. (1997) Transport of [2-^{14}C]-jasmonic acid from leaves to roots mimics wound-induced changes in endogenous jasmonic acid pools in *Nicotiana sylvestris*. *Planta*, **203** 436-41.

Zhang, Z.-P., Krumm, T. and Baldwin, I.T. (1997) Structural requirements of jasmonates and mimics for nicotine induction in *Nicotiana sylvestris*. *J. Chem. Ecol.*, **23** 2777-89.

Ziegler, J., Hamberg, M., Miersch, O. and Parthier, B. (1997) Purification and characterisation of allene oxide cyclase from dry corn seeds. *Plant Physiol.*, **114** 565-73.

5 Plant-microbe interactions and secondary metabolites with antiviral, antibacterial and antifungal properties

Jürgen Reichling

5.1 Introduction

Depending on the estimate, some 2×10^6–2×10^7 different species of bacteria, fungi and animals live on the earth. These require organic carbon from about 3×10^5 species of green, carbon dioxide-assimilating plants in order to live. As a consequence, green plants are subject to constant attack by parasites, symbiotes, phytopathogenic microbes and herbivores. Furthermore, they cannot evade the attacks of the usually mobile herbivores and pathogens. Though tied to a specific location, plants have developed effective defence mechanisms for resistance to microbial infections and to damage caused by herbivores and phytophagous insects. Their success is obvious; disease is more the exception than the rule. For example, of the 10^5 species of fungus that exist, only about 150 species attack apple trees (e.g. *Malus domestica*).

The multicomponent complex of mechanisms for resistance in plants functions by means of different defence strategies, which can be classified as 'preformed' and 'induced defence mechanisms'.

The preformed defence mechanisms include the presence of a variety of antimicrobial agents (e.g. more or less toxic low molecular weight secondary metabolites, such as cyanogenic glycosides, mustard oil glycosides, alkaloids, phenols, essential oils and tannins) and physical barriers (e.g. hairs, spikes, thorns, bark and bud scales), lignification, suberization and the formation of callose, agglutinins and enzyme inhibitors (e.g. extracellular microbial hydrolases). Of course, the maintenance of such a system of defence requires the investment of a considerable amount of energy, which is not then available for other plant functions (Nahrstedt, 1979).

In addition, plants have developed more economic methods of defence which are only preserved in the plant genome. One hypothesis states that all plants contain the genetic potential for 'inducible resistance mechanisms' to fungal and bacterial diseases. Therefore, it is supposed that resistance and susceptibility in plants are not determined by the presence or absence of genetic information for resistance mechanisms but rather by the speed with which the cryptic information is expressed, the activity of the gene products and the magnitude of the resistance response. In some cases of local attack, inducible defence mechanisms offer the whole plant protection ('induced systemic defence'), for example, through the production of proteinase inhibitors (PIs). In other situations, defence is restricted to the site of infection and the immediate vicinity ('induced local defence'), for example, through: wound healing reactions; sealing of the cell walls with callose and lignin deposits (Kauss, 1987); rapid synthesis of hydrolases, which break down the surface molecules of the pathogens, e.g. chitinase, glucanase; and preventive necroses or hypersensitive reactions.

In the present chapter, the phenomenon of induced local defence by means of phytoalexins is examined, together with the antimicrobial effects of preformed secondary metabolites.

5.2 Induced local defence by means of phytoalexins

If plants are attacked by bacteria or obligatory biotrophic fungi, various reactions may occur, depending on the nature of the organisms which are encountering each other:

In 'compatible reactions', the attacking microbe and the plant coexist for a prolonged period of time; the microbe receives nutritive substances from the plant and is not rejected.

In 'incompatible reactions', the attacking microbe and the plant cannot coexist; the microbe is rejected following the development of a hypersensitive reaction. In a hypersensitive reaction, affected cells or tissue are sacrificed by the plant in order to prevent the spread of the attack from one spot to the entire plant. After penetration of an incompatible microbe (for example, a fungus that is pathogenic for the plant), the directly neighbouring cells turn brown and die-off, often within a few hours. Within the resulting area of necrosis, which measures one or only a few millimeters in diameter, the pathogen cannot find any nutrition and dies (Schloesser, 1983).

In such an area of necrosis, substances showing antibiotic effects can be found; namely, the 'phytoalexins'. Evidence exists for the role of phytoalexins in protecting plants against diseases caused by fungi and bacteria.

5.2.1 The phytoalexins

Müller and Boerger (1940) defined the phytoalexins (Greek: phyton, meaning plant; alexis, meaning defence) as low molecular weight, antibiotically effective substances of plant secondary metabolism, the synthesis and accumulation of which is induced by pathogens or herbivores (for comprehensive review, see Cline and Albersheim, 1981; Kombrink and Hahlbrock, 1985; Kuc and Rush, 1985; Ebel *et al.*, 1985; Ebel, 1986; Mayer, 1989; Lamb *et al.*, 1992; Niemann, 1993; Grayer and Harborne, 1994). The majority of phytoalexins are found in, or immediately adjacent to, the browned, necrotic, infected tissues at concentrations that are inhibitory to the development of fungi and bacteria (Hahn *et al.*, 1981, 1985; Smith and Banks, 1986; Keen, 1986). The phytoalexin defence mechanism is not highly specific with regard to its induction, the products produced and the specifity of the products to inhibit the development of pathogens. Therefore, phytoalexin accumulation may be a primitive plant response to metabolic insult, stress or antimicrobial infection, which has persisted and remained effective as part of a multicomponent disease resistance mechanism.

The induction of phytoalexin synthesis in plant tissue has been studied mainly in pathogenic fungi; however, studies of attacks by viruses, bacteria, nematodes, arachnida and insects have also been conducted. Correspondingly, antibacterial, fungistatic and nematostatic phytoalexins have been discovered, as have those which deter insects from feeding. These substances usually demonstrate a biostatic or biocidal effect at relatively high concentrations (10^{-4}–10^{-5} M/L). In healthy plants not under attack, these substances are only found at very low concentrations, if at all. At present, we are aware of over 200 different phytoalexins in more than 25 plant families. Their molecular structures reflect the variation in secondary plant metabolic pathways, since phytoalexins can be found among the alkaloids, coumarins, dihydrophenanthrenes, flavonoids, isoflavonoids, phenols, polyacetylenes, steroids, stilbenes and terpenes (for selected phytoalexins see Figures 5.1 and 5.2).

Phytoalexin production has been reported in Dicotyledones, rarely in Monocotyledones and Gymnosperms, and not at all in nonvascular plants (see Table 5.1). It seems that similarities are evident between phytoalexins from plants within a family. Accumulation of phytoalexins has been studied most carefully in the families of Leguminosae (= Fabaceae) and Solanaceae, where the phytoalexins have chemosystematic properties. For Leguminosae, more than 100 phytoalexins have been reported belonging to isoflavonoids, furanoacetylenes, stilbenes, benzofurans, chromones and flavanones. Plants in the Leguminosae have not been reported to produce sesquiterpenoid phytoalexins. On the other hand, sesquiterpenes are typical phytoalexins for plants in the Solanaceae (Sprecher and Urbasch, 1983; Wolters and Eilert, 1983). Furthermore, exact analysis has shown that an individual plant can often synthesize more than one and varying phytoalexins. In some cases, their chemical

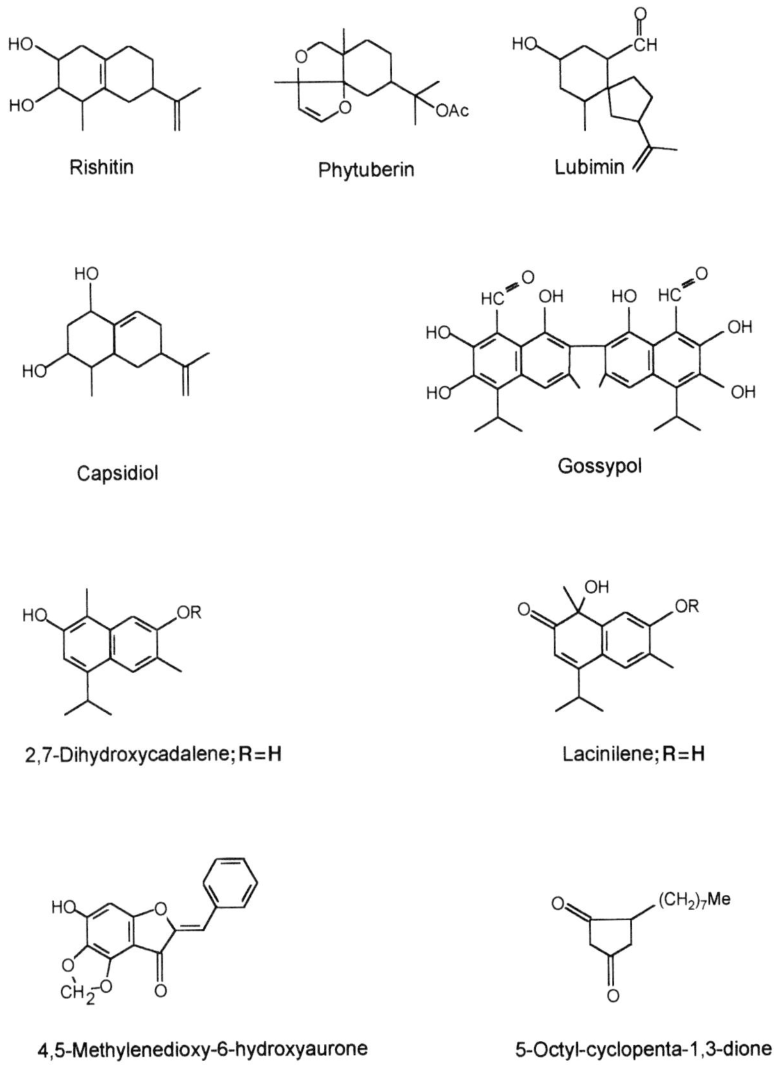

Figure 5.1 Chemical structures of selected phytoalexins.

structures are very similar (e.g. several glyceollins synthesized by soybean), but in others very different (e.g. rishitin and chlorogenic acid synthesized by the potato).

Phytoalexins are synthesized relative quickly after contact with the attacking pathogen. After a lag phase, at a minimum of 2 h, the bioactive substance can be measured and the amounts increase during the

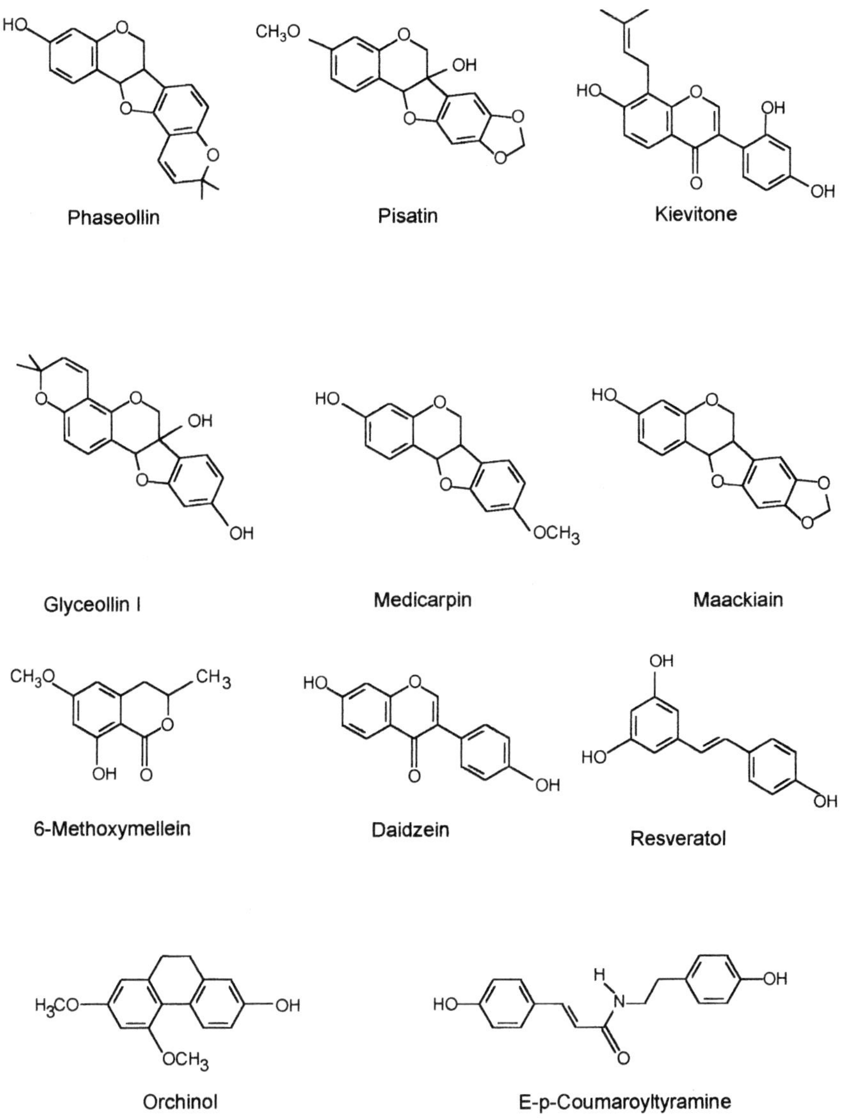

Figure 5.2 Chemical structures of selected phytoalexins.

following hours and days for up to about 96 h, sometimes longer, until maximum accumulation has been achieved. Subsequently, the levels of phytoalexin decrease to those which existed before the attack. This means that high levels of phytoalexin accumulation do not persist in plants once

Table 5.1 Phytoalexin accumulation in plants

Plant source	Phytoalexins	Fungi, bacteria and abiotic elicitors	References
Apiaceae			
Apium graveolens	Furanocoumarins: angelicin; bergapten; columbianetin; isopimpinellin; psoralen; 4,5′,8-trimethylpsoralen	Fungus: *Fusarium oxysporum* f. sp. *apii*	Heath-Pagliuso *et al.*, 1992
Daucus carota	Isocoumarins: 6-methoxymellein	Fungus: *Chaetonium globosum*	Amin *et al.*, 1988
Glehnia littoralis	Furanocoumarins: bergapten; demethylsuberosin; psoralen; xanthotoxin	Bacterium: *Pseudomonas cichorii*	Masuda *et al.*, 1998
Asteraceae			
Carthamus tinctorius	Polyacetylenes: safinol; dehydrosafinol; Coumarins: ayapin	Fungi: *Phytophthora* sp.; *Alternaria carthami*	Sprecher and Urbasch, 1983; Wolters and Eilert, 1983; Nahrstedt, 1979
Cichorium intybus	Sesquiterpenes: cichoralexin	Bacterium: *Pseudomonas cichorii*	Monde *et al.*, 1990b
Lactuca sativa	Sesquiterpenes: lettucenin A; costunolid	Fungi: *Botrytis cinerea; Bremia lactucae*	Sprecher and Urbasch, 1983; Wolters and Eilert, 1983; Bestwick *et al.*, 1995
Helianthus annuus	Coumarins: ayapin; scopoletin	$CuCl_2$	Sprecher and Urbasch, 1983; Wolters and Eilert, 1983; Gutierrez *et al.*, 1995
Hypochoeris radicata	Sesquiterpenes: isohypoglabric acid methyl ester; hypochoeroside K Alkenals: 6-hydroxyhexadienal; mucondialdehyde	$CuCl_2$	Maruta *et al.*, 1995
Polymnia sonchifolia	Acetophenones: 4′-hydroxy-3′-(3-methylbutanoyl)acetophenone; 4′-hydroxy-3′-(3-methyl-2-butenyl)acetophenone; Benzofurans: 5-acetyl-2-(1-hydroxy-1-methylethyl)benzofuran	Bacterium: *Pseudomonas cichorii*	Takasugi and Masuda, 1996
Brassicaceae			
Brassica oleracea	Indole alkaloids: brassinin; cyclo-brassinin; methoxybrassenin A; methoxybrassenin B; methoxybrassitin; spiroprassinin	Bacterium: *Pseudomonas cichorii*	Monde *et al.*, 1991

Caryophyllaceae			
Dianthus caryophyllus	Dianthalexins, dianthramides, dianthramines, dianthanilides: e.g. hydroxydianthalexin B; hydroxydianthramide S methylester; methoxydianthramide S	Fungus: *Fusarium oxysporum* f. sp. *dianthi*	Niemann *et al.*, 1991; Niemann, 1993
Convolvulaceae			
Ipomoea batatas	Furanoterpenoids: ipomoeamaron	Fungus: *Ceratocystis* sp.	Sprecher and Urbasch, 1983; Wolters and Eilert, 1983; Nahrstedt, 1979
Cupressaceae			
Cupressus sempervirens	Sesquiterpenes: 6-isopropyltropolone-β-glucoside; 5-(3-hydroxy-3-methyltrans-1-butenyl)-6-isopropyl-tropolone-β-glucoside	Fungus: *Diplodia pinea* f. sp. *cupressi*	Madar *et al.*, 1995
Dioscoreaceae			
Dioscorea bulbifera	Dihydrostilbenes: demethylbatasin IV; dihydroresveratrol	Fungus: *Botryodiplodia theobromae*	Adesanya *et al.*, 1989
Dioscorea dumentorium	Dihydrostilbenes: demethylbatasin IV; dihydroresveratrol	Fungus: *Botryodiplodia theobromae*	Adesanya *et al.*, 1989
Euphorbiaceae			
Ricinus communis	Diterpenes: casbene	Fungi: *Rhizopus stolonifer*; *Aspergillus* sp.	Lee and West, 1981; Hill *et al.*, 1996; Nahrstedt, 1979
Fabaceae			
Cicer arietinum	Isoflavonoids: maackiain; medicarpin; (-)-vestitone	Fungus: *Ascochyta rabiei*; *Colletotrichum gloeosporioides*	Kessmann *et al.*, 1988; Soby *et al.*, 1997
Glycine max	Isoflavonoids: glyceollin; glycinol; hydroxyphaseollin; phaseollin	Fungus: *Phytophthora megasperma* f. sp. *glycine*	Hahn *et al.*, 1985; Parniske *et al.*, 1991
Medicago sativa	Isoflavonoids: daizein; maackiain; medicarpin; trifolirhizin; (-)-vestitone	Fungus: *Verticillium alboatrium*; $HgCl_2$	Gustine and Moyer, 1982; Walton *et al.*, 1993
Phaseolus vulgaris	Isoflavonoids: phaseollin; phaseollidin; kievitone; wighteone	Fungi: *Fusarium solani* f. sp. *phaseoli; Monilinia fructicola*	Kuc and Rush, 1985; Li D. *et al.*, 1995; Turbek *et al.*, 1992
Pisum sativum	Isoflavonoids: pisatin; (+)-2-hydroxypisatin	$CuCl_2$	Matthews and van Etten; 1983; Miao *et al.*, 1991; Kobayashi *et al.*, 1993

Table 5.1 (Continued)

Plant source	Phytoalexins	Fungi, bacteria and abiotic elicitors	References
Trifolium pratense	Isoflavonoids: daidzein, medicarpin	$HgCl_2$	Gustine and Moyer, 1982
Vigna angularis	Isoflavonoids: daidzein; isoflavone; ligballinol	Actinomycin D	Kobayashi and Otha, 1983
Vicia faba	Polyacetylenes: wyerone; wyerone acid; wyerone epoxide	Fungus: *Botrytis cinerea*	Sprecher and Urbasch, 1983; Wolters and Eilert, 1983; Nahrstedt, 1979
Iridaceae			
Iris pseudacorus	Isoflavones: ayamenin A; ayamenin B; ayamenin C; ayamenin D; biochanin A; irilin A; irilin B; irilin C; iristectorigenin A; iristectorigenin B; genistein, lupinalbin A; pratensein; tectorigenin	$CuCl_2$	Hanawa *et al.*, 1991
Juncaceae			
Juncus roemerianus	Dihydrophenanthrenes: juncusol		Boger *et al.*, 1985
Lactaceae			
Cephalocereus senilis	Aurones: 4,5-methylenedioxy-6-hydroxyaurone	Chitin	Pare *et al.*, 1991
Liliaceae			
Allium cepa	5-octyl-cyclopenta-1,3-dione; 5-hexylcyclopenta-1,3-dione	Fungus: *Botrytis cinerea*	Tverskoy *et al.*, 1991
Veratrum grandiflorum	Stilbenoids: resveratrol; oxyresveratrol; piceid; oxyresveratrol-3-*O*-glucoside	$CuCl_2$	Hanawa *et al.*, 1992
Malvaceae			
Gossypium hirsutum	Sesquiterpenes: gossypol; hemigossypol; deoxyhemigossypol; 2,7-dihydroxycadalene; 2-hydroxy-7-methoxycadalene; lacinilene C; lacinilene C7-methyl ether; 7-hydroxycalamenene; 7-hydroxycalamenen-2-one	Bacterium: *Xanthomonas campestris pv malvacearum*; Fungus: *Aspergillus flavus*	Sprecher and Urbasch, 1983; Wolters and Eilert, 1983; Essenberg *et al.*, 1990; Zeringue, 1990; Davila-Huerta *et al.*, 1995
Moraceae			
Morus alba	Phenylbenzofuran: moracin A	Fungus: *Fusarium solani* f. sp. *mori*	Nahrstedt, 1979

Musaceae			
Musa acuminata	Phenalenones: methyl-2-benzimidazole carbamate; 2-(4′-hydroxyphenyl)-naphthalic anhydride	Fungus: *Colletrotrichum musae*	Hirai *et al.*, 1994
Orchidaceae			
Orchis militaris	Dihydrophenanthrenes: orchinol; hircinol; loroglossol	Fungus: *Rhizoctonia* spp.	Nahrstedt, 1979
Papaveraceae			
Papaver somniferum	Alkaloids: sanguinarine	Fungus: *Sclerotina sclerotiorum*	Eilert *et al.*, 1985
Platanaceae			
Platanus acerifolia	Coumarins: scopoletin; umbelliferone	Fungus: *Ceratocystis fimbriata* f. sp. *platani*	Modafar *et al.*, 1993
Poaceae			
Avena sativa	*N*-cinnamoylanthranilic acids: avenanthramide G	Fungus: *Helminthosporium victoriae*	Miyagawa *et al.*, 1996
Pyricularia oryzae	Diterpenenes: oryzalexin A-F Flavanones: sakuranetin	Fungus: *Pyricularia oryzae* abiotic elicitor: UV-light	Kato *et al.*, 1993, 1994
Saccharum sp.	Stilbenes: piceatannol	Fungus: *Colletotrichum falcatum*	Brinker and Seigler, 1991
Zea mays	Several caffeoyl esters	Fungi: *Colletotrichum graminicola*; *Helminthosporium maydis*	Lyons *et al.*, 1990
Ranunculaceae			
Thalictrum rugosum	Alkaloids: berberine	Fungus: *Saccharomyces cerevisiae*	Funk *et al.*, 1987
Rosaceae			
Mespilus germanica	Dibenzofurans: α-cotonefuran; 6-hydroxy-, 6-methoxy- and 7-hydroxy-6-methoxy-α-pyrufurans	Fungus: *Nectria cinnabarina*	Kokubun *et al.*, 1995a
Photinia davidiana	Dibenzofurans: eriobofuran; 9-hydroxyeriobofuran; 7-methoxyeriobofuran	Fungus: *Nectria cinnabarina*	Kokubun *et al.*, 1995b
Pyracantha coccinea	Dibenzofurans: eriobofuran; 9-hydroxyeriobofuran; 7-methoxyeriobofuran	Fungus: *Nectria cinnabarina*	Kokubun *et al.*, 1995b
Pyrus pyrifolia	Phenols: 3,5-di-*O*-caffeoylquinic acid	Fungus: *Alternaria alternata*	Kodoma *et al.*, 1998
Sanguisorba minor	2′,6′-Dihydroxy-4′-methoxyacetophenone	Fungus: *Botrytis cinerea*	Kokubun *et al.*, 1994

Table 5.1 (Continued)

Plant source	Phytoalexins	Fungi, bacteria and abiotic elicitors	References
Sorbus aucuparia	Biphenyls: aucuparin, 2′-methoxyaucuparin; 4′-methoxyaucuparin; 2′-hydroxyaucuparin; isoaucuparin	Fungus: *Nectria cinnabarina*	Kokubun *et al.*, 1995a
Solanaceae			
Capsicum frutescens	Sesquiterpenes: capsidiol	Fungi: *Botrytis cinerea*; *Fusarium* sp.	Sprecher and Urbasch, 1983; Wolters and Eilert, 1983
Datura stramonium	Sesquiterpenes: lubimin; hydroxylubimin	Fungi: *Monilinia fructicola*; *Phytophthora* sp.	Sprecher and Urbasch, 1983; Wolters and Eilert, 1983; Kuc and Rush, 1985
Lycopersicon esculentum	Sesquiterpenes: rishitin Phenols: *p*- coumaroyltyramine, E-feruloyltyramine Polyacetylenes: falcarinol; falcarindiol	Fungus: *Phytophthora megasperma* Abiotic elicitors: chitosan; mechanical wounding	Sprecher and Urbasch, 1983; Gross, 1982; Pearce *et al.*, 1998
Nicotiana spp.	Sesquiterpenes: capsidiol; glutinosone; oxyglutinosone; 5-epi-aristolochene; phytuberol; phytuberin	Bacteria: *Pseudomonas solanacereum; P. syringae*	Tanaka and Fujimori, 1985;
Solanum tuberosum	Sesquiterpenes: C-1′ epimers of (2*R*,5*S*,10*R*)-2-(1′,2′-dihydroxy-1′-methylethyl)-6,10-dimethyl-spiro[4,5] dec-6-en-8-one and their 2′-*O*-β-D-glucopyranosides	Fungi: *Phoma foveata, Fusarium* spp.	Engström, 1998
Solanum tuberosum	Sesquiterpenes: anhydro-β-rotunol; hydroxylubimin; lubimin; phytuberin; phytuberol; rishitin; rishitinol; solavetivone	Fungus: *Phytophthora infestans*	Bostock *et al.*, 1981; Kuc and Rush, 1985
Vitaceae			
Vitis vinifera	Stilbenes: α-viniferin, ε-viniferin	Fungus: *Botrytis cinerea*	Nahrstedt, 1979

a pathogen or stress has been contained and plant metabolism has returned to normal.

The levels of phytoalexin in the plant tissue are regulated by new synthesis and degradation of secondary metabolites. For example, at the site of the hypersensitive reaction, there is a decreased phytoalexin degradation (Keen, 1986). At the same time, new synthesis is transiently greatly increased. It is set into motion by a chemical signal ('*elicitor*') of the attacking pathogen. In resistant plants, the chemical signal causes a rapid increase in the synthesis of key enzymes of the plant secondary metabolism. It has been shown that, prior to enzyme synthesis, the corresponding messenger ribonucleic acid (mRNA) has developed (Chappell and Hahlbrock, 1984; Cramer *et al.*, 1985; Grab *et al.*, 1985). For example, before the phytoalexins, glyceollin I–III, accumulate in the soybean plant, the activity of a number of metabolic isoflavonoid enzymes greatly increases, namely that of phenyalanine ammonia lyase (PAL), *p*-coumaric acid- coenzyme A (CoA)-ligase, chalcone synthase and chalcone isomerase. Simultaneously, an increase in PAL, *p*-coumaric acid-CoA-ligase and chalcone synthase mRNA is seen (Kombrink and Hahlbrock, 1985; Smith and Banks, 1986; Ebel *et al.*, 1985). While enzymes of secondary metabolism are being newly synthesized in response to the pathogen's signal, the activities of many other enzymes of primary metabolism remain unchanged.

5.2.2 Elicitors

In order to trigger the genes necessary for phytoalexin synthesis and synthesis of the corresponding enzymes, the plant must receive information about the attack. The corresponding signals are called 'elicitors'. These are distinguished as biotic and abiotic elicitors.

5.2.2.1 Biotic elicitors

Biotic elicitors are organic substances, usually containing carbohydrates, which develop their signal effect at very low concentrations (about 10^{-9} M/L). They may originate from the attacking microbe (e.g. fungus) or the attacked plant. Such elicitors were first isolated from microbes in the cell walls of spores and hyphae of pathogenic and nonpathogenic fungi. There were, for example: oligosaccharides composed of the same saccharide units, such as β-1,3-glucan (Kombrink and Hahlbrock, 1985); oligosaccharides composed of various saccharide units (Albersheim and Darvill, 1985; Davis *et al.*, 1986); glycoproteins (West, 1981); and fatty acids, such as eicosapentenoic acid, arachidonic acid (Bostock *et al.*, 1981; Preisig and Kuc, 1985).

At the site of infection, these elicitors are released either spontaneously or under the effect of plant cell wall enzymes from the wall of the fungus. In the interaction between soybean and the fungus, *Phytophthora megasperma* f. sp. *glycinea*, for example, an endoglucanase and a glucosilase from the soybean plant (Keen and Yoshikawa, 1983) induce the release of elicitors, which signal phytoalexin synthesis and a number of other defence reactions, such as: local accumulation of chitinase (Roby *et al.*, 1986); local accumulation of lignin (Dean and Kuc, 1987); and systemic production of proteinase inhibitors (Walker-Simmons *et al.*, 1984).

Elicitors can also be derived from the cell wall of the infected plant. There they are broken down under the effect of enzymes of varying origin ('endogenous elicitors'). The activating enyzmes may originate from the microbe: for example, an endopolygalacturonic acid lyase from the bacterium, *Erwinia carotovora* (Davis *et al.*, 1986), and an endo-1,4-polygalacturonase from the fungus, *Rhizopus stolonifer* (Lee and West, 1981) release cell wall fragments, which serve as elicitors for phytoalexin synthesis in the infected plant. In addition, the plant tissue itself may deliver the enzymes, which then release the elicitors from the plant cell wall (Albersheim and Darvill, 1985).

5.2.2.2 Abiotic elicitors

Chemical or physical factors that put a plant under stress and then also trigger the production of phytoalexins are called abiotic elicitors. These are: heavy metal salts, such as $CuCl_2$ or $HgCl_2$ (Adesanya *et al.*, 1984); agents which interact with deoxyribonucleic acid (DNA), such as actinomycin D or ultraviolet rays (Hardwiger and Schwochau, 1971a,b); and metabolic inhibitors, such as trichloroacetic acid, monoiodoacetate or 2,4-dinitrophenol (Cruickshank, 1966).

Although these elicitors have nothing to do with the defence of a primary attack by a foreign organism, they still provoke suitable defence reactions and protect the plant that has been weakened or injured by abiotic stress from subsequent parasitic attack, which is deterred directly by the high concentrations of phytoalexins. The observation that abiotic stimuli induce the synthesis and accumulation of phytoalexins in plant tissues can easily be explained through the stress-determined release of endogenous, constitutive elicitors from the cell wall (Smith and Banks, 1986; Albersheim and Darwill, 1985).

5.2.3 Specificity of phytoalexin accumulation

The synthesis and accumulation of a certain phytoalexins are induced by very different species of microbes. Cruickshank and Perrin (1961) name, for example, 19 species of fungi that induce the development of pisatin on the pods of peas. Furthermore, different races of the same fungus that are compatible with the host plant induce phytoalexin development, just like those races that are incompatible with the host and cause a hypersensitive reaction. Therefore, the phytoalexins are not a specific means of defence directed toward a certain microbe (as, for example, antibodies are); their effect is nonspecific and directed against numerous microbes.

On the other hand, the difference between compatible and incompatible reactions is interesting. Here, it is seen that various races of a fungus can affect the extent and kinetics of phytoalexin accumulation to completely different degrees. Incompatible races usually cause phytoalexin synthesis to begin very quickly and the amount of phytoalexin accumulated is relatively high. It takes longer for phytoalexin synthesis to commence after an attack by a compatible race and the final levels are considerably lower. As an example, both compatible and incompatible races of *Phytophthora megasperma* f. sp. *glycinea* induce the accumulation of glyceollines in soybean. Twelve hours after infection, the same levels of glyceollines are reached. After this time, the levels of glyceollines continue to increase greatly in the incompatible race, whereas they are likely to remain constant in the compatible race (Hahn *et al.*, 1985).

An exact analysis shows that incompatible and compatible microbes alike deliver elicitor molecules. The varying levels of phytoalexin accumulation that have been observed might be due to the following factors: 1) Various species and races of microbes send differing, specific elicitor signals; 2) Phytoalexins are degraded by enzymes of penetrating microbes (Maloney and van Etten, 1994; Li D. *et al.*, 1995). For example, among the antimicrobial phytoalexins produced by *Phaseolus vulgaris* is the prenylated isoflavonoid, kievitone. The bean pathogen, *Fusarium solani* f. sp. *phaseoli*, secretes the glycoenzyme, kievitone hydrolase, which catalyzes conversion of kievitone to a less toxic metabolite (Li D. *et al.*, 1995); or 3) Compatible species or races of microbes synthesize suppressor molecules, which annul or reduce the effect of the elicitors of different origin. Such suppressor molecules have been found in strains of the pea pathogens, *Mycosphaerella pinodes* and *Erysiphe pisi* (Schloesser, 1983).

5.3 Antibacterial and antifungal agents of higher plants

The indiscriminate use of antibiotics has resulted in the emergence of a number of resistant bacteria and bacterial strains. To overcome the increasing resistance of nosocomial bacteria, more effective antimicrobial agents with novel modes of action must be developed. Medicinal plants used in traditional medicines to treat infectious diseases seem to be an abundant source of new bioactive secondary metabolites. Therefore, in the last few years, a variety of medicinal plants and plant extracts have been screened for their antimicrobial activity. In addition, using the bioassay-guided fractionation of bioactive plant extracts, different secondary metabolites exhibiting antimicrobial activities have been isolated. The results obtained by this method have clearly revealed that the antimicrobial activity is mainly due to alkaloids, flavonoids, phenolic compounds, terpenoids and tannins (see Figures 5.3 and 5.4). Furthermore, essential oils have also been reported to be active against Gram-positive and Gram-negative bacteria, as well as against yeasts and fungi (see Tables 5.2–5.4). Bioactive secondary metabolites and essential oils are considered to be part of the preformed defence system of higher plants. It is assumed that all families of higher plants possess more or less bioactive secondary metabolites involved in the comprehensive plant defence system. For the distribution of antimicrobial secondary metabolites in the plant kingdom, see Grayer and Harborne (1994).

Whilst it is beyond the scope of the present chapter to review this expanding scientific field extensively, its progress will be documented by the most important results published in the last 10 yrs. The essential oils studied and secondary metabolites isolated were preferably tested *in vitro* against some of the following microorganisms: 1) Gram-positive bacteria, e.g. *Bacillus cereus*, *B. subtilis*, *Mycobacterium intracellulare*, *Sarcinia flava*, *S. lutea*, *Staphylococcus aureus*, *S. epidermidis*, *Streptococcus faecalis*, *S. hemolyticus and S. pneumoniae*; 2) Gram-negative bacteria, e.g. *Enterobacter cloacae*, *Escherichia coli*, *Klebsiella oxytoca*, *K. pneumoniae*, *Proteus mirabilis*, *P. morgani*, *P. rettgeri*, *Pseudomonas aeruginosa*. *Salmonella enteritidis*, *S. typhosa*, *S. typhimurium*, *Shigella flexneri and S. sonnei*; 3) Yeasts, e.g. *Candida albicans*, *C. kruzei*, *C. tropicalis*, *Saccharomyces cerevisiae*, *Schizosaccharomyces pombe*, *Torula glabrata*, *Torulopsis utilis*, *T. glabrata and Trichosporon capitatum*; and 4) Fungi, e.g. *Aspergillus fumigatus*, *A. niger*, *A. ochraceus*, *Epidermophyton flocosum*, *Fusarium sporotrichoides*, *F. tricintum*, *Microsporum canis*, *Penicillium rubrum*, *P. spinulosum*, *Trichophyton rubrum and T. mentagrophytes*.

5.3.1 Essential oils with antimicrobial activity

Essential oils are widely distributed in certain plant families, e.g. Alliaceae, Apiaceae, Asteraceae, Brassicaceae, Lamiaceae, Myrtaceae and Rutaceae. They are mixtures of different lipophilic and volatile substances, such as monoterpenes, sesquiterpenes and/or phenylpropa-

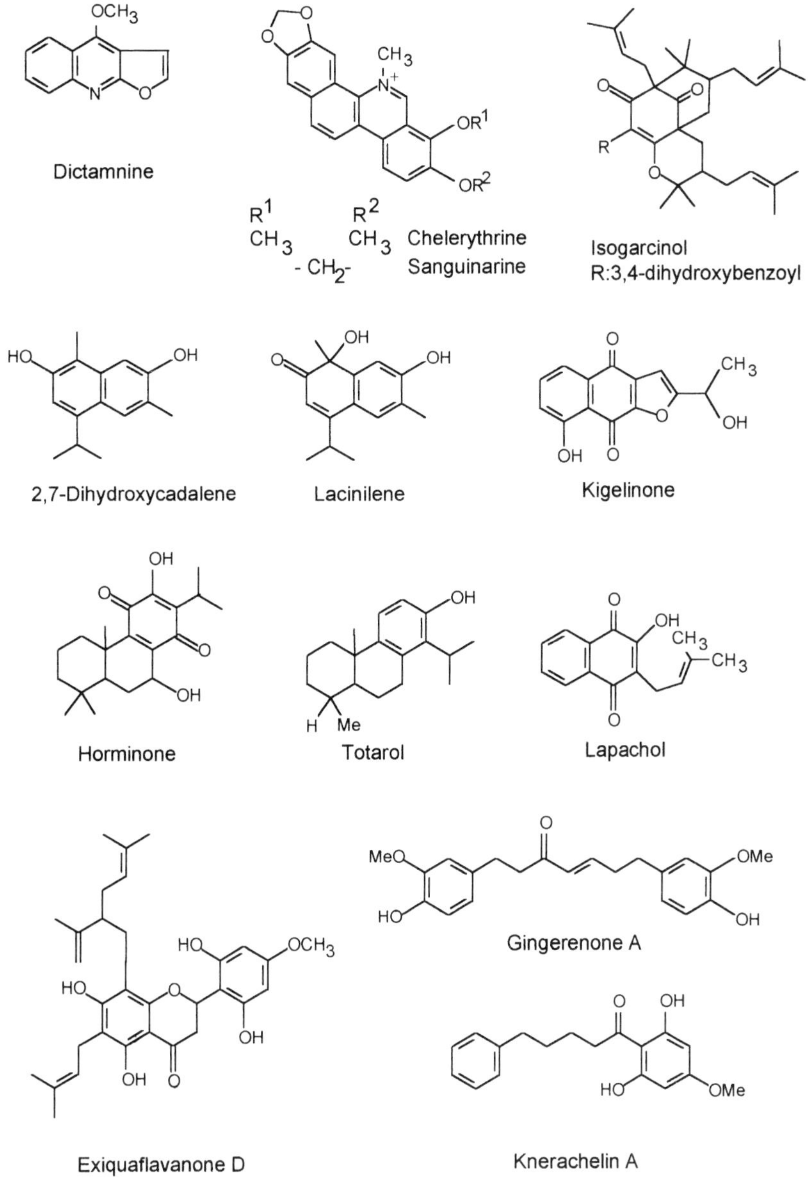

Figure 5.3 Chemical structures of several plant-derived compounds with antimicrobial properties.

noids, and have a pleasant odour. Therefore, testing and evaluation of antimicrobial activity is difficult because of their volatilty and insolubility

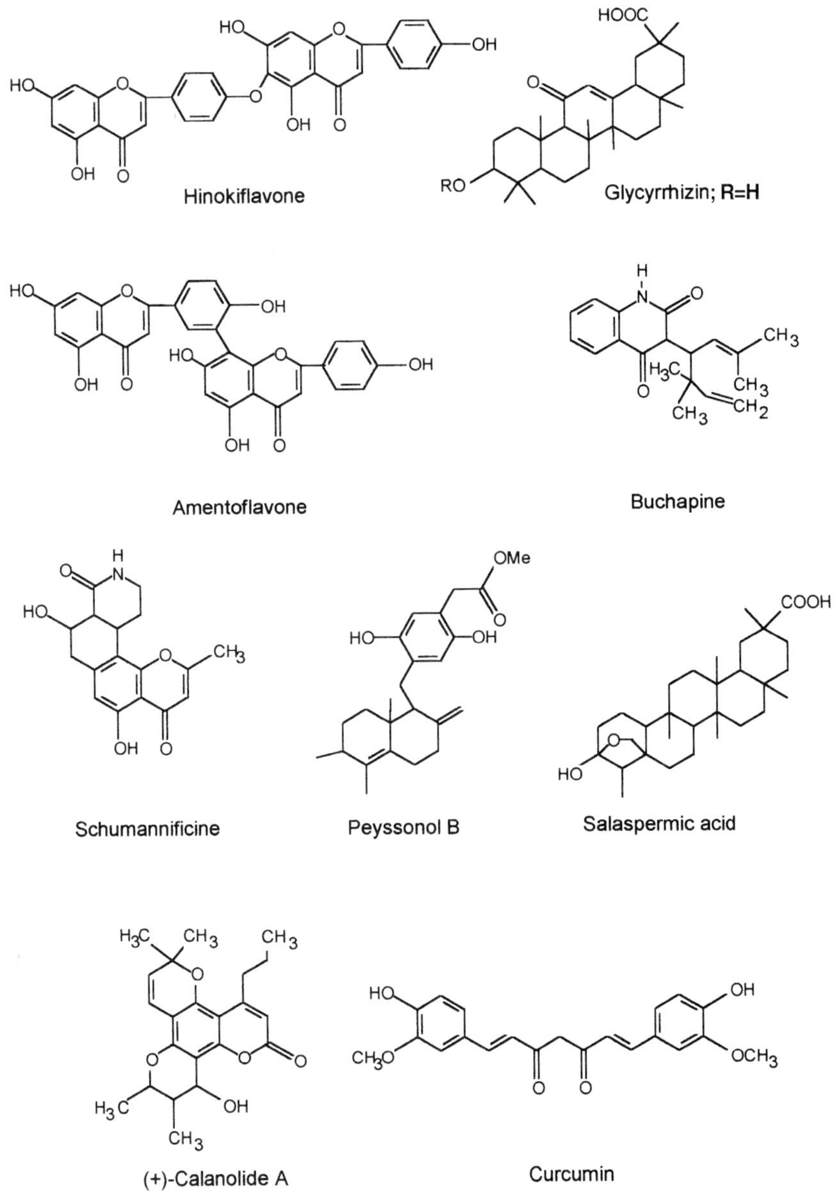

Figure 5.4 Chemical structures of plant-derived anti-human immunodeficiency virus (HIV) compounds.

in water. The assay technique, the growth medium and the microorganisms used are especially important when testing essential oils. Depending

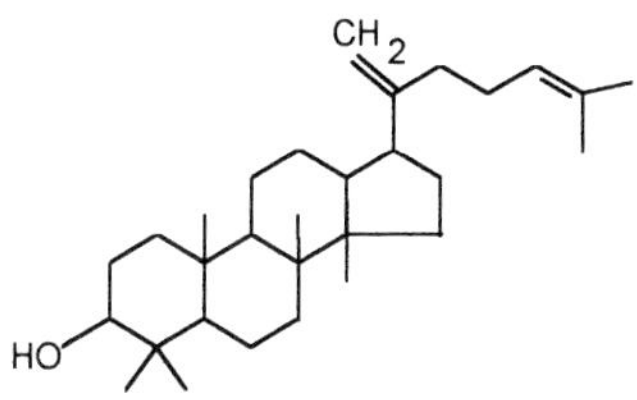

Dammaradienol

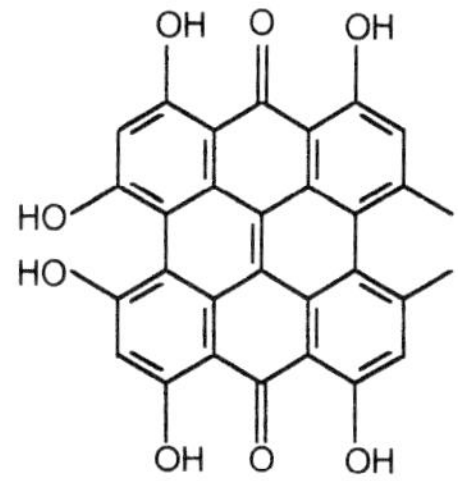

Hypericin

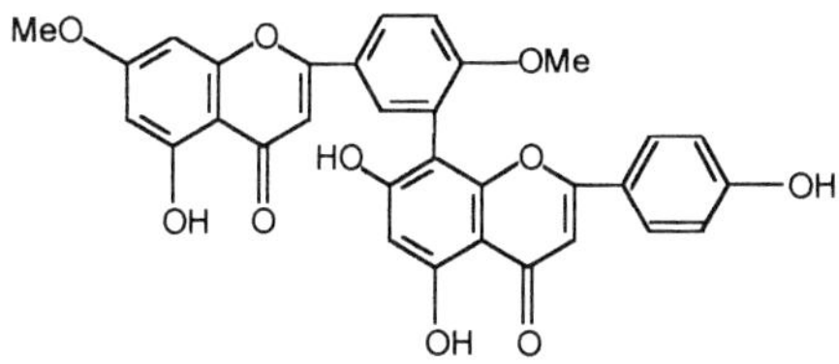

Ginkgetin

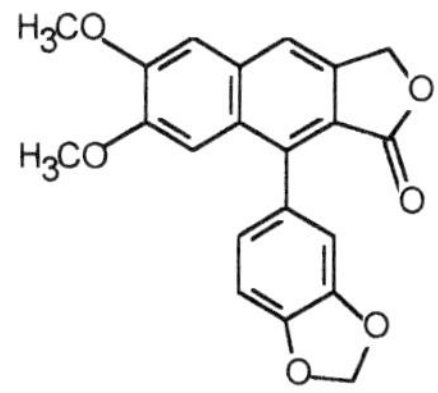

Justicidin B

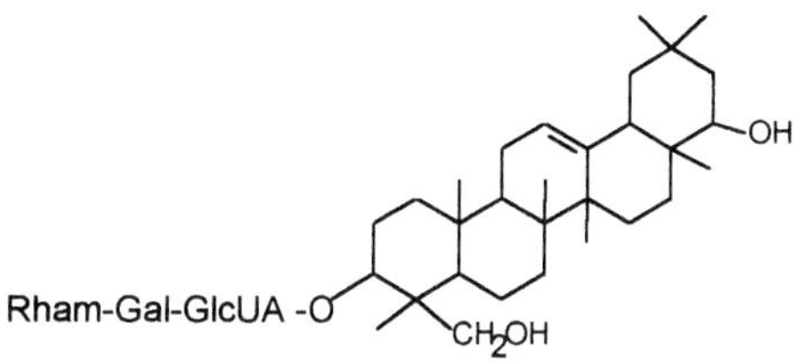

Soyasaponin I

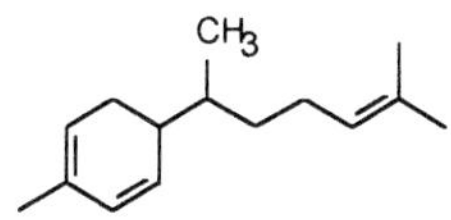

α-Zingiberene

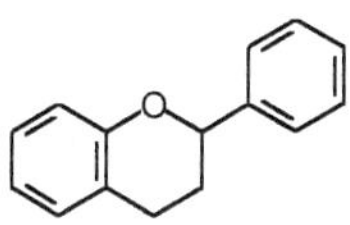

Flavan

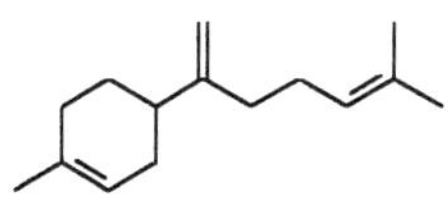

β-Bisabolene

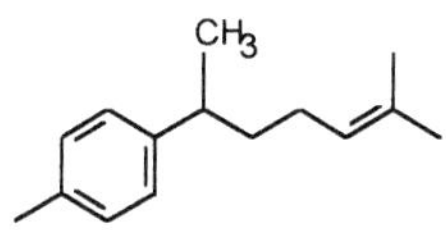

ar-Curcumene

Figure 5.5 Chemical structures of several plant-derived compounds with antiviral properties.

on its insolubility in water, an emulsifying agent, e.g. Tween 80, dimethyl sulphoxide (DMSO), has to be used to disperse the essential oil in broth liquid or agar medium. All these facts should be taken into account when estimating the antimicrobial activity of essential oils examined by different researchers.

During the last decade (1988–1998), a large number of studies have been performed to assess the antimicrobial activity of different essential oils (Jalsenjak *et al.*, 1987; Gergis *et al.*, 1990; Kartnig *et al.*, 1991; Economou and Nahrstedt, 1991; Barel *et al.*, 1991; Hammerschmidt *et al.*, 1993; Schales *et al.*, 1993; Panizzi *et al.*, 1993; Kalodera *et al.*, 1994; Carson and Riley, 1994, 1995; Carson *et al.*, 1995; Pattnaik *et al.*, 1995a,b, 1996; Nenoff *et al.*, 1996; Kilibarda *et al.*, 1996; Tirillini *et al.*, 1996; Hili *et al.*, 1997). All essential oils showed antimicrobial activity against at least one of the microorganisms tested. The most active essential oils were those of *Cymbopogon* spp., *Eucalyptus citriodora*, *Melaleuca alternifolia*, *Mentha piperita* and *Thymus vulgaris*. With regard to the highly resistant Gram-negative bacterium, *Pseudomonas aeruginosa*, only the essential oil from *Calamintha nepeta* exhibited significant activity (see Table 5.2).

While essential oils were extensively tested against a broad spectrum of bacteria, yeasts and fungi, the interaction between the essential oils and the bacteria which ultimately induce the antibacterial activity is not well understood. Recently, Takaisi-Kikuni and co-workers (1996) studied the effect of various amounts of the essential oil of *Cymbopogon densiflorus* (lemongrass oil) on the metabolic activity, growth and morphology of *Staphylococcus aureus*. Relatively high concentrations of the oil impaired staphylococcal growth in a bacteriostatic manner (chloramphenicol-type), and in low doses metabolism became ineffective due to energy losses in the form of heat. Ultrastructural data revealed morphological changes characteristic of the induction of bacteriolysis by bactericidal antibiotics (penicillin-type). Based on these results, it was assumed that the essential oil may have antibacterial activity by influencing bacterial targets involved in cytoplasmatic and cell wall metabolism. In a further study, Pattnaik and co-workers (1995a,b, 1996) showed that palmarosa oil and peppermint oil induced the formation of an elongated, filamentous form of *E.coli* cells at concentrations of 1.66 μg/ml.

5.3.2 *Isolated secondary plant metabolites with antimicrobial properties*

5.3.2.1 *Alkaloids*

The antimicrobial activity of alkaloids has been extensively reviewed (Clark and Hufford, 1992; Wink, 1993). Recently, Verpoorte (1998) published another comprehensive review on this subject. The review considered about 300 alkaloids with antimicrobial activity and reported that bioactive alkaloids could be found within acridone-, aporphine-, benzophenanthridine-, bisbenzylisoquinoline-, indole-, isoquinoline-, piperidine-, protoberberine-, quinoline-, terpenoid- and steroid-type alkaloids. The studies on the antimicrobial activity of alkaloids published in the last decade are summarized in Table 5.3 (for more detail, see Verpoorte, 1998).

Dictamnine, a furoquinoline alkaloid, isolated from the root bark of *Dictamnus dasycarpus* (a traditional Chinese medicine), exhibited strong antifungal activity against the pathogenic fungus, *Cladosporium cucumerium* (minimal concentration required to cause 50% inhibition [MIC_{50}] 25.0 μg/ml). It is an interesting fact that Grayer and Harborne (1994) have previously suggested that furoquinoline alkaloids may play an important role in the defence of plants against potentially pathogenic fungi.

Recently, Colombo and Bosisio (1996) reported on the pharmacological activity of *Chelidonium majus* (Papaveraceae). The plant has a long history of use in the treatment of several diseases in European countries. It contains various isoquinoline alkaloids with protopine, protoberberine and benzophenanthridine structures, e.g. sanguinarine, chelidonine, chelerythrine, berberine and coptisine. *C. majus* extracts and their purified compounds exhibited interesting antiviral, antitumoral and antimicrobial properties both *in vitro* and *in vivo*. Sanguinarine and chelerythrine have been reported to display antibacterial activity, with an MIC value of 6.25 μg/ml. Sanguinarine is used in oral health products, such as mouthwashes and toothpastes. *In vitro* studies have indicated that the antiplaque action of sanguinarine is due to its ability to inhibit the adherence of bacteria to newly formed pellicle. The MIC values of the compound ranged 1–32 μg/ml for most species of plaque bacteria (Grenby, 1995; Godowski, 1989).

Polymorphonuclear neutrophils (PMNs) represent an important defence mechanism against bacterial infection. Superoxide is one of the most important factors released by PMNs following a variety of stimulations, including that of bacteria. It was found that 25–200 μg/ml of ofloxacin (1-[5-isoquinolinesulfonyl]-2-methylpiperazine) augmented superoxide production of PMNs. This augmentation was assumed to be due to the enhancement of leukocyte protein kinase C (Nagafuji *et al.*, 1993).

5.3.2.2 Anthraquinones

Anthraquinonic compounds, traditionally used as laxatives, possess many other pharmacological properties, including microbiological action (see Table 5.4). Of eight pure anthraquinones tested, only rhein showed significant antimicrobial activity. Aloe-emodin, emodin, chrysophanic acid, physcion, aloin, sennoside A and sennoside B were found to be inactive (Didry *et al.*, 1994).

Table 5.2 Essential oils with antimicrobial activity

Essential oil derived from	Bacteria Gram (+)	Bacteria Gram (-)	Yeasts	Fungi	MIC µg/ml	References
Abies alba	+					Schales *et al.*, 1993
Achillea fragrantissima	+	-	y		850–1900	Barel *et al.*, 1991
Aegle marmelos	+	-	y	f		Pattnaik *et al.*, 1996
Ageratum conyzoids	+	-				Pattnaik *et al.*, 1996
Allium sativum			y			Aboul Ela *et al.*, 1996
Artemisia parviflora	+	-	y			Mehrotra *et al.*, 1993
Artemisia rhoxburghiana	+	-	y			Mehrotra *et al.*, 1993
Bystropogon pulmosus	+	-	y	f		Economou and Nahrstedt, 1991
Calamintha nepeta	+	-	y		*5.0–20.0*	Panizzi *et al.*, 1993
Cinnamomum camphora	+	-	y			Hili *et al.*, 1997
Cinnamomum zeylanicum	+	-	y			Hili *et al.*, 1997
Citrus aurantium	+	-	y	f		Pattnaik *et al.*, 1996
Cochleospermum regium	+				*1500–5000*	Brum *et al.*, 1997
Coriandrum sativum	+	-	y			Hili *et al.*, 1997
Cymbopogon citratus	+	-	y			Hili *et al.*, 1997; El-Kamali *et al.*, 1998
Cymbopogon flexuosus	+	-	y	f	*0.16–11.6*	Pattnaik *et al.*, 1995a,b, 1996
Cymbopogon martini	+	-	y	f	*0.5–8.3*	Pattnaik *et al.*, 1995a,b, 1996
Cymbopogon winterianus	+	-	y	f		Pattnaik *et al.*, 1995a,b, 1996
Daucus carota	+	-			120–18500	Kilibarda *et al.*, 1996
Ellateria cardamomum	+	-	y			Hili *et al.*, 1997
Eucalyptus spp.	+	-	y		1000–5000	Hajji and Fkih-Tetouani, 1993; Chand *et al.*, 1994
Eucalyptus citriodora	+	-	y	f	*0.16–10.0*	Pattnaik *et al.*, 1996
Helichrysum amorginum	+	-			750–1250	Chinou *et al.*, 1996
Helichrysum italicum	+	-			3250–3750	Chinou *et al.*, 1996
Jasona candicans	+	-				Hammerschmidt *et al.*, 1993
Jasona montana	+	-				Hammerschmidt *et al.*, 1993

Melaleuca alternifolia	+	-	y	f	500–6000	Chand *et al.*, 1994; Nenoff *et al.*, 1996; Hili *et al.*, 1997
Melaleuca alternifolia	+	-			*0.12–2.0%* of oil	Carson and Riley, 1994, 1995; Carson *et al.*, 1995
Mentha arvensis	+	-		f		Singh *et al.*, 1992
Mentha piperita	+	-	y	f	*0.27–10.0*	Pattnaik *et al.*, 1996; Hill *et al.*, 1997
Micromeria thymifolia	+	-	y	f	*0.1–60.0%* of oil	Kalodera *et al.*, 1994
Nigella sativa	+		y	f	2500	Aboul Ela *et al.*, 1996; El-Kamali *et al.*, 1998
Ocimum gratissimum	+	-	y	f	312.5–625	Ndounga and Quamba, 1997
Ocimum basilicum	+	-	y	f	1250–5000	Ndounga and Quamba, 1997
Pelargonium graveolens	+	-	y	f		Pattnaik *et al.*, 1996
Picea abies	+		y		1.6–100	Kartnig *et al.*, 1991
	+	-				Schales *et al.*, 1993
Pinus sylvestris	+	-				Schales *et al.*, 1993
Piper angustifolia	+	-	y	f	10–100	Tirillini *et al.*, 1996
Pogostemon patchouli	+	-	y	f		Pattnaik *et al.*, 1996
Pulicaria undulata	+	-				El-Kamali *et al.*, 1998
Rosmarinus officinalis	+	-	y		*5.0–40.0*	Panizzi *et al.*, 1993; Hili *et al.*, 1997
Salvia officinalis	+	-	y	f		Hili *et al.*, 1997
Satureja montana	+	-	y		*5.0–40.0*	Panizzi *et al.*, 1993
Sideritis cladestina	+	-	y			Gergis *et al.*, 1990
Sideritis sipylea	+	-	y			Gergis *et al.*, 1990
Syzygium aromatikum	+	-	y			Hili *et al.*, 1997
Thymus vulgaris	+	-	y		*1.0–5.0*	Panizzi *et al.*, 1993
	+	-	y		125–8000	Janssen *et al.*, 1988
Thymus masticina	+	-	y			Hili *et al.*, 1997

Values in italics indicate agar/broth dilution method.
Abbreviations: MIC, minimal inhibitory concentration; Gram(+), Gram-positive; Gram(-), Gram-negative.

Table 5.3 Alkaloids with antimicrobial activity

Alkaloids	Bacteria Gram (+)	Bacteria Gram (-)	Yeasts	Fungi	MIC μg/ml	References
Diterpenoid alkaloids						
8-acetylheterophyllisine				f	100–250	Rahman *et al.*, 1997
panicutin				f	75–200	Rahman *et al.*, 1997
vilmorrianone				f	100–225	Rahman *et al.*, 1997
Terpenoid indole alkaloids						
Yohimbine-heteroyohimbine-type						
Usambarensine and derivatives						Caron *et al.*, 1988
dihydro-(3α)	+				64.0	Caron *et al.*, 1988
tetrahydro-(3α,17α)	+					Caron *et al.*, 1988
tetrahydro-(3α,17β)	+				32.0	Caron *et al.*, 1988
Tetrahydrousambarensine derivatives						
10′-OH-(3α,17α)	+					Caron *et al.*, 1988
10′-OH-(3α,17β)	+					Caron *et al.*, 1988
10,10′-diOHMe-(3α,17α)	+					Caron *et al.*, 1988
10,10′-diOMe,NMe-(3α,17α)	+					Caron *et al.*, 1988
10′-OH,10-OMe,NMe-(3α,17α)	+					Caron *et al.*, 1988
10,10′-diOH,NME-(3α,17α)	+					Caron *et al.*, 1988
10-OH,10′-OMe,NMe-(3α,17α)	+					Caron *et al.*, 1988
cinchophylline (3α,17α)	+					Caron *et al.*, 1988
cinchophylline (3α,17β)	+					Caron *et al.*, 1988
cinchophylline (3β,17α)	+				16.0	Caron *et al.*, 1988
cinchophylline (3β,17β)	+				32.0	Caron *et al.*, 1988
and derivatives						
17,4′didehydro (3α)	+					Caron *et al.*, 1988
18,19-dihydro (3α,17α)	+					Caron *et al.*, 1988
18,19-dihydro (3α,17β)	+					Caron *et al.*, 1988

18,19-dihydro (3β,17α)	+					Caron *et al.*, 1988
18,19-dihydro (3β,17β)	+					Caron *et al.*, 1988
19-OH,18,19-diH (3α,17α)	+					Caron *et al.*, 1988
19-OH,18,19-diH (3α,17β)	+				32.0	Caron *et al.*, 1988
19-OH,18,19-diH (3β,17α)	+					Caron *et al.*, 1988
19-OH,18,19-diH (3β,17β)	+					Caron *et al.*, 1988
ochrolifuanine E	+					Caron *et al.*, 1988
ochrolifuanine F	+	-			32.0	Caron *et al.*, 1988
18,19-dihydroochrolifuanine F	+	-				Caron *et al.*, 1988
Iboga-type						
conoduramine	+	-			15–400	Munoz *et al.*, 1994
conodurine	+	-			4–400	Munoz *et al.*, 1994
Miscellaneous terpenoid indols						
stemmadenine	+	-	y		*1.2–37.5*	Mariee *et al.*, 1988
Various indols						
clausenal	+	-	y	f	*3.0–25.0*	Chakraborty *et al.*, 1995
cryptoheptine	+	-			*6.2–100*	Paulo *et al.*, 1994
cryptolepine	+	-	y		*1.5–500*	Paulo *et al.*, 1994
cryptolepine	+	-	y		*6.3–500*	Cimanga *et al.*, 1996
cryptoquindoline	+	-			*100*	Paulo *et al.*, 1994
harman				f	*25–100*	Quetin-Leclercq *et al.*, 1995
harmaline	+	-	y	f		Ahmad *et al.*, 1992
harmalol	+	-		f		Ahmad *et al.*, 1992
harmine	+	-	y	f		Ahmad *et al.*, 1992
harmine			y	f	*12.5–100*	Quetin-Leclercq *et al.*, 1995
harmol	+	-	y	f		Ahmad *et al.*, 1992
harmol				f	100	Ahmad *et al.*, 1992
norharman			y	f	*12.5–100*	Quetin-Leclercq *et al.*, 1995

Table 5.3 (Continued)

Alkaloids	Bacteria Gram (+)	Bacteria Gram (-)	Yeasts	Fungi	MIC µg/ml	References
N_b-methylharmalan			y		*50.0*	Quetin-Leclercq *et al.*, 1995
melinonine F			y		*12.5–100*	Quetin-Leclerc *et al.*, 1995
hydroxycryptoleptine	+	-			*100*	Paulo *et al.*, 1994
quindoline	+	-			*100*	Paulo *et al.*, 1994
quindoline	+	-	y		*8–500*	Cimanga *et al.*, 1996
quadrigemine B	+	-			*125*	Mahmud *et al.*, 1993
yuehchukene	+				20.0–25.0	Waterman, 1990
Pyrrolidinoindoline-type						
isopsychotridine E	+		y	f	5.0–100	Saad *et al.*, 1995
hodgkinsine A	+		y	f	5.0–100	Saad *et al.*, 1995
quadrigemine C	+		y	f	5.0–50	Saad *et al.*, 1995
quadrigemine H	+		y	f	10.0–75	Saad *et al.*, 1995
psychotridine E	+		y	f	25.0–100	Saad *et al.*, 1995
vatine	+		y	f	10.0–100	Saad *et al.*, 1995
vatine A	+		y	f	25.0–100	Saad *et al.*, 1995
vatamine	+		y	f	10.0–100	Saad *et al.*, 1995
vatamidine	+		y	f	25.0–100	Saad *et al.*, 1995
Isoquinoline alkaloids						
Bisbenzylisoquinoline-type						
dehatrine	+			f	*300–1000*	Tsai *et al.*, 1989
Aporphine-type						
actinodaphnine	+	-	y	f	*50–1000*	Tsai *et al.*, 1989
anhydroushinsunine			y	f	*125–1000*	Tsai *et al.*, 1989
anhydroushinsunine methoiodide			y	f	*62.5–1000*	Tsai *et al.*, 1989
anonaine	+	-	y	f	*3.0–125*	Simeon *et al.*, 1990; Paulo *et al.*, 1992; Tsai *et al.*, 1989

asimilobine	+		y		*12.0–100*	Simeon *et al.*, 1990
bulbocapnine	+	-	y		*1000*	Abbasoglu *et al.*, 1991
O-methylbulbocapnine			y		*500–1000*	Tsai *et al.*, 1989
O-methylbulbocapnine methoiodide	+				*300*	Tsai *et al.*, 1989
dicentrine methoiodide	+				*50.0*	Tsai *et al.*, 1989
glaucine methoiodide	+				*4300*	Tsai *et al.*, 1989
glaziovine	+				*12.0–50.0*	Simeon *et al.*, 1990
isoboldine	+	-		f	*500*	Abbasoglu *et al.*, 1991; Paulo *et al.*, 1992
lanuginosine	+			f	*25.0–100*	Simeon *et al.*, 1990; Ferdous *et al.*, 1992
laurelliptine				f	*500*	Paulo *et al.*, 1992
laurotetanine	+	-	y		*100–1000*	Tsai *et al.*, 1989
liriodenine	+	-	y	f	*0.4–100*	Simeon *et al.*, 1990; Pabuccuoglu *et al.*, 1991
liriodendronine			y		*25.0*	Pabuccuoglu *et al.*, 1991
liriodenine methoiodide	+		y		*0.4–6.0*	Pabuccuoglu *et al.*, 1991
lysicamine	+		y	f	*12.0–26.0*	Pabuccuoglu *et al.*, 1991; Simeon *et al.*, 1990
lysicamine methoiodide			y		*0.8–6.2*	Pabuccuoglu *et al.*, 1991
magnoflorine			y		*250*	Tsai *et al.*, 1989
N-methylxylopine methoiodide	+	-			*50–300*	Tsai *et al.*, 1989
norushinsunine	+		y		*6.0–100*	Simeon *et al.*, 1990
N-methylasimilobine	+				*25.0*	Simeon *et al.*, 1990
nuciferine	+				*50–1000*	Simeon *et al.*, 1990
N-methylactinodaphnine	+	-	y	f	*50–1000*	Tsai *et al.*, 1989
N-methyllaurotetanine	+	-	y		*100–1000*	Tsai *et al.*, 1989
roemerine methoiodide	+				*50–100*	Tsai *et al.*, 1989
2-*O*-methylliriodendronine			y		*50.0*	Pabuccuoglu *et al.*, 1991
oxostephanine	+	-			*50–150*	Ferdous *et al.*, 1992
xylopine	+		y		*25.0–100*	Tsai *et al.*, 1989; Simeon *et al.*, 1990

Table 5.3 (Continued)

Alkaloids	Bacteria Gram (+)	Bacteria Gram (-)	Yeasts	Fungi	MIC μg/ml	References
Benzophenanthridine-type						
chelerythrine	+	-	y	f	*6.3–100*	Abbasoglu *et al.*, 1991
sanguinarine	+	-	y	f	*0.5–100*	Abbasoglu *et al.*, 1991
Protoberberine-type						
β-allocryptopine	+	-				Abbasoglu *et al.*, 1991
berberine	+	-	y	f	*0.31–100*	Okunade *et al.*, 1994
canadine	+	-				Abbasoglu *et al.*, 1991
corydaline	+	-				Abbasoglu *et al.*, 1991
cryptopine	+	-				Abbasoglu *et al.*, 1991
ophiocarpine	+	-				Abbasoglu *et al.*, 1991
palmatine	+	-	y		*1000*	Abbasoglu *et al.*, 1991
protopine	+	-			*100*	Abbasoglu *et al.*, 1991
scoulerine	+	-				Abbasoglu *et al.*, 1991
stylopine	+	-				Abbasoglu *et al.*, 1991
tetrahydropalmatine	+				*50.0*	Simeon *et al.*, 1990
Miscellaneous isoquinoline alkaloids						
adlumidine	+	-				Abbasoglu *et al.*, 1991
bicuculline	+	-				Abbasoglu *et al.*, 1991
corydaldine	+	-				Abbasoglu *et al.*, 1991
crinamine	+				10.0	Adesanya *et al.*, 1992
diacetylcrinamine	+				10.0	Adesanya *et al.*, 1992
diacetalhamayne	+				10.0	Adesanya *et al.*, 1992
eupolauridine			y		*1.56*	Liu *et al.*, 1990
hydrastinine	+	-				Abbasoglu *et al.*, 1991
3-methoxysampangine			y	f	*0.2–3.1*	Liu *et al.*, 1990

Steroidal alkaloids						
α-chaconine				f	*60–100* μM	Fewell and Roddick, 1993
α-solanine				f	*80–100* μM	Fewell and Roddick, 1993
Miscellaneous types of alkaloids						
Furoquinoline-type						
dictamnine				f	25.0	Zhao *et al.*, 1998
Piperidine-type						
julifloricine	+	-	y		0.5–100	Aqeel *et al.*, 1989; Tawara *et al.*, 1993
euphococcinine		-			*1000*	Aqeel *et al.*, 1989; Tawara *et al.*, 1993
Pyrrolizidine-type						
9-angeloylretronecine		-	y	f		Marquina *et al.*, 1989
heliotrine			y			Marquina *et al.*, 1989
lasiocarpine	+		y	f		Marquina *et al.*, 1989
supinine			y			Marquina *et al.*, 1989
Miscellaneous-types						
antofine	+			f		Baumgartner *et al.*, 1990
ficuseptine	+			f		Baumgartner *et al.*, 1990
illukumbin B				f		Greger *et al.*, 1992, 1993
methylillukumbin B				f		Greger *et al.*, 1992, 1993
methylillukumbin A				f		Greger *et al.*, 1992, 1993
N-methylsinharine				f		Greger *et al.*, 1992, 1993
sinharine				f		Greger *et al.*, 1992, 1993

Values in italics indicate agar/broth dilution method. Abbreviations: MIC, minimal inhibitory concentration; Gram (+), Gram-positive bacteria; Gram (-), Gram-negative bacteria.

5.3.2.3 Diterpenoids

Many diterpenoids exhibit antimicrobial activity (see Table 5.4). Some of these compounds may be involved in the resistance of higher plants (e.g. conifers).

Abietane-type: lanigerol and forskalinone, isolated from the roots of *Salvia lanigera* and *Salvia forskahlei*, respectively, demonstrated moderate antibacterial activity against Gram-positive bacteria (Ulubelene *et al.*, 1996; El-Lakany *et al.*, 1995). Additional antibacterial diterpenoids were isolated from *Salvia hypargeia* and *Salvia sclarea*. Hypargenins A, B, C, D and F, as well as 2,3-dehydrosalvipisone, sclareol, manool and 7-oxoroyleanone, were active against *Staphylococcus aureus*, while hypargenin F was also active against *Mycobacterium tuberculosis* (Ulubelen *et al.*, 1988, 1994). Horminone, 11-hydroxy-12-oxo-abietatriene and 7α,11-dihydroxy-12-methoxy-abietatriene, isolated from the roots and aerial parts of *Plectranthus hereroensis* (Lamiaceae), were active against several Gram-positive and Gram-negative bacteria (Batista *et al.*, 1994; Dellar *et al.*, 1996). Totarol, isolated from the bark of *Podocarpus nagi* (Podocarpaceae), exhibited potent bactericidal activity against Gram-positive bacteria, among which *Propionibacterium acnes* was the most sensitive. Totarol also showed strong activity both against penicillin-resistant and penicillin-susceptible strains of *Staphylococcus aureus* (Kubo *et al.*, 1992). Later Haraguchi and co-workers (1996) studied the biological mechanism of totarol in *Pseudomonas aeruginosa*. It was shown that the compound inhibited oxygen consumption and respiratory-driven proton translocation in whole cells and oxidation of nicotinamide adenine dinucleotide (reduced form) (NADH) in membrane preparations. NADH-cytochrome c reductase, NADH-2,6-dichlorophenol indophenol (DCIP) reductase and NADH-coenzyme Q (CoQ) reductase were also inhibited.

Clerodane-type: from the leaves of *Premna schimperi* (Verbenaceae), 12-oxo-ent-3,13-clerodien-15-oic acid and, from the roots of *Croton sonderianus* (Euphorbiaceae), hardwickic acid and 3,4-seco-trachylobanoic acid were obtained. The diterpenes were active against *Staphylococcus aureus*, and the two latter compounds were also active against *Bacillus subtilis* (McChesney and Clark, 1991).

Ent-beyerenane-type: nine of the natural diterpenes, isolated from *Sideritis pusilla*, and five semisynthetic compounds were tested for antimicrobial activity. The natural compound, 1-acetyljativatriol, as well as the semisynthetic substances showed antimicrobial activity against Gram-positive bacteria. All compounds were inactive against Gram-negative bacteria. Studies on the structure-activity relationship revealed that the 12,17-dihydroxy group is responsible for the antimicrobial

Table 5.4 Secondary metabolites with antimicrobial activity

Compounds	Bacteria Gram (+)	Bacteria Gram (-)	Yeasts	Fungi	MIC µg/ml	References
Acetophenones						
xanthoxyline	+	-				De Godoy *et al.*, 1991
Anthraquinones						
rhein	+				*15.6–125*	Didry *et al.*, 1994
Apocarotenoids						
cochloxanthin		-	y	f	*500* (total inhibition)	Diallo *et al.*, 1991
dihydrocochloxanthin		-	y	f	*500* (total inhibition)	Diallo *et al.*, 1991
Benzophenones						
garcinol	+				*6.3–25.0*	Iinuma *et al.*, 1996
isogarcinol	+				*12.5–25.0*	Iinuma *et al.*, 1996
xanthochymol	+				*3.1–12.5*	Iinuma *et al.*, 1996
Benzoquinones						
2-hydroxy-5-methoxy-3-(8′*Z*,11,14′-pentadecatrienyl)-1,4-benzoquinone				f	6.0	Suzuki *et al.*, 1998
Chromenes						
methylripariochromene A				f		Bandara *et al.*, 1992
Coumarins						
angelicin				f	4160–6030	Afek *et al.*, 1995
bergapten				f	4090–5950	Afek *et al.*, 1995
columbianetin				f	25.0–48.0	Afek *et al.*, 1995
herniarin	+	-	y	f		Ceska *et al.*, 1992
8-methoxyporalen	+	-	y	f		Ceska *et al.*, 1992
psoralen				f	3850–5160	Afek *et al.*, 1995

Table 5.4 (Continued)

Compounds	Bacteria Gram (+)	Bacteria Gram (-)	Yeasts	Fungi	MIC μg/ml	References
umbelliferone	+	-	y	f		Ceska *et al.*, 1992
xanthotoxin				f	3920–4510	Afek *et al.*, 1995
Furanocoumarin-type						
oxypeucedanin				f	1 μg (minimum quantity for inhibition)	Marston *et al.*, 1995
oxypeucedanin hydrate				f	10 μg (minimum quantity for inhibition)	Marston *et al.*, 1995
Dihydrochalcones						
asebogenin	+	-			0.1–5.0 μg/TLC	Orjala *et al.*, 1994
2′, 6′-dihydroxy-4′-methoxy-dihydrochalcone	+	-			0.3–5.0 μg/TLC	Orjala *et al.*, 1994
2′,4′,6′-trihydroxydihydrochalcone				f		Miles *et al.*, 1991
piperaduncin A	+	-			1.6–5.0 μg/TLC	Orjala *et al.*, 1994
piperaduncin B	+	-			0.2–5.0 μg/TLC	Orjala *et al.*, 1994
piperaduncin C	+	-			1.5–5.0 μg/TLC	Orjala *et al.*, 1994
Diterpenes						
Abietane-type						
2,3-dehydrosalvipisone	+				10.5	Ulubelen *et al.*, 1994
7α,12-dihydroxy-17(15→16)-abeo-abieta-8,12,16-trien-11,14-dione	+	-	y		*15.6–31.2*	Batista *et al.*, 1994
7,11-dihydroxy-12-methoxy-abietatriene	+				*10.0–60.0*	Dellar *et al.*, 1996
forskalinone	+				168–670	Ulubelen *et al.*, 1996
horminone	+	-	y		*7.8–250*	Batista *et al.*, 1994
11-hydroxy-12-oxo-abietatriene	+				*10.0–40.0*	Dellar *et al.*, 1996; Batista *et al.*, 1994
hypargenin A	+				15.6	Ulubelen *et al.*, 1988
hypargenin B	+				125.0	Ulubelen *et al.*, 1988
hypargenin C	+				15.6–125	Ulubelen *et al.*, 1988

hypargenin D	+				62.5	Ulubelen *et al.*, 1988
hypargenin F	+	-			62.5–125	Ulubelen *et al.*, 1988
lanigerol	+				1000	El-Lakany *et al.*, 1995
manool	+				13.8	Ulubelen *et al.*, 1994
sclareol	+				48.3	Ulubelen *et al.*, 1994
totarol	+				*0.39–1.56*	Kubo *et al.*, 1992
totaradiol	+				*25–200*	Kubo *et al.*, 1992
Labdane-type						
labdane-14-ene-8,13-diol	+	-	y			Chinou *et al.*, 1994
labdane-13(*E*)-ene-8a,15-diol	+	-	y			Chinou *et al.*, 1994
labdane-13(*E*)-ene-8a-ol-15-yl acetate	+	-	y			Chinou *et al.*, 1994
labdane-7,13(*E*)-dien-15-ol	+	-	y			Chinou *et al.*, 1994
13-episclareol	+	-	y			Chinou *et al.*, 1994
8,13-epoxylabdan-14-en	+	-	y			Chinou *et al.*, 1994
8,13-epoxy-13-*epi*-labdan-14-en	+	-	y			Chinou *et al.*, 1994
Beyerenane-type						
1-acetyljativatriol	+				25.0–50.0	Diaz *et al.*, 1988
Clerodane-type						
clerodin				f		Cole *et al.*, 1991
hardwickic acid	+		y	f		McChesney and Clark, 1991
jodrellin A				f		Cole *et al.*, 1991
jodrellin B				f		Cole *et al.*, 1991
12-oxo-ent-3,13(16)-clerodien-15-oic acid	+				6.3–25.0	Habtemariam *et al.*, 1990
3,4-secotrachylobanoic acid	+		y	f		McChesney and Clark, 1991
Seco-kaurane-type						
trichorabdal A	+				*12.5–100*	Osawa *et al.*, 1994
trichorabdal B	+				*12.5–100*	Osawa *et al.*, 1994
trichorabdal C	+				*100–> 200*	Osawa *et al.*, 1994
trichorabdal H	+				*50–200*	Osawa *et al.*, 1994

Table 5.4 (Continued)

Compounds	Bacteria Gram (+)	Bacteria Gram (-)	Yeasts	Fungi	MIC μg/ml	References
Miscellaneous-type						
pseudolaric acid B			y		*0.78–12.5*	Li E. *et al.*, 1995
Flavonoids						
Catechin-type						
(-)-epigallocatechin	+	-			*50–100*	Mori *et al.*, 1987
Flavonol-type						
5,7-dihydroxy-3,8-dimethoxyflavone	+	-			20–50	Tomas-Lorente *et al.*, 1991
3,5-dimethoxy-6,7-methylene-dioxyflavone	+				20.0	Pomilio *et al.*, 1992
3,6-dimethoxy-5,7-dihydroxyflavone	+				15.0	Pomilio *et al.*, 1992
3′-7-di-*O*-methylquercetin				f		Miles *et al.*, 1993
5-hydoxy-3-methoxy-6,7-methylenedioxyflavone	+				75.0	Pomilio *et al.*, 1992
3,5,6,7,8-pentamethoxyflavone				f	5.0 per TLC	Tomas-Barberan *et al.*, 1988
3,5,6,7-tetramethoxyflavone	+			f	20.0	Tomas-Barberan *et al.*, 1988; Pomilio *et al.*, 1992
datisetin	+	-			*100*	Mori *et al.*, 1987
galangin	+				50.0	Pomilio *et al.*, 1992
isorhamnetin-3-*O*-robinobioside	+				50.0	Pomilio *et al.*, 1992
kaempferol	+				*50.0*	Pomilio *et al.*, 1992
3-*O*-methylquercetin	+	-	y	f	*6.3–100*	van Puyvelde *et al.*, 1989
morin	+	-			*100*	Mori *et al.*, 1987
myricetin	+	-			*50–100*	Mori *et al.*, 1987
platanoside	+	-				Mitrokotsa *et al.*, 1993
quercetagetin	+	-			*100*	Mori *et al.*, 1987
quercetagetin-7-arabinosyl-galactoside	+	-			38–130	Tereschuk *et al.*, 1997

quercetin 3-*O*-rhamnoside	+	-			Hasan and Ahmad, 1996
quercetin 3-*O*-glucosyl-(1→ 4)-galactoside	+	-			Hasan and Ahmad, 1996
robinetin	+	-		*100*	Mori *et al.*, 1987
rutin		-		32.0	Bernard *et al.*, 1997
tiliroside	+	-			Mitrokotsa *et al.*, 1993
Flavone-type					
7,8-dihydroxyflavone	+	-		*100*	Mori *et al.*, 1987
5,7-dihydroxy-6-methoxyflavone	+			15.0	Pomilio *et al.*, 1992
5,6-dimethoxy-7-hydroxyflavone	+			75.0	Pomilio *et al.*, 1992
5,6,7,8-tetramethoxyflavone			f	2.0 per TLC plate	Tomas-Barberan *et al.*, 1988
dimethylchrysin			f	1.0 per TLC plate	Tomas-Barberan *et al.*, 1988
flavone			f		Weidenbörner *et al.*, 1990
trimethylgalangin			f	1.0 per TLC plate	Tomas-Barberan *et al.*, 1988
Flavanone-type					
abyssinone-V	+	-		2.9–26.4	Ratsimamanga-Urverg *et al.*, 1994
7,4′-dihydroxyflavan			f		Achenbach *et al.*, 1988
euchrestaflavanone A	+	-		5.58–23.53	Ratsimamanga-Urverg *et al.*, 1994
exiguaflavanone B	+			50.0	Iinuma *et al.*, 1994
exiguaflavanone D	+			1.56–6.25	Iinuma *et al.*, 1994
flavanone			f		Weidenbörner *et al.*, 1990
naringenin	+				Osawa *et al.*, 1992
Flavanonol-type					
(+)-dihydrorobinetin	+	-		*200*	Mori *et al.*, 1987
3′,4′-dihydroxy-7-methoxyflavan			f		Achenbach *et al.*, 1988
7,4′-dihydroxy-3′-methoxyflavan			f		Achenbach *et al.*, 1988
7,3′-dimethoxy-4′-hydroxyflavan			f		Achenbach *et al.*, 1988
obtustyrene			f		Achenbach *et al.*, 1988

Table 5.4 (Continued)

Compounds	Bacteria Gram (+)	Bacteria Gram (-)	Yeasts	Fungi	MIC μg/ml	References
Isoflavan-type						
5,7-dihydroxy-4′-hydroxyisoflavan				f		Weidenbörner *et al.*, 1990
6,7-dihydroxy-4′-methoxyisoflavan				f		Weidenbörner *et al.*, 1990a,b
5,7-dihydroxy-4′-methoxyisoflavan				f		Weidenbörner *et al.*, 1990a,b
Isoflavanone-type						
biochanin A				f		Weidenbörner *et al.*, 1990a,b
darbergioidin	+					Osawa *et al.*, 1992
desmodianone A	+		y		*1.0–100*	Delle Monache *et al.*, 1996
desmodianone B	+	-	y		*1.0–100*	Delle Monache *et al.*, 1996
dihydrobiochanin A	+					Osawa *et al.*, 1992
dihydrogenistein	+					Osawa *et al.*, 1992
dihydrocajanin	+					Osawa *et al.*, 1992
2,7-dihydroxy-3(3′-methoxy-4′-hydroxy)-5-methoxy-isoflavone				f		Miles *et al.*, 1993
ferreirin	+					Osawa *et al.*, 1992
Lignans						
Aryltetralin-type						
4′-*O*-demethyldehydropodophyllotoxin				f		Rahman *et al.*, 1995
picropodophyllone				f		Rahman *et al.*, 1995
Cyclolignan-type						
galbulin				f		Sartorelli *et al.*, 1998
oleiferin-B				f		Sartorelli *et al.*, 1998
oleiferin-G				f		Sartorelli *et al.*, 1998
oleiferin-F				f		Sartorelli *et al.*, 1998
oleiferin-H				f		Sartorelli *et al.*, 1998
verrucosin				f		Sartorelli *et al.*, 1998

Limonoids						
limonoid glycoside	+	-				Srivastava, 1986
mahmoodin	+	-				Siddiqui *et al.*, 1992
naheedin	+	-				Siddiqui *et al.*, 1992
Monoterpenes						
camphene	+		y	f		Tirillini *et al.*, 1996
carvacrol	+	-			*1.0–5.0* mM	Lucchini *et al.*, 1990; Didry *et al.*, 1993
carvone	+	-	y	f	*1.56–100* µl/ml	Hinou *et al.*, 1989; Naigre *et al.*, 1996
carvone	+	-	y	f	2500–12500	Aboul Ela *et al.*, 1996
1,8-cineole	+	-	y	f	*0.4–6.6*	Pattnaik *et al.*, 1997
1,8-cineole	+	-	y		6250–25000	Aboul Ela *et al.*, 1996
citral	+	-	y	f	*0.1–6.6*	Pattnaik *et al.*, 1997
citral	+	-				Hinou *et al.*, 1989
geraniol	+	-			*125–375*	Hinou *et al.*, 1989; Chinou *et al.*, 1996
geraniol	+	-	y	f	*0.3–5.0*	Pattnaik *et al.*, 1997
geranyl acetate	+				*250–750*	Hinou *et al.*, 1989; Chinou *et al.*, 1996
isobornyl acetate	+					Hinou *et al.*, 1989
isolimonene	+	-	y	f	*0.78–100* µl/ml	Naigre *et al.*, 1996
isomenthone	+	-				Economou and Nahrstedt, 1991
isopulegol	+	-	y	f	*0.78–12.5* µl/ml	Naigre *et al.*, 1996
limonene			y		*100* µl/ml	Naigre *et al.*, 1996
linalool	+	-				Hinou *et al*, 1989
linalool	+	-	y	f	2500–6250	Aboul Ela *et al.*, 1996
linalool	+	-	y	f	*0.2–6.6*	Pattnaik *et al.*, 1997
linalyl acetate	+	-				Hinou *et al.*, 1989
menthene	+	-				Hinou *et al.*, 1989
menthenefuran	+					Hinou *et al.*, 1989

Table 5.4 (Continued)

Compounds	Bacteria Gram (+)	Bacteria Gram (-)	Yeasts	Fungi	MIC μg/ml	References
menthone	+	-				Hinou *et al.*, 1989; Economou and Nahrstedt, 1991
menthyl acetate	+					Hinou *et al.*, 1989
myrtenal	+					Hinou *et al.*, 1989
neryl acetate	+	-			*250–750*	Hinou *et al.*, 1989; Chinou *et al.*, 1996
β-pinene	+					Hinou *et al.*, 1989
piperitone	+					Hinou *et al.*, 1989
pulegone	+	-	y	f	*0.4–85.0%* of oil	Hinou *et al.*, 1989; Economou and Nahrstedt, 1991; Kalodera *et al.*, 1994
sabinene	+					Hinou *et al.*, 1989
terpinolene	+	-				Hinou *et al.*, 1989
terpineol	+	-				Hinou *et al.*, 1989
terpinyl acetate	+	-				Hinou *et al.*, 1989
tetrahydrogeraniol	+					Hinou *et al.*, 1989
thymol	+	-			*1.0–4.0* mM	Lucchini *et al.*, 1990; Didry *et al.*, 1993
thymol	+				*50–100*	Osawa *et al.*, 1994
Naphthoquinones						
cassiaside B	+				10.0	Messana *et al.*, 1991
dehydro-α-lapachone	+				100–200	Binutu *et al.*, 1996
dehydro-α-lapachone			y	f	200–400	Binutu *et al.*, 1996
isopinnatal	+				100–200	Binutu *et al.*, 1996
isopinnatal			y	f	200–400	Binutu *et al.*, 1996
juglone	+				*125–500*	Didry *et al.*, 1994
kigelinone	+				100	Binutu *et al.*, 1996

kigelinone			y	f	100–200	Binutu *et al.*, 1996
lapachol	+				100–200	Binutu *et al.*, 1996
			y	f	200–400	Binutu *et al.*, 1996
lawsone	+				*125–500*	Didry *et al.*, 1994
plumbagin	+				*1.0–250*	Didry *et al.*, 1994
quinquangulin-6-*O*-apiofuranosyl-(1→6)-*O*-glucopyranoside	+		y		100	Messana *et al.*, 1991
rubrofusarin-6-*O*-glucopyranoside	+				10.0	Messana *et al.*, 1991
Phenanthrenes						
9,10-Dihydrophenanthrene type						
desvinyljuncusol	+		y		*6.25–50.0*	Boger *et al.*, 1985
2-hydroxy-3-methyl-9,10-dihydrophenanthrene	+		y		*3.12–25.0*	Boger *et al.*, 1985
juncusol	+		y		*12.5–25.0*	Boger *et al.*, 1985
Phenolic compounds						
caffeic acid	+		y	f	100–200	Binutu *et al.*, 1996
ferulic acid	+		y	f	100–200	Binutu *et al.*, 1996
gingerenone A				f	10.0 ppm	Endo *et al.*, 1990
4-hydroxystyrene	+	-		f	32–500	Kobayashi *et al.*, 1996
knerachelin A	+				*8.0–32.0*	Zahir *et al.*, 1993
knerachelin B	+				*4.0–16.0*	Zahir *et al.*, 1993
3-methoxy-4-acetoxystyrene	+	-		f	32–500	Kobayashi *et al.*, 1996
	+	-		f	65–1000	Kobayashi *et al.*, 1996
p-coumaric acid	+		y	f	200–400	Binutu *et al.*, 1996
Phenylpropanoids						
1-allyl-2,6-dimethoxy-3,4-methylenedioxybenzene	+	-			100 ppm	Masuda *et al.*, 1991
colenemal	+					Brader *et al.*, 1997
elemicin				f	20 μg (minimum quantity for inhibition)	Marston *et al.*, 1995

Table 5.4 (Continued)

Compounds	Bacteria Gram (+)	Bacteria Gram (-)	Yeasts	Fungi	MIC μg/ml	References
honokiol	+	-			*20.0–80.0*	Chang *et al.*, 1998
trans-isoelemicin				f	8.0 μg (minimum quantity for inhibition)	Marston *et al.*, 1995
magnolol	+	-			*20–160*	Chang *et al.*, 1998
myristicin				f	8.0 μg (minimum quantity for inhibition)	Marston *et al.*, 1995
plicatin B	+		y		12.5–50.0	Schmitt *et al.*, 1991
precolpuchol	+			f		Brader *et al.*, 1997
Phloroglucinols						
hyperbrasilol A	+					Rocha *et al.*, 1995
isouliginosin B	+					Rocha *et al.*, 1995
japonicine A	+					Rocha *et al.*, 1995
uliginosin A	+					Rocha *et al.*, 1995
Resorcinol						
malabaricone B	+		y		1.0–16.0	Orabi *et al.*, 1991
malabaricone C	+		y		2.0–32.0	Orabi *et al.*, 1991
Sesquiterpenoids						
allo-aromadendrane-10β,14-diol	+	-				De Siqueira *et al.*, 1997
argentone				f		Maatooq *et al.*, 1996
carisone				f		Maatooq *et al.*, 1996
caryophyllene oxide	+				13.8	Ulubelen *et al.*, 1994
cernuol	+	-	y	f	*5.0–200*	Smirnov *et al.*, 1998
dentatin A	+	-				Gören *et al.*, 1990
glaucolide A	+					Montanaro *et al.*, 1996
8α-hydroxyanhydroverlotorin	+	-				Gören *et al.*, 1990
15-hydroxyargentone				f		Maatooq *et al.*, 1996

hymenin	+	-				Montanaro *et al.*, 1996
6-*O*-isobutyroylplenolin	+				*38–300*	Taylor and Towers, 1998
isospeciformin	+	-				Gören *et al.*, 1990
6-*O*-methylacrylylplenolin	+				*75–300*	Taylor and Towers, 1998
15-nor-argentone				f		Maatooq *et al.*, 1996
8-oxo-argentone				f		Maatooq *et al.*, 1996
8-oxo-15-nor-argentone				f		Maatooq *et al.*, 1996
6-*O*-angeloylplenolin	+				*75–300*	Taylor and Towers, 1998
spathulenol	+				136.0	Ulubelen *et al.*, 1994
tabulin	+	-				Gören *et al.*, 1990
tanachin	+	-				Gören *et al.*, 1990
vernodalin	+	-	y	f		Al Magboul *et al.*, 1997
vernoleptin	+	-	y	f		Al Magboul *et al.*, 1997
Steroids						
cryptanoside A	+					Vasanth *et al.*, 1997
cryptanoside C	+	-				Vasanth *et al.*, 1997
20-hydroxyecdysone	+	-		f		Ahmad *et al.*, 1996
stigmasterol	+	-				Srivastava *et al.*, 1997
Stilbenes						
(*E*)-3-chloro-4-stilbenol				f	40–43 ppm	Schultz *et al.*, 1992
(*E*)-3,5-dimethoxy-4-stilbenol				f	49–60 ppm	Schultz *et al.*, 1992
(*E*)-3,5-dimethoxystilbene				f	90 ppm	Schultz *et al.*, 1992
(*E*)-3-methoxy-4-stilbenol				f	40 ppm	Schultz *et al.*, 1992
(*Z*)-4-methoxy-3-stilbenol				f	8–64 ppm	Schultz *et al.*, 1992
(*E*)-5-methoxy-3-stilbenol				f	42–163 ppm	Schultz *et al.*, 1992
(*E*)-4-stilbenol				f	12–31 ppm	Schultz *et al.*, 1992
(*E*)-3-stilbenol				f	35–54 ppm	Schultz *et al.*, 1992
(*Z*)-3-stilbenol				f	25–75 ppm	Schultz *et al.*, 1992
(*E*)-3,4-stilbenediol				f	33–34 ppm	Schultz *et al.*, 1992
(*E*)-3,5-stilbenediol				f	29–140 ppm	Schultz *et al.*, 1992

Table 5.4 (Continued)

Compounds	Bacteria Gram (+)	Bacteria Gram (-)	Yeasts	Fungi	MIC μg/ml	References
Tannins						
1,2,3,4,6-penta-galloylglucose	+	-			*256–1024*	Burapadaja and Bunchoo, 1995
1,2,3,4,6-penta-galloylglucose			y		*512*	Burapadaja and Bunchoo, 1995
1,3,6-tri-galloylglucose	+	-			*460–1024*	Burapadaja and Bunchoo, 1995
chebulagic acid	+	-			*256–1024*	Burapadaja and Bunchoo, 1995
corilagin	+	-			*128–1024*	Burapadaja and Bunchoo, 1995
punicalagin	+	-			*256–1024*	Burapadaja and Bunchoo, 1995
Thiophenes						
5′-methyl-5-[4-(3-methyl-1-oxobutoxy)-1-butynyl]-2,2′-bithiophene				f	*200*	Ahmad and Alam, 1995
5′-hydroxymethyl-5-[butyl-3-en-1-yn]-2,2′-bithiophene isovaleroxy ester				f	*200*	Ahmad and Alam, 1995
Triterpenes						
Cucurbitane-type						
mormodicine I				f		Chandravadana *et al.*, 1997
mormodicine II				f		Chandravadana *et al.*, 1997
Cycloartane-type						
astrasieversianin II	+	-			2.0–10.0	Calis *et al.*, 1997a,b
astragaloside I	+	-			20–30	Calis *et al.*, 1997a,b
astragaloside II	+	-			20.0	Calis *et al.*, 1997a,b
astrasieversianin X	+	-			20–50	Calis *et al.*, 1997
astragaloside VI		-			50.0	Calis *et al.*, 1997
cyclocanthoside G		-			10.0	Calis *et al.*, 1997
(24*R*)-24,25-epoxycycloartan-3-one	+				*8.0*	Cantrell *et al.*, 1996
(3β,24*R*)-24,25-epoxycycloartan-3-ol	+				*8.0*	Cantrell *et al.*, 1996

Lupane-type						
ceanothic acid	+	-			*42–250*	Li *et al.*, 1997
27-hydroxy ceanothic acid	+	-			*562–875*	Li *et al.*, 1997
ceanothetric acid	+	-			*406–625*	Li *et al.*, 1997
Miscellaneous-tritepenes						
argentatine A	+	-	y			Martinez-Vazquez *et al.*, 1994
Oleanane-type						
arjungenin	+	-			*17.5*	Nandy *et al.*, 1997
arjungenin methylester	+	-	y		*3.0–6.4*	Nandy *et al.*, 1997
arjunglucoside	+	-			*38.0*	Nandy *et al.*, 1997
belleric acid	+	-			*3.0–10.0*	Nandy *et al.*, 1997
belleric acid methylester	+	-	y		*6.4–33.3*	Nandy *et al.*, 1997
bellericagenin A	+	-	y		*6.0–40.0*	Nandy *et al.*, 1997
bellericagenin B	+	-	y	f	*6.2–17.0*	Nandy *et al.*, 1997
bellericaside A	+	-	y		*7.0–25.0*	Nandy *et al.*, 1997
bellericaside B	+	-	y	f	*1.5–10.0*	Nandy *et al.*, 1997
bellericoside	+	-	y		*12.5–22.0*	Nandy *et al.*, 1997
cyclamin			y		*80–160*	Calis *et al.*, 1997b
cyclaminorin			y		*80–160*	Calis *et al.*, 1997b
deglucocyclamin			y		*80–160*	Calis *et al.*, 1997b
dillinic acid A	+	-				Nick *et al.*, 1994
dillinic acid B	+	-				Nick *et al.*, 1994
dillinic acid C	+	-				Nick *et al.*, 1994
3β-hydroxyolean-12-en-28-oic acid				f	500	Verma *et al.*, 1998
3β-hydroxyolean-12-en-28-oic acid methyl ester				f	600	Verma *et al.*, 1998
3β-24-dihydroxyolean-12-en-28-oic acid				f	500–600	Verma *et al.*, 1998
3β-24-dihydroxyolean-12-en-28-oic acid methyl ester				f	600–1000	Verma *et al.*, 1998

Table 5.4 (Continued)

Compounds	Bacteria Gram (+)	Bacteria Gram (-)	Yeasts	Fungi	MIC µg/ml	References
pulsatilla saponin D			y	f	*8.0–250*	Ekabo and Farnsworth, 1996
salzmannianoside A			y	f	*8.0–125*	Ekabo and Farnsworth, 1996
salzmannianoside B			y	f	*8.0–250*	Ekabo and Farnsworth, 1996
tomentosic acid methylester	+	-			*8.3–36.5*	Nandy *et al.*, 1997
Phenol nor-triterpene-type						
3-*O*-methyl-6-oxo-tingenol	+				*35–39*	Gonzalez *et al.*, 1996
6-oxo-iguesterol	+				*25.0*	Gonzalez *et al.*, 1996
6-oxo-tingenol	+				*12.0–14.0*	Gonzalez *et al.*, 1996
Ursane-type						
22β-acetoxylantic acid	+	-	y	f		Barre *et al.*, 1997
rubrinol	+	-			*130–200*	Akhtar *et al.*, 1994
Xanthones						
BR-xanthone A				f		Gopalakrishnan *et al.*, 1997
garcinone D				f		Gopalakrishnan *et al.*, 1997
gartanin				f		Gopalakrishnan *et al.*, 1997
mangostin				f		Gopalakrishnan *et al.*, 1997
γ-mangostin				f		Gopalakrishnan *et al.*, 1997

Values in italics indicate agar/broth dilution method; the others using disc/agar diffusion method. Abbreviations: MIC, minimal inhibitory concentration; Gram (+), Gram-positive bacteria; Gram (-), Gram-negative bacteria; TLC, thin layer chromatographic plate; inhibitory concentration; Gram.

activity of these compounds. In contrast, if a hydroxy group was present at C-1, no antimicrobial activity was observed (Diaz *et al.*, 1988).

Labdane-type: some labdane-type diterpenes, isolated from *Aframomum aulacocarpos* and *Cistus incanus*, demonstrated only weak activity against Gram-positive and Gram-negative bacteria (Chinou *et al.*, 1994).

Seco-kaurane-type: trichorabdal A and B, isolated from the leaves of *Rabdosia trichocarpa*, have antimicrobial effects against Gram-positive and Gram-negative periodontopathic bacteria. Both compounds completely inhibited the growth of the Gram-negative bacterium, *Porphyromonas gingivalis*, at 12.5 μg/ml (Osawa *et al.*, 1994).

5.3.2.4 Flavonoids

Flavonoids are known to demonstrate a variety of biological activities, including antithrombic effects, anti-inflammatory and antispasmodic actions, antiviral, antifungal, antibacterial and antitumoral activities, and diuretic properties. They also play an important role in normal plant growth and development and defence against infection and injury (Cody *et al.*, 1986, 1988). Whereas the antibacterial and antifungal activities of the flavonoids have been reported repeatedly, little is known about the mode of action of the bioactive flavonoids. Mori and co-workers (1987) reported antibacterial activity, a structure-activity relationship, and the effects of several flavonoids (e.g. flavones, flavonols, flavanones, flavanonols and catechins) on DNA and ribonucleic acid (RNA) synthesis in *Proteus vulgaris* (Gram-negative) and *Staphylococcus aureus* (Gram-positive). Certain flavonols (e.g. myricetin, robinetin) were the most effective antibacterial agents. Flavanones and flavanonols were not at all effective. A free 3′,4′,5′-trihydroxy B-ring and a free 3-OH group are necessary for antibacterial activity (Table 5.4).

The antimicrobial activities of extracts and constituents of *Gomphrena martiona* and *Gomphrena boliviana* (Amaranthaceae) were determined, in order to identify the compounds responsible for the traditional medicinal use of these plants. A bioassay-guided fractionation of a petroleum ether extract yielded five 5,6,7-trisubstituted flavones showing high activity against *Mycobacterium phlei*, MIC_{50} 15.0–75.0 μg/ml, which approaches that of the commercial antibiotic streptomycin sulfate (MIC_{50} 10.0 μg/ml). Other natural and synthetic flavonoids with diverse structures were tested to define structure-activity relationships. For example, 5,6,7-trisubstituted flavones exhibited minor MIC_{50} values as compared to the 5,7-disubstituted flavones. The occurrence of the rare 5,6,7-trisubstituted flavones in both *Gomphrena* species may explain the medicinal use of these plants against bacterial diseases. Furthermore, the lipophilic flavonoids, located in the epidermis and/or cuticula of the leaves, may be responsible for the resistance of these species to plant diseases (Pomilio *et al.*, 1992).

Recently, Bernard *et al.* (1997) described, for the first time, a DNA topoisomerase inhibitor specific for topoisomerase IV. Three flavonoids were isolated from cottenseed flour, which promoted *E. coli* topoisomerase IV-dependent DNA cleavage. Rutin, the most active, inhibited topoisomerase IV-dependent decatenation activity (concentration required to produce half maximum inhibition [IC_{50}] 64 μg/ml). None of the flavonoids isolated had any stimulatory activity on *E. coli* DNA gyrase-dependent or calf thymus topoisomerase II-dependent DNA cleavage.

The isoflavanones, dihydrobiochanin A, ferreirin, darbergioidin and dihydrocajanin, isolated from *Swartzia polyphylla* (Leguminosae), had potent antibacterial activity against cariogenic bacteria (e.g. *Streptococcus cricetus, S. rattus, S. mutans*) (Osawa *et al.*, 1992).

5.3.2.5 Monoterpenoids

Several monoterpenes, all constituents of essential oils, were tested for antimicrobial activity against Gram-positive and Gram-negative bacteria, yeasts and filamentous fungi (e.g. Carson and Riley, 1995; Chinou *et al.*, 1996; Pattnaik *et al.*, 1997). Citral, geraniol, linalool, terpinen-4-ol and α-terpineol were the most effective antibacterial agents (see Table 5.4).

5.3.2.6 Miscellaneous phenolic compounds

Plants release a variety of organic compounds into the environment by leaf leachates and root exudates and through decomposition of litter. The growth of microorganisms in the rhizosphere may be profoundly controlled by these compounds. Certain classes of these compounds, predominantly phenols, phenolic acids and aromatic alcohols, have been reported in the literature to exert an antibacterial effect (see also Table 5.4). Ferulic, isovanillic, *p*-hydroxycinnamic, *p*-hydroxybenzoic, syringic, caffeic, gentisic, protocatechuic, *p*-coumaric, vanillic and *p*-hydroxybenzoic acid, isolated from different plant sources, have exhibited potent antibacterial activities. The mechanism of action of phenolic compounds is described as being nonspecific and resulting in alterations of the cytoplasmic membrane (Lucchini *et al.*, 1990; Binutu *et al.*, 1996; Fernandez *et al.*, 1996). Caffeic acid inhibited aflatoxin production of *Aspergillus flavus* without inhibiting the fungal growth. The compound also showed bactericidal activity towards *Pseudomonas aeruginosa* and *Staphylococcus aureus* (Paster *et al.*, 1988).

In a further study, Kobayashi *et al.* (1996) identified 4-hydroxystyrene, 3-methoxy-4-hydroxystyrene and 3-methoxy-4-acetoxystyrene as exudate components from wheat roots in sterile hydroponic culture. This indicates that these antimicrobial components may play a significant role in the defence system as allelochemicals for the rhizosphere.

The benzophenone derivatives, garcinol and isogarcinol, both isolated from the pericarps of *Garcinia purpurea* (Guttiferae), as well as xanthochymol, isolated from the pericarp of *G. subelliptica*, exhibited antibacterial properties against methicillin-resistant and methicillin-sensitive *Staphylococcus aureus*. Xanthochymol inhibited the methicillin-resistant bacterium in the concentration range 3.1–12.5 μg/ml. This MIC_{50} value was almost equal to that of the antibiotic vancomycin (6.26 μg/ml), which is currently used to treat methicillin-resistant infections (Iinuma *et al.*, 1996).

The naphthoquinones, plumbagin, juglone, lawsone, kigelinone, isopinnatal, dehydro-α-lapachone and lapachol, displayed a wide spectrum of antimicrobial activity (Didry *et al.*, 1994; Binutu *et al.*, 1996).

The dimeric phenylpropanoids, magnolol and honokiol, both isolated from stem bark of *Magnolia obovata*, displayed strong antimicrobial activity against the periodontopathic microorganisms, *Porphyromonas gingivalis*, *Prevotella gingivalis*, *Actinobacillus actinomycetemcomitans*, *Capnocytophaga gingivalis* and *Veillonella disper*. Their activities were similar to that of listerine (MIC: 10–80 μg/ml), a known oral antiseptic (Chang *et al.*, 1998).

5.3.2.7 Triterpenoids

During the search for antimicrobial compounds from higher plant sources, several bioactive triterpenes were isolated by bioassay-guided fractionation and purification of aqueous or alcohol plant extracts (see Table 5.4). A methanol extract of *Ceanothus americanus* demonstrated antimicrobial activity against selected oral pathogens. Ceanothic acid and ceanothetric acid isolated from the extract demonstrated growth inhibitory effects against *Streptococcus mutans*, *Actinomyces viscosus*, *Porphyromonas gingivalis* and *Prevotella intermedia*, with MIC values ranging 42.0 – 625.0 μg/ml (Li *et al.*, 1997). In a further study, Perese and co-workers (1997) isolated acetyl aleuritolic acid from the aqueous-ethanolic extract of *Croton urucurana* (Euphorbiaceae), a triterpene exhibiting antibacterial activity both against *Staphylococcus aureus* and *Salmonella typhimurium*. Furthermore, 22β-acetoxylantic acid, a triterpene derived from *Lantana camara*, was active against *Staphylococcus aureus* and *Salmonella typhyi* (Barre *et al.*, 1997).

5.4 Secondary metabolites from higher plants with antiviral properties

The search for selective antiviral agents, focused mainly on anti-human immunodeficiency virus (HIV) agents, has been vigorous in recent years but progress in the development of useful new antivirals has been slow. Meanwhile, the frequency of viral resistance to the relatively few antiviral drugs currently used is increasing (Mohrig, 1996). Furthermore, the treatment of viral infections is often unsatisfactory and new viral pathogens are likely to be discovered. There is a need to find new substances with not only intracellular but also extracellular virucidal properties. Most of the known antiseptics and disinfectants fail to kill all pathogens in a given time at room temperature (Vlietinck *et al.*, 1991). The methods commonly used for the evaluation of *in vitro* antiviral activities of synthetic and natural substances are based mainly on the inhibition of cytopathic effects, the reduction or inhibition of plaque formation and reduction in the virus yield, but also on other viral functions in selected host cells cultures (Vlietinck *et al.*, 1991).

Using bioassay-guided fractionation of bioactive plant extracts, it was found that various proteins, glycoproteins, polysaccharides and secondary metabolites are the most active antiviral substances of plant extracts. Secondary metabolites with antiviral properties originate in a whole range of substance classes, such as alkaloids, lignans, phenols, phenolic glycosides, quinones, flavonoids, coumarins, tannins, sesquiterpenes and saponins (Van den Berghe *et al.*, 1986; Vlietinck *et al.*, 1998). These compounds inhibit different stages in the replication of various viruses, e.g. virus adsorption, virus-cell fusion, translation, virus assembly and virus release. In the following review, the progress on this expanding scientific field is documented by the most important results published in the last decade (see Tables 5.7 and 5.8). Tables 5.5 and 5.6 include the abbreviations of the viruses and host cell systems reviewed. In the following sections, those substances whose mode of action is known are discussed in more detail.

5.4.1 Isolated secondary metabolites with anti-HIV properties

In the last 10 years, much effort has been made to find effective anti-HIV agents. The acquired immunodeficiency syndrome (AIDS) is a pandemic immunosuppressive disease, which results in life-threatening, opportunistic infections. Human immunodeficiency virus, a retrovirus, has been identified as the aetiological agent of AIDS. HIV infects human CD4+ lymphocytes, monocytes and macrophages. Direct binding between the viral envelope gp 120 glycoprotein and the CD4+ receptor is necessary to initiate a productive infection. The HIV-1 infected cells express viral gp 120 glycoproteins on their surface, which bind with the CD4+ receptors of infected cells, leading to the formation of giant cell complexes (syncytia). Subsequently, the syncytia undergo cytolysis and cells die within a few days.

Any novel synthetic or plant-derived anti-HIV compound that selectively interferes with HIV-1 replication, virus adsorption, syncytium

Table 5.5 Abbreviations of viruses reviewed

Double-stranded deoxyribonucleic acid (dsDNA) viruses		
Adenoviridae		
ADV	=	adenovirus
ADV31	=	adenovirus type 31
Herpesviridae		
BHV 1	=	bovine herpes virus type 1
BMV	=	bovine mammalitis virus
CMV	=	cytomegalovirus
EHV-1	=	equine herpes virus type 1
HCMV	=	human cytomegalovirus
HSV-1	=	herpes simplex virus type 1
HSV-2	=	herpes simplex virus type 2
MCMV	=	murine cytomegalovirus
PRV	=	pseudorabies virus
VZV	=	varicella-zoster virus
Poxviridae		
VV	=	vaccinia virus
Double-stranded ribonucleic acid (dsRNA) viruses		
Birnaviridae		
IPNV	=	infectious pancreatic necrosis virus
Reoviridae		
BRV	=	bovine rotavirus
(-)-Single-stranded DNA (ssDNA)/(+)-ssDNA		
Hepadnaviridae		
HBV	=	hepatitis B virus
(+)-Single-stranded ribonucleic acid (ssRNA) viruses		
Coronaviridae		
BCV	=	bovine coronavirus
Flaviviridae		
DV-4	=	dengue virus type 4
JEV	=	Japanese encephalitis virus
YFV	=	yellow fever virus
Picornaviridae		
COXB1	=	coxsackievirus B type 1
COXB2	=	coxsackievirus B type 2
EMCV	=	encephalomyocarditis virus
HRV-1B	=	human rhinovirus type 1B
HRV-2B	=	human rhinovirus type 2B
PCV	=	picornavirus
PMV	=	poliomyelitis virus
POLIO 1	=	poliovirus 1
RV-2	=	rhinovirus type 2
RV-1B	=	rhinovirus type 1B

Table 5.5 (Continued)

Retroviridae		
AMV	=	avian myeloblastosis virus
EAV	=	equine anaemia virus
HIV-1	=	human immunodeficiency virus type 1
HIV-2	=	human immunodeficiency virus type 2
MMLV	=	Moloney murine leukaemia virus
Mo-MuLV	=	Moloney murine leukaemia virus
Togaviridae		
SINV	=	Sindbis virus
SFVL10	=	Semliki forest L10 virus
SV	=	Sindbis virus
VEEV	=	Venezuelan equine encephalomyelitis virus
(-)-Single-stranded ribonucleic acid (ssRNA) viruses		
Bunyaviridae		
PTV	=	Punta Toro virus
RVFV	=	Rift Valley fever virus
SFV	=	sandfly fever virus
Orthomyxoviridae		
FPVA	=	avian influenza (fowl plague) virus A
INFA	=	influenza A virus
INFB	=	influenza B virus
Paramyxoviridae		
BPIV 3	=	bovine parainfluenza virus type 3
BRSV	=	bovine respiratory syncytial virus
MEVA	=	measles Edmonston A virus
Para-3	=	parainfluenza virus type 3
RSV	=	respiratory syncytial virus
Rhabdoviridae		
IHNV	=	infectious haematopoietic necrosis virus
VSV	=	vesicular stomatitis virus
OMV	=	Onorhynchus masou virus

formation, inhibition of HIV-1 reverse transcriptase (HIV-1 RT) or viral protease should be considered relevant in the treatment of AIDS. In a recently published review, Vlietinck and co-workers (1998) summarized the most important results of the research conducted in the last 10 yrs. It could be demonstrated that many compounds of plant origin inhibit different stages in the replication cycle of HIV: virus adsorption, e.g. chromone alkaloids, isoquinoline alkaloids, polyphenolics, flavonoids, coumarins, phenols, tannins and triterpenes; virus-cell fusion, e.g.

Table 5.6 Abbreviations of host cells reviewed

A-549	=	human adenocarcinoma A-549 cells
BHK-21	=	baby hamster kidney cells
BSC-1	=	monkey kidney epithelial cells
C-8166	=	human lymphoblastic cells
CEM-SS	=	human lymphoblastoid cells
CER	=	chicken embryo related cells
CV-1	=	African green monkey kidney cells
HEF	=	human embryonic fibroblast cells
HeLa	=	(Helen Lake) human epithelial cervical carcinoma cells
HEL	=	human embryonic lung cells
MA 104	=	monkey kidney cells
MDBK	=	Madin-Darby bovine kidney cells
MDCK	=	Madin-Darby canine kidney cells
MOLT-4	=	human leukaemic T-cells
MRC-5	=	human embryo lung cells
MT-4	=	human T-cell leukaemia virus-I (HTLV-I)-carrying cells
PLC/PRF/5	=	human hepatoma cells
PBM	=	primary human peripheral blood mononuclear cells
SC-1	=	mouse cells
3T3-L1	=	mouse cells
Vero	=	African green monkey kidney cells
VK	=	primary vervet monkey kidney cells

triterpenes; reverse transcription, e.g. benzophenanthridine alkaloids, protoberberine alkaloids, isoquinoline and quinoline alkaloids, coumarins, flavonoids, lactones, tannins, iridoids and triterpenes; proteolytic cleavage (protease inhibition), e.g. triterpenes, xanthones and coumarins; glycosylation, e.g. indolizidine alkaloids, piperidine alkaloids and pyrrolizidine alkaloids; integrase inhibition, e.g. flavonoids, tannins; and virus assembly and release, e.g. hypericin and pseudohypericin. The most important secondary metabolites with antiviral properties against HIV are summarized in Table 5.7 (see also Fig. 5.4).

5.4.1.1 Alkaloids

A number of chromone alkaloids were isolated from the rootbark of *Schumanniophyton magnificum*. Of all the compounds tested, schumannificine exhibited the strongest activity against HIV-1. Potent anti-herpes simplex virus (HSV)-1 activity was also observed for a number of its derivatives. It was assumed that the presence of a piperidine ring and unsubstituted hydroxy groups on the molecules is necessary for the anti-HIV activity, which is considered to be due to irreversible binding to gp

Table 5.7 Secondary metabolites with anti-human immunodeficiency virus (HIV) activity

Secondary metabolites	Viruses	Host cells/ biological targets	Evaluation of antiviral activity	EC_{50}/IC_{50} μM	References
Alkaloids					
Colchicine-type					
colchicine	HIV-1	H9 lymphocytes	p24	0.01	Tatematsu *et al.*, 1991
Flavonoid-type					
O-demethylbuchena-vianine	HIV-1	CEM-SS	XTT assay	0.26	Beutler *et al.*, 1992
N,O-didemethylbuchena-vianine	HIV-1	CEM-SS	XTT assay	142	Beutler *et al.*, 1992
Piperidone-type					
N-methylschummannificine	HIV-1	C8166	gp120	5.0	Houghton *et al.*, 1994
anhydroschummannificine	HIV-1	C8166	gp120	0.4	Houghton *et al.*, 1994
N-methylanhydroschummannificine	HIV-1	C8166	gp120	20.0	Houghton *et al.*, 1994
rohitukine	HIV-1	C8166	gp120	30.0	Houghton *et al.*, 1994
schummanificine	HIV-1	C8166	gp120	1.6	Houghton *et al.*, 1994
Pyridino-type					
schummanniophytine	HIV-1	C8166	gp120	8.0	Houghton *et al.*, 1994
isoschummanniophytine	HIV-1	C8166	gp120	80.0	Houghton *et al.*, 1994
N-methylschumman-niophytine	HIV-1	C8166	gp120	80.0	Houghton *et al.*, 1994
Quinoline-type					
buchapine	HIV-1	CEM-SS	XTT assay	0.94	McCormick *et al.*, 1996
buchapine	HIV-1	RT		12.0	McCormick *et al.*, 1996
3-(3-methyl-2-butenyl)-4-(3-methyl-2-butenyl)-oxyl-2-quinolinone	HIV-1	CEM-SS	XTT assay	1.64	McCormick *et al.*, 1996
	HIV-1	RT		8.0	McCormick *et al.*, 1996

Coumarins					
calanolide A	HIV-1	CEM-SS	XTT assay	0.1	Currens *et al.*, 1996
	HIV-1	C8166	XTT assay	2.0	Currens *et al.*, 1996
inophyllum B	HIV-1	RT		0.042	Taylor *et al.*, 1994
Diterpenes					
Abietane-type					
carnosolic acid	HIV-1	protease		0.08 μg/ml (IC_{90})	Paris *et al.*, 1993
rosmanol	HIV-1	protease		0.60 μg/ml (IC_{90})	Paris *et al.*, 1993
7-*O*-methylrosmanol	HIV-1	protease		1.5 μg/ml (IC_{90})	Paris *et al.*, 1993
7-*O*-ethylrosmanol	HIV-1	protease		1.7 μg/ml (IC_{90})	Paris *et al.*, 1993
Kauren-type					
tripterifordin	HIV-1	H9 lymphocytes	p24	1.0	Chen *et al.*, 1992a,b
Flavonoids					
Flavonol-type					
fisetin	HIV-1	H9 lymphocytes	p24	122	Hu *et al.*, 1994
galangin	HIV-1	H9 lymphocytes	p24	28.0	Hu *et al.*, 1994
hesperidin	HIV-1	H9 lymphocytes	p24	> 164	Hu *et al.*, 1994
kämpferol	HIV-1	protease		2.0 μg/ml	Mahmood *et al.*, 1996
	HIV-1	C8166	gp120	4.0 μg/ml	Mahmood *et al.*, 1996
kämpferol-3-glucoside	HIV-1	C8166	gp120	8.0 μg/ml	Mahmood *et al.*, 1996
kämpferol-3-(6-*p*-coumaroyl)-β-glucoside	HIV-1	C8166	gp120	8.0 μg/ml	Mahmood *et al.*, 1996
morin	HIV-1	H9 lymphocytes	p24	> 331	Hu *et al.*, 1994
myricetin	HIV-1	H9 lymphocytes	p24	35.0	Hu *et al.*, 1994
quercetin	HIV-1	protease		20.0 μg/ml	Mahmood *et al.*, 1996
	HIV-1	H9 lymphocytes	p24	132	Hu *et al.*, 1994
	HIV-1	C8166	gp120	20.0 μg/ml	Mahmood *et al.*, 1996
quercetin-3-*O*-β-galactopyranoside	HIV-1	integrase		64.6 μg/ml	Kim *et al.*, 1998

Table 5.7 (Continued)

Secondary metabolites	Viruses	Host cells/ biological targets	Evaluation of antiviral activity	EC_{50}/IC_{50} μM	References
quercetin-3-*O*-α-rhamnopyranoside	HIV-1	integrase		75.2 μg/ml	Kim *et al.*, 1998
quercetin-3-*O*-(2″-galloyl)-α-arabinopyranoside	HIV-1	integrase		18.1 μg/ml	Kim *et al.*, 1998
quercetin-3-*O*-(2″-galloyl)-β-galactopyranoside	HIV-1	integrase		27.9 μg/ml	Kim *et al.*, 1998
quercetin-3-*O*-(2″,6″-digalloyl)-β-galactopyranoside	HIV-1	integrase		24.2 μg/ml	Kim *et al.*, 1998
Flavone-type					
acacetinrhamnoglucosid	HIV-1	H9 lymphocytes	p24	> 231	Hu *et al.*, 1994
acacetingalactoside	HIV-1	H9 lymphocytes	p24	8.0	Hu *et al.*, 1994
apigenin	HIV-1	H9 lymphocytes	p24	9.0	Hu *et al.*, 1994
	HIV-1	RT		443	Lin *et al.*, 1997
apigeningalactosid	HIV-1	H9 lymphocytes	p24	61.0	Hu *et al.*, 1994
baicalin	HIV-1	CEM-SS	focal syncytium formation	25.0	Li B.-Q. *et al.*, 1993
	HIV-1	RT		0.015	Li B.-Q. *et al.*, 1993
chrysin	HIV-1	H9 lymphocytes	p24	5.0	Hu *et al.*, 1994
7,8-dihydroxyflavone	HIV-1	H9 lymphocytes	p24	10.0	Hu *et al.*, 1994
3-hydroxyflavone	HIV-1	H9 lymphocytes	p24	13.0	Hu *et al.*, 1994
4′,5,7-trihydroxyflavone	HIV-1	H9 lymphocytes	p24	92.0	Hu *et al.*, 1994
flavone	HIV-1	H9 lymphocytes	p24	50.0	Hu *et al.*, 1994
luteolin	HIV-1	H9 lymphocytes	p24	10.0	Hu *et al.*, 1994
Flavanone-type					
(+)-catechin	HIV-1	H9 lymphocytes	p24	> 345	Hu *et al.*, 1994
flavanone	HIV-1	H9 lymphocytes	p24	45.0	Hu *et al.*, 1994
Biflavonoid-type					
agathisflavone	HIV-1	RT		100	Lin *et al.*, 1997

agathisflavone	HIV-1	PBM		33.6	Lin *et al.*, 1997
amentoflavone	HIV-1	RT		119	Lin *et al.*, 1997
	HIV-1	PBM		94.0	Lin *et al.*, 1997
hinokiflavone	HIV-1	RT		62.0	Lin *et al.*, 1997
	HIV-1	PBM		4.1	Lin *et al.*, 1997
morelloflavone	HIV-1	RT		116.0	Lin *et al.*, 1997
	HIV-1	PBM		6.9	Lin *et al.*, 1997
robustaflavone	HIV-1	RT		65.0	Lin *et al.*, 1997
	HIV-1	PBM		> 100	Lin *et al.*, 1997
swertifrancheside	HIV-1	nucleic acid polymerase		42.9	Pengsuparp *et al.*, 1995
	HIV-2	nucleic acid polymerase		56.6	Pengsuparp *et al.*, 1995
Isoflavonoids					
Pterocarpan-type					
3-*O*-methylcalopocarpin	HIV-1	CEM-SS	pra	0.2 μg/ml	McKee *et al.*, 1997
sandwicensin	HIV-1	CEM-SS	pra	2.0 μg/ml	McKee *et al.*, 1997
Isoflavan-type					
5-deoxyglyasperin F	HIV-1	NIC primary anti-HIV screen	XTT assay		McKee *et al.*, 1997
5-hydroxyneobavaisoflavanone	HIV-1	NIC primary anti-HIV screen	XTT assay		McKee *et al.*, 1997
Lignans					
anolignan A	HIV-1	RT		60.4 μg/ml	Rimando *et al.*, 1994
anolignan B	HIV-1	RT		1072.0 μg/ml	Rimando *et al.*, 1994
interiotherin A	HIV-1	H9 lymphocytes	p24	3.1 μg/ml	Chen *et al.*, 1996
mal.4	HIV-1	PBM	p24	1 - 10 μg/ml	Gnabre *et al.*, 1995, 1996
phyllamycin B	HIV-1	RT		3.5	Chang *et al.*, 1995
	HIV-1	human DNA polymerase-α		289	Chang *et al.*, 1995
retrojusticin B	HIV-1	RT		5.49	Chang *et al.*, 1995
	HIV-1	human DNA polymerase-α		989	Chang *et al.*, 1995
schisantherin D	HIV-1	H9 lymphocytes	p24	0.5 μg/ml	Chen *et al.*, 1996

Table 5.7 (Continued)

Secondary metabolites	Viruses	Host cells/ biological targets	Evaluation of antiviral activity	EC_{50}/IC_{50} μM	References
Phenolics					
curcumin	HIV-1	integrase		40.0	Mazumder *et al.*, 1995
protolichesterinic acid	HIV-1	nucleic acid polymerase		24.3	Pengsuparp *et al.*, 1995
	HIV-2	nucleic acid polymerase		29.6	Pengsuparp *et al.*, 1995
putranjivain	HIV-1	RT		3.9	El-Mekkawy *et al.*, 1995
Sesquiterpenes					
peyssonol A	HIV-1	RT		38.7	Loya *et al.*, 1995
	HIV-2	RT		23.7	Loya *et al.*, 1995
peyssonol B	HIV-1	RT		34.5	Loya *et al.*, 1995
	HIV-2	RT		28.0	Loya *et al.*, 1995
Tannins					
gallic acid methyl ester	HIV-1	integrase		35.8 μg/ml	Kim *et al.*, 1998
1.6-digalloyl-glucose	HIV-1	RT		270	El-Mekkawy *et al.*, 1995
digallic acid	HIV-1	RT		200	El-Mekkawy *et al.*, 1995
punicalin	HIV-1	RT		8.0	Nonaka *et al.*, 1990
punicacortein	HIV-1	RT		5.0	Nonaka *et al.*, 1990
sanguin H-11	HIV-1	RT		20.0	Nonaka *et al.*, 1990
Caffeoylquinate-type					
3,4-di-caffeoylquinic acid	HIV-1	RT		19.4	Chang *et al.*, 1995
	HIV-1	human DNA polymerase-α		11.6	Chang *et al.*, 1995
3,5-di-caffeoylquinic acid	HIV-1	RT		1.16	Chang *et al.*, 1995
	HIV-1	human DNA polymerase-α		2.32	Chang *et al.*, 1995
methyl 3,4-di-caffeoyl-quinic acid	HIV-1	RT		94.0	Chang *et al.*, 1995
	HIV-1	human DNA polymerase-α		45.0	Chang *et al.*, 1995
methyl 3,5-di-caffeoyl-quinic acid	HIV-1	RT		1.7	Chang *et al.*, 1995

methyl 3,5-di-caffeoyl-quinic acid	HIV-1	human DNA polymerase-α		3.77	Chang *et al.*, 1995
Galloylglucose-type					
1,2,6-tri-*O*-galloyl-β-glucose	HIV-1	integrase		28.3 μg/ml	Kim *et al.*, 1998
1,2,3,4,6-penta-*O*-galloyl-β-glucose	HIV-1	integrase		28.0 μg/ml	Kim *et al.*, 1998
Galloylquinate-type					
3-galloylquinic acid	HIV-1	RT		72.6	Chang *et al.*, 1995
	HIV-1	human DNA polymerase-α		8.72	Chang *et al.*, 1995
3,4-di-galloylquinic acid	HIV-1	RT		7.81	Chang *et al.*, 1995
	HIV-1	human DNA polymerase-α		0.61	Chang *et al.*, 1995
3,5-di-galloylquinic acid	HIV-1	RT		1.31	Chang *et al.*, 1995
	HIV-1	human DNA polymerase-α		0.48	Chang *et al.*, 1995
3,5-di-galloyl-4-di-galloylquinic acid	HIV-1	RT		10.0	Nishizawa *et al.*, 1989
	HIV-1	H9 lymphocytes	p24	6.25	Nishizawa *et al.*, 1989
3,4-di-galloyl-5-di-galloylquinic acid	HIV-1	RT		10.0	Nishizawa *et al.*, 1989
	HIV-1	H9 lymphocytes	p24	6.25	Nishizawa *et al.*, 1989
3-di-galloyl-4,5-di-galloylquinic acid	HIV-1	RT		10.0	Nishizawa *et al.*, 1989
	HIV-1	H9 lymphocytes	p24	6.25	Nishizawa *et al.*, 1989
1,3,4,5-tetra-galloylquinic acid	HIV-1	RT		10.0	Nishizawa *et al.*, 1989
	HIV-1	H9 lymphocyte cells	p24	6.25	Nishizawa *et al.*, 1989
3,4,5-tri-galloylquinic acid	HIV-1	H9 lymphocyte cells	p24	6.52	Nishizawa *et al.*, 1989
	HIV-1	RT		0.08	Chang *et al.*, 1995
	HIV-1	human DNA polymerase-α		0.17	Chang *et al.*, 1995
Galloylshikimate-type					
3,4-di-galloylshikimic acid	HIV-1	RT		1.77	Chang *et al.*, 1995
	HIV-1	human DNA polymerase-α		1.05	Chang *et al.*, 1995
3,5-di-galloylshikimic acid	HIV-1	RT		0.52	Chang *et al.*, 1995
	HIV-1	human DNA polymerase-α		0.31	Chang *et al.*, 1995
3,4,5-tri-galloylshikimic acid	HIV-1	RT		0.10	Chang *et al.*, 1995
	HIV-1	human DNA polymerase-α		0.22	Chang *et al.*, 1995

Table 5.7 (Continued)

Secondary metabolites	Viruses	Host cells/ biological targets	Evaluation of antiviral activity	EC_{50}/IC_{50} μM	References
2-phenylethanol-(6-galloyl)-glucoside	HIV-1	C8166	gp120	40.0 μg/ml	Mahmood *et al.*, 1996
Triterpenes					
1β-hydroxyaleuteriolic acid 3-*p*-hydroxybenzoate	HIV-1	RT		3.7	Pengsuparp *et al.*, 1995
	HIV-2	RT		59.0	Pengsuparp *et al.*, 1995
betulinic acid	HIV-1	H9 lymphocytes	p24	1.4	Fujioka and Kashiwada, 1994
celasdine-B	HIV-1	H9 lymphocytes	p24	0.8 μg/ml	Kuo and Kuo, 1997
epipomolic acid	HIV-1	protease		17.9 μg/ml (inhibition: 42%)	Xu *et al.*, 1996
gleditsia saponin C	HIV-1	H9 lymphocytes	p24	1.1	Konoshima *et al.*, 1995
glycyrrhizin	HIV-1	MT-4	pra	150.0	Ito *et al.*, 1987
gymnocladus saponin G	HIV-1	H9 lymphocytes	p24	2.7	Konoshima *et al.*, 1995
maslinic acid	HIV-1	protease		17.9 μg/ml (inhibition: 100%)	Xu *et al.*, 1996
paltanic acid	HIV-1	H9 lymphocytes	p24	6.5	Fujioka and Kashiwada, 1994
salaspermic acid	HIV-1	H9 lymphocytes	p24	10.0	Chen *et al.*, 1992
salispermic acid	HIV-1	RT		16.0 μg/ml	Chen *et al.*, 1992
soyasaponin II	HIV-1	MT-4	pra	112.0	Hayashi *et al.*, 1997
suberosol	HIV-1	H9 lymphocytes	p24	3.0 μg/ml	Li *et al.*, 1993
tormetic acid	HIV-1	protease		17.9 μg/ml (inhibition: 49%)	Xu *et al.*, 1996
ursolic acid	HIV-1	protease		17.9 μg/ml (inhibition: 85%)	Xu *et al.*, 1996

Abbreviations: EC_{50}/IC_{50}, effective/inhibition concentration causing a 50% reduction in the viral cytopathic effect or in viral replication; p24, p24 antigen capture assay; gp 120, reduction of gp 120 production in cells; RT, reverse transcriptase; pra, plaque reduction assay; XTT, tetrazolium metabolic assay; NCI, National Cancer Institute. For abbreviations of the viruses and host cell systems reviewed see Tables 5.5 and 5.6.

Table 5.8 Secondary metabolites with antiviral activity

Secondary metabolites	Viruses	Host cells / biological targets	Evaluation of antiviral activity	EC_{50}/IC_{50} µg/ml	References
Alkaloids					
Acridone-type					
1,3-*O*-methyl-*N*-methyl-acridone	HSV-2	HEF	pra	4.9	Yamamoto *et al.*, 1989
acronycine	HSV-2	HEF	pra	3.3	Yamamoto *et al.*, 1989
5-methoxyacronycine	HSV-2	HEF	pra	5.5	Yamamoto *et al.*, 1989
dimethoxyacronycine	HSV-2	HEF	pra	6.5	Yamamoto *et al.*, 1989
atalaphillidine	HSV-2	HEF	pra	0.73	Yamamoto *et al.*, 1989
	HSV-2	HEF	pra	0.82	Yamamoto *et al.*, 1989
N-methylatalaphilline	HSV-2	HEF	pra	8.4	Yamamoto *et al.*, 1989
citracridone-I	HSV-2	HEF	pra	1.3	Yamamoto *et al.*, 1989
citrusinine-I	HSV-2	HEF	pra	0.74	Yamamoto *et al.*, 1989
	HSV-1	HEF	pra	0.56	Yamamoto *et al.*, 1989
	HSV-2	HEF	pra	0.74	Yamamoto *et al.*, 1989
	HCMV	HEF	pra	1.5	Yamamoto *et al.*, 1989
5-hydroxy-*N*-methyl-severifoline	HSV-2	HEF	pra	2.0	Yamamoto *et al.*, 1989
Isoquinoline-type					
deoxypancratistatin	JEV	Vero	mtt	0.48	Gabrielsen *et al.*, 1992
	YFV	Vero	mtt	0.4	Gabrielsen *et al.*, 1992
	DV-4	LLCMK2	pra	0.69	Gabrielsen *et al.*, 1992
	PTV	Vero	mtt	0.66	Gabrielsen *et al.*, 1992
	RVFV	Vero	pra	5.1	Gabrielsen *et al.*, 1992
	SFV	Vero	mtt	1.7	Gabrielsen *et al.*, 1992
cis-dihydronarciclasine	JEV	Vero	mtt	0.96	Gabrielsen *et al.*, 1992
	YFV	Vero	mtt	1.3	Gabrielsen *et al.*, 1992
	DV-4	LLCMK2	pra	2.5	Gabrielsen *et al.*, 1992
	PTV	Vero	mtt	2.2	Gabrielsen *et al.*, 1992
	RVFV	Vero	pra	1.4	Gabrielsen *et al.*, 1992
trans-dihydronarciclasine	JEV	Vero	mtt	0.004	Gabrielsen *et al.*, 1992
	YFV	Vero	mtt	0.003	Gabrielsen *et al.*, 1992
	DV-4	LLCMK2	pra	0.015	Gabrielsen *et al.*, 1992
	PTV	Vero	mtt	0.008	Gabrielsen *et al.*, 1992

Table 5.8 (Continued)

Secondary metabolites	Viruses	Host cells / Biological targets	Evaluation of antiviral activity	EC_{50}/IC_{50} μg/ml	References
isonarciclasine	JEV	Vero	mtt	0.72	Gabrielsen *et al.*, 1992
	YFV	Vero	mtt	0.22	Gabrielsen *et al.*, 1992
	DV-4	LLCMK2	pra	0.27	Gabrielsen *et al.*, 1992
	PTV	Vero	mtt	0.28	Gabrielsen *et al.*, 1992
	RVFV	Vero	pra	3.3	Gabrielsen *et al.*, 1992
lycorine	JEV	Vero	mtt	0.33	Gabrielsen *et al.*, 1992
	YFV	Vero	mtt	0.28	Gabrielsen *et al.*, 1992
	DV-4	LLCMK2	pra	0.24	Gabrielsen *et al.*, 1992
	PTV	Vero	mtt	0.50	Gabrielsen *et al.*, 1992
	RVFV	Vero	pra	0.93	Gabrielsen *et al.*, 1992
lycoricidine	JEV	Vero	mtt	0.056	Gabrielsen *et al.*, 1992
	YFV	Vero	mtt	0.053	Gabrielsen *et al.*, 1992
	DV-4	LLCMK2	pra	0.059	Gabrielsen *et al.*, 1992
	PTV	Vero	mtt	0.042	Gabrielsen *et al.*, 1992
	RVFV	Vero	pra	0.15	Gabrielsen *et al.*, 1992
	SFV	Vero	mtt	0.058	Gabrielsen *et al.*, 1992
narciclasine	JEV	Vero	mtt	0.008	Gabrielsen *et al.*, 1992
	YFV	Vero	mtt	0.006	Gabrielsen *et al.*, 1992
	DV-4	LLCMK2	pra	0.015	Gabrielsen *et al.*, 1992
	PTV	Vero	mtt	0.0074	Gabrielsen *et al.*, 1992
pancratistatin	JEV	Vero	mtt	0.022	Gabrielsen *et al.*, 1992
	YFV	Vero	mtt	0.016	Gabrielsen *et al.*, 1992
	DV-4	LLCMK2	pra	0.063	Gabrielsen *et al.*, 1992
	RVFV	Vero	pra	0.16	Gabrielsen *et al.*, 1992
pretazettine	JEV	Vero	mtt	0.60	Gabrielsen *et al.*, 1992
	YFV	Vero	mtt	0.50	Gabrielsen *et al.*, 1992
	PTV	Vero	mtt	0.61	Gabrielsen *et al.*, 1992
	RVFV	Vero	pra	2.9	Gabrielsen *et al.*, 1992
	SFV	Vero	mtt	0.82	Gabrielsen *et al.*, 1992

Miscellaneous alkaloids					
hippeastrine	HSV	Vero	pra	25.0 μg (100% plaque inhibition)	Renard-Nozaki *et al.*, 1989
Piperidiono-type					
rohitukine	HSV-1	Vero	vap	1.6 μM	Houghton *et al.*, 1994
schummannificine	HSV-1	Vero	vap	0.5 μM	Houghton *et al.*, 1994
N-methylschummannificine	HSV-1	Vero	vap	0.5 μM	Houghton *et al.*, 1994
anhydroschummannificine	HSV-1	Vero	vap	0.06 μM	Houghton *et al.*, 1994
N-methylanhydro-schummannificine	HSV-1	Vero	vap	0.5 μM	Houghton *et al.*, 1994
Pyridino-type					
schummanniophytine	HSV-1	Vero	vap	40.0 μM	Houghton *et al.*, 1994
isoschummanniophytine	HSV-1	Vero	vap	50.0 μM	Houghton *et al.*, 1994
N-methylschummanniophytine	HSV-1	Vero	vap	50.0 μM	Houghton *et al.*, 1994
Pyrrolidinoindoline-type					
hodgkinsine A	VSV	Vero	pra	10.0	Saad *et al.*, 1995
	HSV-1	Vero	pra	30.0	Saad *et al.*, 1995
Coumarins					
collinin	HBV	DNA replication		68.3	Chang *et al.*, 1997
oxynitidine	HBV	DNA replication		200	Chang *et al.*, 1997
Diterpenoids					
scopadulciol	HSV-1	HeLa	pra	0.016 μM	Hayashi and Hayashi, 1996
Flavonoids					
Flavan-type					
flavan	HRV-1B	HeLa	pra	0.27 μg/plate	Denyer *et al.*, 1994
4′,6-dichloroflavan	HRV-1B	HeLa	pra	0.02 μg/plate	Denyer *et al.*, 1994
Flavone-type					
isoscutellarein	INFA	sialidase		20.0 μM	Nagai and Miyaichi, 1992
	INFA	MDBK	mtt	16.0 nmol/well	Nagai and Miyaichi, 1992
isoscutellarein-8-methylether	INFA	sialidase		55.0 μM	Nagai and Miyaichi, 1992
	INFA/B	MDCK	mtt	20.0 μM	Nagai and Miyaichi, 1995

Table 5.8 (Continued)

Secondary metabolites	Viruses	Host cells / Biological targets	Evaluation of antiviral activity	EC_{50}/IC_{50} μg/ml	References
3-Methoxyflavone-type					
ternatin	ADV	Vero	pra	3.74	Simeos *et al.*, 1990
	HSV-1	Vero	pra	16.46	Simeos *et al.*, 1990
	HSV-2	Vero	pra	5.23–37.4	Simeos *et al.*, 1990
	VSV	Vero	pra	14.21–17.95	Simeos *et al.*, 1990
	POLIO	Vero	pra	1.5	Simeos *et al.*, 1990
Flavonol-type					
galangin	HSV-1	VK	pra	12–47	Meyer *et al.*, 1997
	COX B1	VK	pra	12–47	Meyer *et al.*, 1997
	ADV 31	PLC/PRF/5	pra	12–47	Meyer *et al.*, 1997
isoquercitrin	HSV-1	Vero	pra	40.0 μg (inhibition: 100%)	Abou-Karam and Shier, 1992
quercetin	potato virus X	plant tissue		1.0	French and Towers, 1992
Biflavonoid-type					
ginkgetin	HSV-1	HeLa	pra	0.91	Hayashi *et al.*, 1992
	HSV-1	Vero	pra	0.76	Hayashi *et al.*, 1992
	HSV-2	Vero	pra	0.83	Hayashi *et al.*, 1992
	HCMV	HEL	pra	1.75	Hayashi *et al.*, 1992
Lignans					
Lactone-type					
4′-demethylpodophyllotoxin	HSV-1	CV-1	pra	0.04	San Feliciano *et al.*, 1993
	VSV	BHK	pra	0.1	San Feliciano *et al.*, 1993
deoxypodophyllotoxin	HSV-1	CV-1	pra	0.01	San Feliciano *et al.*, 1993
	VSV	BHK	pra	< 0.01	San Feliciano *et al.*, 1993
deoxypicropodophyllotoxin	HSV-1	CV-1	pra	0.8	San Feliciano *et al.*, 1993
	VSV	BHK	pra	0.08	San Feliciano *et al.*, 1993
diphyllin	VSV	HEL	eptt	0.25	Asano *et al.*, 1996
	SINV	3T3-L1	pra	1.0	MacRae *et al.*, 1989
diphyllin apioside	VSV	HEL	eptt	0.25	Asano *et al.*, 1996
diphyllin apioside-acetate	VSV	HEL	eptt	0.13	Asano *et al.*, 1996
epipodophyllotoxin acetate	HSV-1	CV-1	pra	0.2	San Feliciano *et al.*, 1993
	VSV	BHK	pra	0.1	San Feliciano *et al.*, 1993

epipicropodophyllotoxin acetate	HSV	CV-1	pra	0.8	San Feliciano *et al.*, 1993
epipicropodophyllotoxin	VSV	BHK	pra	2.0	San Feliciano *et al.*, 1993
justicidin A	VSV	HEL	eptt	0.13	Asano *et al.*, 1996
justicidin B	VSV	HEL	eptt	0.06	Asano *et al.*, 1996
justicidin C	VSV	HEL	eptt	16.0	Asano *et al.*, 1996
justicidin D	VSV	HEL	eptt	16.0	Asano *et al.*, 1996
justicidin B	SINV	3T3-L1	pra	0.01	MacRae *et al.*, 1989
justicidinoside A	VSV	HEL	eptt	16.0	Asano *et al.*, 1996
justicidinoside B	VSV	HEL	eptt	125	Asano *et al.*, 1996
justicidinoside C	VSV	HEL	eptt	125	Asano *et al.*, 1996
α-peltatine	MCMV	3T3-L1	pra	0.01	MacRae *et al.*, 1989
β-peltatine A methyl ether	HSV-1	CV-1	pra	0.01	San Feliciano *et al.*, 1993
β-peltatine A methyl ether	VSV	BHK	pra	0.01	San Feliciano *et al.*, 1993
podophyllotoxin	MCMV	3T3-L1	pra	0.01	MacRae *et al.*, 1989
podophyllotoxione	HSV-1	CV-1	pra	0.4	San Feliciano *et al.*, 1993
podophyllotoxin	VSV	BHK	pra	1.0	San Feliciano *et al.*, 1993
Non-lactone-type					
acetyljunaphthoic acid	HSV-1	CV-1	pra	< 20.0	San Feliciano *et al.*, 1993
	VSV	BHK	pra	> 40.0	San Feliciano *et al.*, 1993
methylacetyljunaphthoate	HSV-1	CV-1	pra	20.0	San Feliciano *et al.*, 1993
	VSV	BHK	pra	20.0	San Feliciano *et al.*, 1993
methyljunaphthoate	HSV-1	CV-1	pra	< 20.0	San Feliciano *et al.*, 1993
	VSV	BHK	pra	10.0	San Feliciano *et al.*, 1993
rhinacanthin E	INFA	MDCK	pra	7.4	Kernan *et al.*, 1997
rhinacanthin F	INFA	MDCK	pra	3.1	Kernan *et al.*, 1997
Naphthodianthrones					
hypericin	HSV-1	BSC-1	pra	20	Tang *et al.*, 1990
	INFA	MDCK	pra	> 100	Tang *et al.*, 1990
	ADV	HeLa	pra	> 100	Tang *et al.*, 1990
	POLIO 1	BSC-1	pra	> 100	Tang *et al.*, 1990
	Mo-MuLV	SC-1	XC cell assay	6.0	Tang *et al.*, 1990

Table 5.8 (Continued)

Secondary metabolites	Viruses	Host cells / Biological targets	Evaluation of antiviral activity	EC_{50}/IC_{50} μg/ml	References
Phenolic compounds					
salicin	PMV	Vero	eptt	25.0	Van Hoof *et al.*, 1989
salireposide	PMV	Vero	eptt	25.0	Van Hoof *et al.*, 1989
	SFVL10	Vero	eptt	50.0	Van Hoof *et al.*, 1989
woodorien	HSV-1	Vero	pra	100 (plaque formation: 16%)	Xu *et al.*, 1993
Sesquiterpenes					
arvoside B	VSV	CER	pra	14.0	De Tommasi *et al.*, 1990
β-bisabolene	HRV-1B	HeLa	pra	14.4 μg/plate	Denyer *et al.*, 1994
ar-curcumene	HRV-1B	HeLa	pra	20.4 μg/plate	Denyer *et al.*, 1994
epicubebol glycoside	VSV	CER	pra	36.0	De Tommasi *et al.*, 1990
	HRV-1B	HeLa	pra	25.0	De Tommasi *et al.*, 1990
β-sesquiphellandrene	HRV-IB	HeLa	pra	0.90 μg/plate	Denyer *et al.*, 1994
α-zingiberene	HRV-1B	HeLa	pra	1.9 μg/plate	Denyer *et al.*, 1994
Tannins					
casuaricitin	HSV-1	CV-1	pra	0.044	Fukuchi *et al.*, 1989
coriariin A	HSV-1	CV-1	pra	0.038	Fukuchi *et al.*, 1989
cornusiin A	HSV-1	CV-1	pra	0.039	Fukuchi *et al.*, 1989
geraniin	HSV-1	CV-1	pra	0.093	Fukuchi *et al.*, 1989
oenothein B	HSV-1	CV-1	pra	0.036	Fukuchi *et al.*, 1989
penta-galloylglucose	HSV-1	CV-1	pra	0.047	Fukuchi *et al.*, 1989
rugosin D	HSV-1	CV-1	pra	0.034	Fukuchi *et al.*, 1989
tellimagrandin I	HSV-1	CV-1	pra	0.036	Fukuchi *et al.*, 1989
4,8-tetramer of epicatechin gallate	HSV-1	CV-1	pra	0.14	Fukuchi *et al.*, 1989
tannic acid	HSV-1	CV-1	pra	0.034	Fukuchi *et al.*, 1989
	HSV-1	CV-1	pra	0.15	Fukuchi *et al.*, 1989
	HSV-2	CV-1	pra	0.15	Fukuchi *et al.*, 1989
	HSV-1	Vero	pra	0.086	Fukuchi *et al.*, 1989
	HSV-1	Vero	pra	0.043	Fukuchi *et al.*, 1989

	HSV-2	Vero	pra	0.041	Fukuchi *et al.*, 1989
	HSV-1	A-549	pra	0.061	Fukuchi *et al.*, 1989
Triterpenes					
Oleanane-type					
glycyrrhizin	VZV	HEF	immunoperoxidase method	710 μM	Baba and Shigeta, 1987
hydroxyoleanonic lactone	HSV-1	Vero	pra	5.0	Poehland *et al.*, 1987
	HSV-2	Vero	pra	5.0	Poehland *et al.*, 1987
soyasaponin II	HSV-1	HeLa	pra	54 μM	Hayashi *et al.*, 1997
	HCMV	HEL	pra	104 μM	Hayashi *et al.*, 1997
	POLIO 1	Vero	pra	> 1000 μM	Hayashi *et al.*, 1997
	INF	MDCK	pra	88 μM	Hayashi *et al.*, 1997
Ursane-type					
ursonic acid	HSV-1	Vero	pra	2.5	Poehland *et al.*, 1987
	HSV-2	Vero	pra	8.0	Poehland *et al.*, 1987
quinovic acid					
-3β-*O*-(β-fucopyranosyl)-4-β-D-glucopyranosid	VSV	CER	pra	20.0	Aquino *et al.*, 1989
-3β-*O*-(β-quinovopyranosyl)-(28-→1)-β-glucopyranosyl ester	VSV	CER	pra	22.4	Aquino *et al.*, 1989
-3β-*O*-(β-fucopyranosyl)-(28→1)-β-glucopyranosyl ester	VSV	CER	pra	31.6	Aquino *et al.*, 1989
-3β-*O*-(β-glucopyranosyl)-(28→1)-β-glucopyranosyl ester	VSV	CER	pra	31.0	Aquino *et al.*, 1989
-3β-*O*-glucopyranoside	VSV	CER	pra	33.1	Aquino *et al.*, 1989
-3β-*O*-(β-glucopyranosyl)-(27→1)-β-glucopyranosyl ester	VSV	CER	pra	70.8	Aquino *et al.*, 1989
-3β-*O*-(β-glucopyranosyl)-(28→1)-β-glucopyranosyl ester	VSV	CER	pra	33.1	Aquino *et al.*, 1989
Dammarane-type					
dammaradienol	HSV-1	Vero	pra	2.5	Poehland *et al.*, 1987
	HSV-2	Vero	pra	3.0	Poehland *et al.*, 1987

Table 5.8 (Continued)

Secondary metabolites	Viruses	Host cells / Biological targets	Evaluation of antiviral activity	EC_{50}/IC_{50} µg/ml	References
dammarenediol-II	HSV-1	Vero	pra	7.0	Poehland *et al.*, 1987
	HSV-2	Vero	pra	7.0	Poehland *et al.*, 1987
dammarenolic acid	HSV-1	Vero	pra	3.0	Poehland *et al.*, 1987
	HSV-2	Vero	pra	2.0	Poehland *et al.*, 1987
hydroxydammarenone	HSV-1	Vero	pra	2.0	Poehland *et al.*, 1987
	HSV-2	Vero	pra	5.0	Poehland *et al.*, 1987
Hopanane-type					
hydroxyhopanone	HSV-1	Vero	pra	7.0	Poehland *et al.*, 1987
	HSV-2	Vero	pra	5.0	Poehland *et al.*, 1987
Miscellaneous triterpenes					
eichlerianic acid	HSV-1	Vero	pra	7.0	Poehland *et al.*, 1987
	HSV-2	Vero	pra	8.0	Poehland *et al.*, 1987
shoreic acid	HSV-1	Vero	pra	7.0	Poehland *et al.*, 1987
	HSV-2	Vero	pra	8.0	Poehland *et al.*, 1987

Abbreviations: EC_{50}/IC_{50}, effective/inhibition concentration for a 50% reduction in the viral cytopathic effect or in viral replication. pra, plaque reduction assay; mtt, colorimetric method to determine viable cells; vap, viral antigen production in cells; eptt, endpoint titration technique. For abbreviations of the viruses and host all systems reviewed see Tables 5.5 and 5.6.

120 rather than to inhibition of reverse transcriptase or protease (Houghton *et al.*, 1994).

5.4.1.2 Coumarins

Novel HIV-1-RT inhibitory coumarin derivatives were recently isolated from the tropical rainforest trees, *Calaphyllum lanigerum* and *Calaphyllum inophyllum*. Calanolides A and B, as well as inophyllum B, have been established as non-nucleoside-specific inhibitors of HIV-1-RT. It has been shown that non-nucleoside-specific HIV-1-RT inhibitors act as noncompetitive inhibitors of this enzyme. Calanolide A inhibited the *in vitro* replication of HIV-1 in human lymphoblastoid cells (CEM-SS), but was inactive against HIV-2 and avian myeloblastosis virus (AMV) (Currens *et al.*, 1996). Boyer and co-workers (1993) showed that both calanolide A and B effectively blocked the DNA-dependent DNA polymerase as well as the RNA-dependent DNA-polymerase activity of HIV-1-RT. HIV-1-RT activity was also inactivated by inophyllum B (Taylor *et al.*, 1994). In further studies, calanolide A similarly inhibited promonocytotropic isolates (effective concentration for a 50% reduction in the viral cytopathic effect or viral replication [EC_{50}] 0.12–0.18 μM) and lymphotrophic isolates (EC_{50} 0.15–0.47 μM) from patients with various stages of HIV-1 disease, as well as drug-resistant strains. In enzyme inhibition assays, calanolide A potently and selectively inhibited HIV-1-RT but not cellular DNA polymerase or HIV-2-RT within the concentration range tested (Currens *et al.*, 1996).

5.4.1.3 Flavonoids

Plant flavonoids are a large group of naturally-occurring phenylchromones found in the leaves, stems, flowers and fruits of most of the higher plants. A variety of *in vitro* and *in vivo* experiments have shown that selected flavonoids exhibit antiallergic, anti-inflammatory, antiviral, and antioxidant activities (Kaul *et al.*, 1985). Several known flavones were examined for their anti-HIV activity. Chrysine was found to be the most active compound in this series (EC_{50} 5.0 μM). Flavonoids with hydroxy groups at C5 and C7 and with a C2 $\rightarrow$ C3 double bond were more active than the others. Based on the results obtained, it was postulated that the presence of substituents (hydroxyl and halogen groups) in the B-ring leads to an increased toxicity and/or decreased activity of flavonoids in general (Hu *et al.*, 1994).

Two other flavones, baicalin and isoscutellarein-8-methylether, were isolated from *Scutellaria baicalensis*, a plant used as a traditional Chinese herbal medicine (Li B.-Q. *et al.*, 1993; Nagai *et al.*, 1995). Baicalin affected syncytium formation on CEM-SS monolayer cells, expression of the HIV-1-specific core antigen p24, and reverse transcriptase activity in the HIV-1-infected H9 lymphocytes (Li B.-Q. *et al.*, 1993). Mahmood *et al.* (1996) tested nine flavonoids, isolated from *Rosa damascena* (Rosaceae), on anti-HIV-1 activity. Kaempferol and its 3-*O*-β-glucopyranoside exhibited the greatest activity against HIV-1 infection of human lymphoblastic (C8166) cells. Kaempferol was effective in reducing maturation of the infectious viral progeny, apparently by selectively inhibiting the viral protease. In contrast, quercetin and two 3-substituted derivatives of kaempferol appeared to inhibit HIV-1 infection by preventing the binding of gp 120 to CD4+.

HIV-1 integrase mediates the insertion of viral DNA into host cellular DNA that is essential for viral replication and virion production. Recently, Kim *et al.* (1998) isolated two flavonol glycoside gallate esters with anti-HIV integrase activity from the leaves of *Acer okamotoarum*. Swertifrancheside, a flavonone-xanthone glucoside isolated from *Swertia franchetiana*, was found to be a potent inhibitor of the HIV-1-RT (IC_{50} 43 μM). The compound inhibited enzyme activity by binding to the template-primer (Pengsuparp *et al.*, 1995). Eleven biflavonoids and their methyl ethers, isolated from *Rhus succedanea* and *Garcinia multiflora*, were evaluated for their anti-HIV-1-RT activity. Robustaflavone and hinokiflavone demonstrated the highest activity against HIV-1-RT; the two compounds displayed similar activity, with IC_{50} values of 65 and 62 μM, respectively (Lin *et al.*, 1997).

5.4.1.4 Lignans

Lignans exhibit a wide range of biological properties, including antifungal, antimicrobial, antiviral and antitumorigenic activities, and inhibition of many enzyme systems. The antiviral activities of lignans have been reviewed by McRae and co-workers (1989). Gnabre and co-workers (1995, 1996) isolated several lignans from the desert plant, *Larrea tridentata* (Zygophyllaceae), by bioassay-guided chromatography. The most predominant anti-HIV-1 compound was 3′-*O*-methyl-nordihydroguaiaretic acid, denoted as mal.4. Mal.4 was shown to exert its inhibitory activity by: interfering with the binding of Sp1 protein to HIV long terminal repeat (LTR) promotor, thus blocking the proviral transcription; HIV Tat (= a potent transcription activator encoded by the HIV-1) transactivation; and suppression of viral replication. Two other lignans, phyllamycin B and retrojusticidin B, isolated from *Phyllanthus myrtifolius* (Euphorbiaceae), have been demonstrated to have a strong inhibitory effect on HIV-1-RT activity but far less inhibitory effect on human DNA polymerase-α activity (Chang *et al.*, 1995). Anolignan A and anolignan B, two dibenzylbutadiene lignans isolated from *Anogeissus acuminata* (Combretaceae), were also identified as active inhibitors of HIV-1-RT activity (Rimando *et al.*, 1994).

5.4.1.5 Tannins

Tannins, such as oligomeric hydrolyzable tannins, complex tannins and other metabolites and condensates, have repeatedly been isolated from medicinal plants. Most of them have exhibited antiviral activities. It was supposed that tannins, like a number of polyanionic compounds, including polyhydroxy carboxylates derived from phenolic compounds, selectively inhibit HIV replication by interacting with the surface glycoprotein, gp 120, to irreversibly prevent the binding of virus to CD4+ receptor. It was also found that reverse transcriptase represents another target for tannins.

It has been shown by several researchers that Chinese herbs represent an important potential source of reverse transcriptase inhibitors. For example, several tannins isolated from Chinese galls exhibit strong inhibitory effects against HIV-1-RT (Takechi *et al.*, 1985; Nishizawa *et al.*, 1989; Nonaka *et al.*, 1990; Okuda *et al.*, 1992; Chang *et al.*, 1994; Büechi, 1998). Recently, El-Mekkawy and co-workers (1995) screened methanolic extracts of 41 medicinal plants used in traditional Egyptian medicine for anti-HIV-1-RT activity. The fruit extracts from *Phyllanthus emblica*, *Quercus pedunculata*, *Rumex cyprius*, *Terminalia bellerica*, *Terminalia chebula* and *Terminalia horrida* revealed significant antiviral activity, with an IC_{50} of 50 μg/ml. Through bioassay-guided fractionation of the active extract of *Phyllanthus emblica*, several tannins, e.g. 1,6-di-*O*-galloyl-β-D-glucose, 1-*O*-galloyl-β-D-glucose, digallic acid and putranjivain A, were isolated as the potent inhibitory substances. The most active compound was putranjivain A, with an IC_{50} of 3.9 μM. The inhibitory mode of action was noncompetitive with respect to the substrate but competitive with respect to a template-primer.

5.4.1.6 Triterpenes

Antiviral saponins have been isolated from various plants. For example, aescine from *Aesculus hippocastanum*, primula saponins from *Primula veris* and *Anagallis arvensis*, saikosaponin A from *Bupleurum falcatum*, theasaponins from *Thea sinencis*, and gymnemic acid from *Gymnema sylvestris*. As yet, the mechanisms of inhibition of virus replication by saponins are not understood in detail.

Glycyrrhizin, the main saponin of *Glycyrrhiza glabra*, inhibits the growth of a number of DNA and RNA viruses, including HIV-1, *in vitro*. It was found that glycyrrhizin interferes with virus adsorption, which was further complemented by an inhibitory effect on protein kinase C. This enzyme seems to be required for the binding of HIV-1 particles to the cellular CD4+ receptors (Ito, 1987). 1-β-Hydroxyaleuritolic acid 3-*p*-hydroxybenzoate, isolated from the roots of *Maprounea africana*, and salaspermic acid, isolated from the roots of the liana *Tripterygium wilfordii* (Celastraceae), were shown to be inhibitors of HIV-1-RT. Salaspermic acid also inhibited HIV-1 replication in H9 lymphocytes. A structure-activity correlation of salaspermic acid with ten related compounds indicated that the acetal linkage of ring A and the carboxyl group in ring E may be required for the anti-HIV-1 activity (Chen *et al.*, 1992b; Pengsuparp *et al.*, 1995).

Celasdin-B, isolated from *Celastrus hindsii*, betulinic acid and platanic acid, both isolated from the leaves of *Syzygium claviforum*, were found to be inhibitors of HIV-1 replication in H9 lymphocytes. Evaluation of anti-HIV activity with structurally-related triterpenoids revealed that the C-3 hydroxy group the C-17 carboxylic acid group and the C-19 substituents contribute to enhanced anti-HIV-1 activity (Fujioka and Kashiwada, 1994; Kuo and Kuo, 1997).

5.4.2 Isolated secondary metabolites with further antiviral properties

5.4.2.1 Alkaloids

Several Amaryllidaceae isoquinoline alkaloids were tested against HSV-1. Alkaloids which may eventually prove to be antiviral agents had a hexahydroindole ring with two functional hydroxyl groups. It was established that the antiviral activity of alkaloids was due to the inhibition of multiplication and not to the direct inactivation of extracellular viruses. The mechanism of the antiviral effect could be explained, in part, as a blocking of viral DNA polymerase activity (Renard-Nozaki *et al.*, 1989).

5.4.2.2 Essential oils

While many essential oils possess high levels of antifungal and antibacterial activity, there is little information concerning the effects of essential oils on viruses or viral infections in either animal or plant systems. Shukla *et al.* (1989) tested the antiviral activity of the essential oils of *Foeniculum vulgare* and *Pimpinella anisum* against potato virus X, tobacco mosaic virus and tobacco ring spot virus on the host, *Chenopodium amaranticolor*. Both essential oils totally inhibited the formation of local lesions at concentrations of 3000 ppm, when the viruses were pretreated with essential oils 30 min before inoculation. Recently, Bishop (1995) tested the essential oil of *Melaleuca alternifolia*

(Myrtaceae) for antiviral activity against tobacco mosaic virus. When applied to *Nicotiana glutinosa* (Solanaceae) plants as a preinoculation spray at 100, 250 and 500 ppm, the essential oil was effective in significantly decreasing lesion numbers for at least 10 days postinoculation.

5.4.2.3 Flavonoids

Galangin, a 3,5,7-trihydroxyflavone isolated from *Helichrysum aureonitens* and *Callicarpa japonica*, exhibited anti-HSV-1 activity at concentrations ranging 12–47 μg/ml. In the same concentration range, galangin also significantly inhibited the cytopathic effect of coxsackie virus B type 1 (COX B 1). It was assumed that the anti-HSV-1 action resulted in a suppression of viral binding to host cells at an early stage of replication (Meyer *et al.*, 1997; Hayashi *et al.*, 1997). The effect of isoscutellarein-8-methylether was investigated on the single-cycle replication of mouse-adapted influenza A and B virus (INFA and INFB) in Madin-Darby canine kidney (MDCK) cells. The compound suppressed the replication of these viruses 6–12 h after incubation, in a dose-dependent manner, by 50% at 20 μM and 90% at 40 μM. In contrast, the agent had only a slight effect on the haemagglutination and RNA-dependent RNA polymerase activities of these viruses *in vitro*. The same compound completely prevented the proliferation of mouse-adapted INFA in mouse lung by intranasal (0.5 mg/kg body weight) and intraperitoneal (4.0 mg/kg body weight) administration, and it was more potent than the known anti-influenza virus substance, amantadine (Nagai *et al.*, 1995). Isoscutellarein, extracted from the leaves of the same plant, showed significant anti-influenza virus activity, similar to that of isoscutellarein-8-methylether. The agent inhibited the replication of INFA in Madin-Darby bovine kidney (MDBK) cells, with an IC_{50} value of 16 μM; and it noncompetitively blocked (IC_{50} 20.0 μM) the hydrolysis of sodium-*p*-nitrophenyl-*N*-acetyl-α-D-neuraminate by influenza virus sialidase (Nagai and Miyaichi, 1992).

Ginkgetin, a biflavone isolated from *Cephalotaxus drupacea*, caused dose-dependent inhibition of the replication of HSV-1, HSV-2 and human cytomegalovirus (HCMV). Adsorption of HSV-1 to host cells and virus penetration into cells were unaffected by this agent. On the other hand, ginkgetin suppressed viral protein synthesis and exerted strong inhibition of transcription of immediate-early genes (Hayashi *et al.*, 1992).

5.4.2.4 Lignans

San Feliciano and co-workers (1993) isolated 19 cyclolignans from *Juniperus sabina* (Cupressaceae). These compounds exhibited antiviral activity against HSV-1 and vesicular stomatitis virus (VSV); the antiviral activity depended on structural variations. The *trans* and *cis* configurations of tetralinelactones were far more active than those of naphthalene and non-lactonic cyclolignan classes.

5.4.2.5 *Miscellaneous phenolic compounds*

A polyphenolic complex (PC), isolated from *Geranium sanguineum* (Geraniaceae), inhibited the reproduction of INFA and INFB *in vitro* and *in ovo*. When influenza viruses were treated with 1 mg/ml of PC, their haemagglutination, neuraminidase and infective activities were reduced completely. Moreover, the polyphenolic complex protected white mice in an experimental influenza infection (Serkedjieva and Manolova, 1992).

Curcumin, a typical yellow pigment of *Curcuma longa*, is widely used as a spice and food colouring (curry). In the past, curcumin exhibited a variety of pharmacological effects, such as antitumour, anti-inflammatory and anti-infectious activity. Recently, Mazumder and co-workers (1995) showed that the agent also inhibits purified HIV-1 integrase *in vitro*.

5.4.2.6 *Thiosulfinates*

Extracts of *Allium sativum* (Alliaceae) have been used traditionally to treat a number of infectious diseases caused by bacteria, fungi, protozoa and viruses. Recently, Weber and co-workers (1992) isolated diallyl thiosulfinate (allicin), allyl methyl thiosulfinate, methyl allyl sulfinate, ajoene, alliin, deoxyalliin, diallyl disulfide and diallyl trisulfide from garlic cloves. The compounds were tested against HSV-1, HSV-2, parainfluenza virus type 3 (Para-3), vaccinia virus (VV), VSV, and human rhinovirus type 2B (HRV-2B) (host cells: Vero cells, HeLa cells; evaluation of antiviral activity: plaque reduction assay). At the highest concentration tested (1.000 mg/ml), the infectivity of all viruses was substantially reduced by a fresh garlic extract. The thiosulfinates appeared to be the active components; the predominant agent was allicin. At a concentration of 25 μg/ml, VSV (reduction of virus titre: 1.3 $\log_{10}$), HSV-1 (reduction of virus titre: 0.5 $\log_{10}$) and Para-3 (reduction of virus titre: 0.2 $\log_{10}$) were sensitive to allicin. The experimental results suggest that fresh garlic extracts, as well as allicin, have direct virucidal activity but no intracellular antiviral properties.

5.5 Conclusions

During the course of evolution, plants have developed effective defence strategies to protect themselves from phytopathogenic microbes and herbivores in their environment. Disease resistance in plants depends upon the activation of coordinated, multicomponent defence mechanisms. A single mechanism or compound with a single metabolic site of action is thought to be unsatisfactory because of the rapid development of resistant strains of pathogens to one selected substance. One mechanism for disease resistance in plants is their ability to accumulate low molecular weight antimicrobial substances (phytoalexins) as a result of infection. In the last decade, many new phytoalexins with antibacterial and antifungal properties have been described. All these compounds belong to the secondary metabolites, including coumarins, isoflavonoids, other phenolic compounds, polyacetylenes and sesquiterpenes. Other components of the disease resistance complex include the presence of preformed antimicrobial agents and physical barriers, lignification, suberization and the formation of callose.

The importance of the plant kingdom as a source of new antimicrobial substances is illustrated by the present review on plant-derived antibacterial, antifungal and antiviral agents. During the last decade, hundreds of different new secondary metabolites with antimicrobial activity have been isolated, e.g. alkaloids, coumarins, flavonoids, lignans, quinones, miscellaneous phenolic compounds, miscellaneous terpenes and tannins. Unfortunately, little is known about the mode of action for most of the natural antibacterial and antifungal agents. In contrast to plant-derived antibacterial and antifungal substances, several secondary metabolites with antiviral properties have exhibited competitive *in vitro* and *in vivo* activities with those found for synthetic antiviral drugs. It has been shown that plant-derived antiviral secondary metabolites interfere with many viral targets, ranging from adsorption of the virus to the host cell via the inhibition of virus-specific enzymes (e.g. reverse transcriptase, protease) to release of the virus from the cells.

From the results so far achieved it is anticipated that bioactive plant-derived secondary metabolites will be used as leads to synthesize new and more active antimicrobial agents as well as substances with new pharmacological effects by repeated structural modification. It is expected that structurally modified natural products will exhibit increased potency, selectivity, duration of action and bioavailability and reduced toxicity. The terpenoid alkaloid, taxol, isolated from the bark of *Taxus brevifolia*, provides an example of this new strategy; taxol and some of its derivatives show an anticancerous activity against ovarian and mammary carcinomas.

References

Abbasoglu, U., Sener, B., Gunay, Y. and Temizer, H. (1991) Antimicrobial activity of some isoquinoline alkaloids. *Arch. Pharm.*, **324** 379-80.

Abou-Karam, M. and Shier, W.T. (1992) Isolation and characterization of an antiviral flavonoid from *Waldsteinia fragarioides*. *J. Nat. Prod.*, **55** 1525-27.

Aboul Ela, M.A., El-Shaer, N.S. and Ghanem, N.B. (1996) Antimicrobial evaluation and chromatographic analysis of some essential and fixed oils. *Pharmazie*, **51** 993-94.

Achenbach, H., Stöcker, M. and Constenla, A. (1988) Flavonoid and other constituents of *Bauhinia manca*. *Phytochemistry*, **27** 1835-41.

Adesanya, S.A., O'Neill, M.J. and Roberts, M.F. (1984) Induced and constitutive isoflavonoids in *Phaseolus mungo* (L.) (Leguminosae). *Z. Naturforsch.*, **39c** 888-93.

Adesanya, S.A., Ogundana, S.K. and Roberts, M.F. (1989) Dihydrostilbene phytoalexins from *Dioscorea bulbifera* and *D. dumentorium*. *Phytochemistry*, **28** 773-74.

Adesanya, S.A., Olugbade, T.A., Odebiyi, O.O. and Aladesammi, J.A. (1992) Antibacterial alkaloids in *Crinum jagus*. *Int. J. Pharmacol.*, **30** 303-307.

Afek, U., Carmeli, S. and Aharoni, N. (1995) Columbianetin, a phytoalexin associated with celery resistance to pathogens during storage. *Phytochemistry*, **39** 1347-50.

Ahmad, A., Kahn, K.A., Sultana, S., Siddiqui, B.S., Begum, S., Faizi, S. and Siddiqui, S. (1992) Study on the *in vitro* antimicrobial activity of harmine, harmaline and their derivatives. *J. Ethnopharmacol.*, **35** 289-94.

Ahmad, A., Ahmad, V.U. and Alam, N. (1995) New antifungal bithienylacetylenes from *Blumea obliqua*. *J. Nat. Prod.*, **58** 1426-29.

Ahmad, V.U., Khaliq-UZ-Zaman, S.M., Ali, M.S., Perveen, S. and Ahmed, W. (1996) An antimicrobial ecdysone from *Asparagus dumosus*. *Fitoterapia*, **LXVII** 88-91.

Akhtar, N., Malik, A., Ali, S.N. and Kazmi, S.U. (1994) Rubrinol, a new antibacterial triterpenoid from *Plumeria rubra*. *Fitoterapia*, **LXV** 162-66.

Al Magboul, A.Z., Bashir, A.K., Khalid, S.A. and Farouk, A. (1997) Antimicrobial activity of vernolepin and vernodalin. *Fitoterapia*, **LXVIII** 83-84.

Albersheim, P. and Darvill, A.G. (1985) Oligosaccharine: Zucker als Pflanzenhormone. *Spektrum der Wissenschaft*, **11** 86-93.

Amin, M., Kurosaki, F. and Nishi, A. (1988) Carrot phytoalexin alters the membrane permeability of *Candida albicans* and multilamellar liposomes. *J. Genet. Microbiol.*, **134** 241-46.

Amoros, M., Lurton, E., Boustie, J. and Girre, L. (1994) Comparison of the anti-herpes simplex virus activities of propolis and 3-methyl-but-2-enyl caffeate. *J. Nat. Prod.*, **57** 644-47.

Aqeel, A., Khursheed, A.K., Viqaruddin, A. and Sabiha, Q. (1989) Antimicrobial activity of julifloricine isolated from *Prosopis juliflora*. *Arzneimittelforschung*, **39** 652-55.

Aquino, R., De Simone, F. and Pizza, C. (1989) Plant metabolites: structure and *in vitro* antiviral activity of quinovic acid glycosides from *Uncaria tomentosa* and *Guettarda platypoda*. *J. Nat. Prod.*, **52** 679-85.

Asano, J., Chiba, K., Tada, M. and Yoshii, T. (1996) Antiviral activity of lignans and their glycosides from *Justicia procumbens*. *Phytochemistry*, **42** 713-17.

Baba, M. and Shigeta, S. (1987) Antiviral activity of glycyrrhizin against varicella-zoster virus *in vitro*. *Antiviral Res.*, **7** 99-107.

Bandara, B.M., Hewage, C.M., Karunaratne, V., Wannigama, C.P. and Adikaram, N.K. (1992) An antifungal chromene from *Eupatorium riparium*. *Phytochemistry*, **31** 1983-85.

Barel, S., Segal, R. and Yashphe, J. (1991) The antimicrobial activity of the essential oil from *Achillea fragrantissima*. *J. Ethnopharmacol.*, **33** 187-91.

Barre, J.T., Bowden, B.F., Coll, J.C., de Jesus, J., de la Fuente, V.E., Janairo, G.C. and Ragasa, Y. (1997) A bioactive triterpene from *Lantana camara*. *Phytochemistry*, **45** 321-24.

Batista, O., Duarte, A., Nascimento, J. and Simeos, M.F. (1994) Structure and antimicrobial activity of diterpenes from the roots of *Plectranthus hereroensis*. *J. Nat. Prod.*, **57** 858-61.

Baumgartner, B., Erdelmeier, C.A., Wright, A.D., Ralli, T. and Sticher, O. (1990) An antimicrobial alkaloid from *Ficus septica*. *Phytochemistry*, **29** 3327-30.

Bernard, F.X., Sable, S., Cameron, B., Provost, J., Desnottes, J.F., Crouzet, J. and Blanche, F. (1997) Glycosylated flavones as selective inhibitors of topoisomerase IV. *Antimicrob. Agents Chemother.*, **41** 992-98.

Bestwick, L., Bennett, H., Mansfield, J.W. and Rossiter, J.T. (1995) Accumulation of the phytoalexin, lettucenin A, and changes in 3-hydroxy-3-methylglutaryl coenzyme A reductase activity in lettuce seedlings with the red spot disorder. *Phytochemistry*, **39** 775-77.

Beutler, J.A., Cardellina II, J.H., McMahon, B. and Boyd, M.R. (1992) Anti-HIV and cytotoxic alkaloids from *Buchenavia capitata*. *J. Nat. Prod.*, **55** 207-13.

Binutu, O.A., Adesogan, K.E. and Okogun, J.I. (1996) Antibacterial and antifungal compounds from *Kigelia pinnata*. *Planta Med.*, **62** 352-53.

Bishop, C.D. (1995) Antiviral activity of the essential oil of *Melaleuca alternifolia* (tea tree) against tobacco mosaic virus. *J. Essen. Oil Res.*, **7** 641-44.

Boger, D.L., Mitscher, L.A., Mullican, M.D., Drake, S.D. and Kitos, P. (1985) Antimicrobial and cytotoxic properties of 9,10-dihydrophenanthrenes: structure-activity studies on juncsol. *J. Med. Chem.*, **28** 1543-47.

Bostock, R.M., Kuc, J.A., Laine, R.A. (1981) Eicosapentenoic and arachidonic acids from *Phytophthora infestans* elicit fungitoxic sesquiterpenes in the potato. *Science*, **212** 67-69.

Boyer, P.L., Currens, M.J., McMahon, J.B., Boyd, M.R. and Hughes, S.H. (1993) Analysis of nonnucleoside drug-resistant variants of human immunodeficiency virus type 1 reverse transcriptase. *J. Virol.*, **67** 2412-20.

Brader, G., Bacher, M., Hofer, O. and Greger, H. (1997) Prenylated phenylpropenes from *Coleonema pulchellum* with antimicrobial activity. *Phytochemistry*, **45** 1207-12.

Brinker, A.M. and Seigler, D.S. (1991) Isolation and identification of piceatannol as a phytoalexin from sugarcane. *Phytochemistry*, **30** 3229-32.

Brum, R.L., Honda, N.K., Hess, S.C., Cruz, A.B. and Moretto, E. (1997) Antibacterial activity of *Cochleospermum regium* essential oil. *Fitoterapia*, **LXVIII** 79-80.

Büechi, S. (1998) Antivirale Gerbstoffe: pharmakologische und klinische Untersuchungen. *Deutsche Apotheker Zeitung*, **138** 1269-74.

Burapadaja, S. and Bunchoo, A. (1995) Antimicrobial activity of tannins from *Terminalia citrina*. *Planta Med.*, **61** 365.

Calis, I., Yürüker, A., Tasdemir, D., Wright, A.D., Sticher, O., Luo, Y.D. and Pezzuto, J.M. (1997a) Cycloartane triterpene glycosides from roots of *Astragalus melanophrurius*. *Planta Med.*, **63** 183-86.

Calis, I., Satana, M.E., Yürüker, A., Kelican, P., Demirdamar, R., Alacam, R., Tanker, N., Rüegger, H. and Sticher, O. (1997b) Triterpene saponins from *Cyclamen mirabile* and their biological activities. *J. Nat. Prod.*, **60** 315-18.

Cantrell, C.L., Lu, T., Fronczek, R. and Fischer, N.H. (1996) Antimycobacterial cycloartanes from *Borrichia frutescens*. *J. Nat. Prod.*, **59** 1131-36.

Caron, C., Hoizey, M.J., Le-Men-Olivier, L., Massiot, G., Zeches, M., Choisy, C., Le-Magrex, E. and Verpoorte, R. (1988) Antimicrobial and antifungal activities of quasi-dimeric and related alkaloids. *Planta Med.*, **54** 409-12.

Carson, C.F. and Riley, T.V. (1994) Susceptibility of *Propionibacterium acnes* to the essential oil of *Melaleuca alternifolia*. *Lett. Appl. Microbiol.*, **19** 24-25.

Carson, C.F. and Riley, T.V. (1995) Antimicrobial activity of the major components of the essential oil of *Melaleuca alternifolia*. *J. Appl. Bacteriol.*, **78** 264-69.

Carson, C.F., Hammer, K.A. and Riley, T.V. (1995) Broth microdilution method for determining the susceptibility of *Escherichia coli* and *Staphylococcus aureus* to the essential oil of *Melaleuca alternifolia* (tea tree oil). *Microbios*, **82** 181-85.

Ceska, O., Chaudhary, S.K., Warrington, P.J. and Ashwood-Smith, M.J. (1992) Coumarins of chamomile, *Chamomilla recutita*. *Fitoterapia*, **LXIII** 387-94.

Chakraborty, A., Saha, C., Podder, G., Chowdhury, B.K. and Bhattacharyya, P. (1995) Carbazole alkaloid with antimicrobial activity from *Clausena heptaphylla*. *Phytochemistry*, **38** 787-89.

Chand, S., Lusunzi, I., Veal, D.A., Williams, L.R. and Karuso, P. (1994) Rapid screening of the antimicrobial activity of extracts and natural products. *J. Antibiot.*, **47** 1295-304.

Chandravadana, M.V., Nidiry, E.S. and Venkateshwarlu, G. (1997) Antifungal activity of momordicines from *Momordica charantia*. *Fitoterapia*, **LXVIII** 383-84.
Chang, C.-W., Lin, M.-T., Lee, S.-S., Chen-Liu, K.C.S., Hsu, F.-L. and Lin, J.-Y. (1995) Differential inhibition of reverse transcriptase and cellular DNA polymerase-α activities by lignans isolated from Chinese herbs, *Phyllanthus myrtifolius* MOON, and tannins from *Lonicera japonica* THUNB and *Castanopsis hystrix*. *Antiviral Res.*, **27** 367-74.
Chang, C.T., Doong, S.L., Tsai, I.L. and Chen, I.S. (1997) Coumarins and anti-HBV constituents from *Zanthoxylum schinifolium*. *Phytochemistry*, **45** 1419-22.
Chang, B.S., Lee, Y.M., Ku, Y., Bae, K. and Chung, C.P. (1998) Antimicrobial activity of magnolol and honokiol against periodontopathic microorganisms. *Planta Med.*, **64** 367-69.
Chappell, J. and Hahlbrock, K. (1984) Transcription of plant defence genes in response to UV light or fungal elicitor. *Nature*, **311** 76-78.
Chen, K., Shi, Q. and Fujioka, T. (1992a) Anti-AIDS agents. 4. Tripterifordin, a novel anti-HIV principle from *Tripterygium wilfordii*: isolation and structural elucidation. *J. Nat. Prod.*, **55** 88-92.
Chen, K., Shi, Q. and Kashiwada, Y. (1992b) Anti-AIDS agents. 6. Salaspermic acid, an anti-HIV principle from *Tripterygium wilfordii*, and the structure-activity correlation with its related compounds. *J. Nat. Prod.*, **55** 340-46.
Chen, D.F., Zhang, S.X., Chen, K., Zhou, B.N., Wang, P., Cosentino, L.M. and Lee, K.H. (1996) Two new lignans, interiotherins A and B, as anti-HIV principles from *Kadsura interior*. *J. Nat. Prod.*, **59** 1066-68.
Chinou, I., Demetzos, C., Harvala, C., Roussakis, C. and Verbist, J.F. (1994) Cytotoxic and antibacterial labdane-type diterpenes from the aerial parts of *Cistus incanus* subsp. *creticus*. *Planta Med.*, **60** 34-36.
Chinou, I.B., Roussis, V., Perdetzoglou, D. and Loukis, A. (1996) Chemical and biological studies on two *Helichrysum* species of Greek origin. *Planta Med.*, **62** 377-79.
Cimanga, K., De Bruyne, T., Lasure, A., Van Poel, B., Pieters, L., Claeys, M., Vanden Berghe, D., Kambu, K., Pona, L. and Vlietinck, A.J. (1996) *In vitro* biological activities of alkaloids from *Cryptolepis sanguinolente*. *Planta Med.*, **62** 22-27.
Clark, A.M. and Hufford, C.D. (1992) *Antifungal Alkaloids*. I. The Alkaloids. Vol. 42 (ed. G.A. Cordell) Academic Press, San Diego, pp. 117-50.
Cline, K. and Albersheim, P. (1981) Host-pathogen interactions. XVII. *Plant Physiol.*, **68** 221-28.
Cody, V., Middleton, E., Harborne, J.B. and Beretz, A. (1986) Plant flavonoids in biology and medicine: biochemical, pharmacological and structure-activity relationships. Alan R. Liss Inc., New York, pp. 67-569.
Cody, V., Middleton, E., Harborne, J.B. and Beretz, A. (1988) Plant flavonoids in biology and medicine. II. Biochemical, cellular and medicinal properties. Alan R. Liss Inc., pp. 1-27.
Cole, M.D., Bridge, P.D., Dellar, J.E., Fellows, L.E., Cornish, M.C. and Anderson, J.C. (1991) Antifungal activity of neo-clerodane diterpenoids from *Scutellaria*. *Phytochemistry*, **30** 1125-27.
Colombo, M.L. and Bosisio, E. (1996) Pharmacological activities of *Chelidonium majus* (L.) (Papaveraceae). *Pharmacol. Res.*, **33** 127-34.
Cimanga, K., De-Bruyne, T., Lasure, A., Van-Poel, B., Pieters, L., Claeys, M., Berghe, D.V., Kambu, K., Tona, L. and Vlietinck, A.J. (1996) *In vitro* biological activities from *Cryptolepsis sanguinolenta*. *Planta Med.*, **62** 22-27.
Cramer, C.L., Ryder, T.B., Bell, J.N. and Lamb, C.J. (1985) Rapid switching of plant gene expression induced by fungal elicitor. *Science*, **277** 1240-43.
Cruickshank, I.A.M. (1966) Defence mechanism in plants. *World Rev. Pest Control*, **5** 161-73.
Cruickshank, I.A.M. and Perrin, D.R. (1961) Studies on phytoalexins. III. *Aust. J. Biol. Sci.*, **14** 336-48.

Currens, M.J., Gulakowski, R.J., Mariner, J.M., Moran, R.A., Buckheit, R.W., Gustafson, K.R., McMahon, J.B. and Boyd, M.R. (1996) Antiviral activity and mechanisms of calanolide A against the human immunodeficiency virus type-1. *J. Pharmacol. Exp. Therapeut.*, **279** 645-51.

Davila-Huerta, G., Hamada, H., Davis, G.D., Stipanovic, R.D., Adams, C.M. and Essenberg, M. (1995) Cadinane-type sesquiterpenes induced in *Gossypium* cotyledons by bacterial inoculation. *Phytochemistry*, **39** 531-36.

Davis, K.R., Darvill, A.G., Albersheim, P. and Dell, A. (1986) Host pathogen interactions XXIX. *Plant Physiol.*, **80** 568-77.

De Godoy, G.F., Miguel, O.G. and Moreira, E.A. (1991) Antibacterial activity of xanthoxyline, constituents of *Sebastiania schottiana*. *Fitoterapia*, **LXII** 269-70.

De Rodriguez, D.J. and Chulia, J. (1990) Search for *in vitro* antiviral activity of a new isoflavonic glycoside from *Ulex europaeus*. *Planta Med.*, **56** 59-62.

De Siqueira, J.M., De Oliveira, C.C. and Diamantino Boaventura, M.A. (1997) Bioactive sesquiterpenoids from *Duguetia grabriuscula*. *Fitoterapia*, **LXVIII** 89-90.

De Tommasi, N., Pizza, C., Conti, C., Orsi, N. and Stein, M.L. (1990) Structure and *in vitro* antiviral activity of sesquiterpene glycosides from *Calendula arvensis*. *J. Nat. Prod.*, **53** 830-35.

Dean, R.A. and Kuc, J. (1987) Rapid lignification in response to wounding and infection as a mechanism for induced systemic protection in cucumber. *Physiol. Mol. Plant Pathol.*, **31** 69-81.

Dellar, J.E., Cole, M.D. and Waterman, P.G. (1996) Antimicrobial abietane diterpenoids from *Plectranthus elegans*. *Phytochemistry*, **41** 735-38.

Delle Monache, G., Botta, B., Vinciguerra, V., de Mello, J.F. and de Andrade Chiappeta, A. (1996) Antimicrobial isoflavones from *Desmodium canum*. *Phytochemistry*, **41** 537-44.

Denyer, C.V., Jackson, P., Loakes, D.M., Ellis, M.R. and Young, D.A. (1994) Isolation of antirhinoviral sesquiterpenes from ginger (*Zingiber officinale*). *J. Nat. Prod.*, **57** 658-62.

Diallo, B., Vanhaelen-Fastre, R. and Vanhaelen, M. (1991) Antimicrobial activity of two apocarotenoids isolated from *Cochlospermum tinctorium* rhizome. *Fitoterapia*, **LXII** 144-45.

Diaz, R.M., Garcia-Granados, A., Moreno, E., Parra, A., Quevedo-Sarmiento, J., de Buruaga, A. and de Buruaga, J.M. (1988) Studies on the relationship of structure to antimicrobial properties of diterpenoid compounds from *Sideritis*. *Planta Med.*, **54** 301-304.

Didry, N., Dubreuil, L. and Pinkas, M. (1993) Activité antibactérienne du thymol, du carvacrol et de l'aldéhyde cinnamique seuls ou associés. *Pharmazie*, **48** 301-304.

Didry, N., Dubreuil, L. and Pinkas, M. (1994) Activity of anthraquinonic and naphthoquinonic compounds on oral bacteria. *Pharmazie*, **49** 681-83.

Ebel, J. (1986) Phytoalexin synthesis: the biochemical analysis of the induction process. *Annu. Rev. Plant Physiol.*, **24** 235-64.

Ebel, J., Staeb, M.R. and Schmidt, W.E. (1985) Induction of enzymes of phytoalexin synthesis in soybean cells by fungal elicitor, in primary and secondary metabolism of plant cell culture (eds. K.H. Neumann, W. Barz and E. Reinhard), Springer, Berlin, pp. 247-54.

Economou, D. and Nahrstedt, A. (1991) Chemical, physiological and toxicological aspects of the essential oil of some species of the genus *Bystropogon*. *Planta Med.*, **57** 347-51.

Eilert, U., Kurz, W.G. and Constabel, F. (1985) Stimulation of sanguinarine accumulation in *Papaver somniferum* cell cultures by fungal elicitors. *J. Plant Physiol.*, **119** 65-76.

Ekabo, O.A. and Farnsworth, N.R. (1996) Antifungal and molluscicidal saponins from *Serjania salzmanniana*. *J. Nat. Prod.*, **59** 431-35.

El-Kamali, H.H., Ahmed, A.H., Mohammed, A.S., Yahia, A.A., El-Tayeb, I.H. and Ali, A.A. (1998) Antibacterial properties of essential oils from *Nigella sativa*, *Cymbopogon citratus* leaves and *Pulicaria undulata* aerial parts. *Fitoterapia*, **LXIX** 77-78.

El-Lakany, A.E., Abdel-Kader, M.S., Sabri, N.N. and Stermitz, F.R. (1995) Lanigerol: a new antimicrobial icetexane diterpene from *Salvia lanigera*. *Planta Med.*, **61** 559-60.

El-Mekkawy, S., Meselhy, M.R., Kusumoto, I.T., Kadota, S., Hattori, M. and Namba, T. (1995) Inhibitory effects of Egyptian folk medicines on human immunodeficiency virus (HIV) reverse transcriptase. *Chem. Pharmaceut. Bull.*, **43** 641-48.

Endo, K., Kanno, E. and Oshima, Y. (1990) Structures of antifungal diarylheptenones, gingerenones A, B, C and isogingerenone B, isolated from the rhizome of *Zingiber officinale*. *Phytochemistry*, **29** 797-99.

Engström, K. (1998) Sesquiterpenoid spiro compounds from potato tubers infected with *Phoma foveata* and *Fusarium* spp. *Phytochemistry*, **47** 985-90.

Essenberg, M., Grover, P.B. and Cover, E.C. (1990) Accumulation of antibacterial sesquiterpenoids in bacterially-inoculated *Gossypium* leaves and cotyledons. *Phytochemistry*, **29** 3107-13.

Ferdous, A.J., Islam, M.O., Hasan, C.M. and Islam, S.N. (1992) *In vitro* antimicrobial activity of lanuginosine and oxostephanine. *Fitoterapia*, **63** 549-50.

Fernandez, M.A., Garcia, M.D. and Saenz, M.T. (1996) Antibacterial activity of the phenolic acid fractions of *Scrophularia frutescens* and *Scrophularia sambucifolia*. *J. Ethnopharmacol.*, **53** 11-14.

Fewell, A.M. and Roddick, J.G. (1993) Interactive antifungal activity of the glycoalkaloids, α-solanine and α-chaconine. *Phytochemistry*, **33** 323-28.

French, C. and Towers, G.H. (1992) Inhibition of infectivity of potato virus X by flavonoids. *Phytochemistry*, **31** 3017-20.

Fujioka, T. and Kashiwada, Y. (1994) Anti-AIDS agents. 11. Betulinic acid and platanic acid as anti-HIV principles from *Syzigium claviflorum*, and the anti-HIV activity of structurally-related triterpenoids. *J. Nat. Prod.*, **57** 243-47.

Fukuchi, K., Sakagami, H., Okuda, T., Hatano, T., Tanuma, S., Kitajima, K., Inoue, Y., Inoue, S., Ichikawa, S., Nonoyama, M. and Konno, K. (1989) Inhibition of herpes simplex virus infection by tannins and related compounds. *Antiviral Res.*, **11** 285-98.

Funk, C., Gügler, K. and Brodelius, P. (1987) Increased secondary product formation in plant cell suspension cultures after treatment with a yeast carbohydrate preparation (elicitor). *Phytochemistry*, **26** 401-405.

Gabrielsen, B., Monath, T.P., Huggins, J.W., Kefauver, D.F., Pettit, G.R., Grszek, G., Hollingshead, M., Kirsi, J.J., Shannon, W.M., Schubert, E.M., Dare, Y., Ugarkar, B., Ussery, M.A. and Phelan, M.J. (1992) Antiviral (RNA) activity of selected Amaryllidaceae isoquinoline constituents and synthesis of related substances. *J. Nat. Prod.*, **55** 1569-81.

Gergis, V., Spiliotis, V. and Poulos, C. (1990) Antimicrobial activity of essential oils from greek *Sideritis* species. *Pharmazie*, **45** 70.

Gnabre, J.N., Brady, J.N., Clanton, D.J., Ito, Y., Dittmer, J., Bates, R.B. and Huang, R.C. (1995) Inhibition of human immunodeficiency virus type 1 transcription and replication by DNA sequence-selective plant lignans. *Biochemistry*, **92** 11239-43.

Gnabre, J.N., Ito, Y., Ma, Y. and Huang, R.C. (1996) Isolation of anti-HIV-1 lignans from *Larrea tridentata* by countercurrent chromatography. *J. Chromatogr. A*, **719** 353-64.

Godowski, K.C. (1989) Antimicrobial action of sanguinarine. *J. Clin. Dent.*, **1** 96-101.

Gonzalez, A.G., Alvarenga, N.L., Ravelo, A.G., Jimenez, I.A., Bazzocchi, I.L., Canela, N.J. and Moujir, L.M. (1996) Antibiotic phenol nor-triterpenes from *Maytenus canariensis*. *Phytochemistry*, **43** 129-32.

Gopalakrishnan, G., Banumathi, B. and Suresh, G. (1997) Evaluation of the antifungal activity of natural xanthones from *Garcinia mangostana* and their synthetic derivatives. *J. Nat. Prod.*, **60** 519-24.

Gören, N., Jakupovic, J. and Topal, S. (1990) Sesquiterpene lactones with antibacterial activity from *Tanacetum argyrophyllum* var. *argyrophyllum*. *Phytochemistry*, **29** 1467-69.

Grab, D., Loyal, R. and Ebel, J. (1985) Elicitor-induced phytoalexin synthesis in soybean cells: changes in the activity of chalcone synthase mRNA and the total population of transplantable mRNA. *Arch. Biochem. Biophys.*, **243** 523-29.

Grayer, R.J. and Harborne, J.B. (1994) A survey of antifungal compounds from higher plants, 1982–1993. *Phytochemistry*, **37** 19-42.

Greger, H., Hofer, O., Kählig, H. and Wurz, G. (1992) Sulfur-containing cinnamides with antifungal activity from *Glycosmis cyanocarpa*. *Tetrahedron*, **48** 1209-18.

Greger, H., Zechner, G., Hadacek, F. and Wurz, G. (1993) Sulphur-containing amides from *Glycosmis* species with different antifungal activity. *Phytochemistry*, **34** 175-79.

Grenby, T.H. (1995) The use of sanguinarine in mouthwashes and toothpaste compared with some other antimicrobial agents. *Br. Dent. J.*, **178** 254-58.

Gross, D. (1987) Chemische Abwehrstoffe der Pflanze. *Biologische Rundschau*, **25** 225-37.

Gustine, D.L. and Moyer, B. (1982) Retention of phytoalexin regulation in legume callus cultures. *Plant Cell Tiss. Org. Cult.*, **1** 255-63.

Gutierrez, M.C., Parry, A., Tena, M., Jorrin, J. and Edwards, R. (1995) Abiotic elicitation of coumarin phytoalexins in sunflower. *Phytochemistry*, **38** 1185-91.

Habtemariam, S., Gray, A.I., Halbert, G.W. and Waterman, P.G. (1990) A novel antibacterial diterpene from *Premna schimperi*. *Planta Med.*, **56** 187-89.

Hahn, M.G., Darvill, A.G. and Albersheim, P. (1981) Host-pathogen interactions XIX. The endogenous elicitor, a fragment of a plant cell wall polysaccharide that elicits phytoalexin accumulation in soybeans. *Plant Physiol.*, **68** 1161-69.

Hahn, M.G., Bonhoff, A. and Grisebach, H. (1985) Quantitative localization of the phytoalexin glyceollin I in relation to fungal hyphae in soybean roots infected with *Phytophthora megasperma* f. sp. *glycinea*. *Plant Physiol.*, **77** 591-601.

Hajji, F. and Fkih-Tetouani, S. (1993) Antimicrobial activity of twenty one *Eucalyptus* essential oils. *Fitoterapia*, **LXIV** 71-77.

Hammerschmidt, F.J., Clark, A.M., Soliman, F.M., El-Kashoury, E.A., El-Kawy, M.M. and El-Fishawy, A.M. (1993) Chemical composition and antimicrobial activity of essential oils of *Jasonic candicans* and *J. montana*. *Planta Med.*, **59** 68-70.

Hanawa, F., Tahara, S. and Mizutani, J. (1991) Isoflavonoids produced by *Iris pseudacorus* leaves treated with cupric chloride. *Phytochemistry*, **30** 157-63.

Hanawa, F., Tahara, S. and Mizutani, J. (1992) Antifungal stress compounds from *Veratrum grandiflorum* leaves treated with cupric chloride. *Phytochemistry*, **31** 3005-3007.

Haragüchi, H., Oika, S., Hüroi, H. and Kübo, F. (1996) Mode of antibacterial action of totarol, a diterpene from *Podocarpus nagi*. *Planta Medica*, **62** 122-25.

Hardwiger, L.A. and Schwochau, M.E. (1971a) Specifity of DNA intercalating compounds in the control of PAL and pisatin levels. *Plant Physiol.*, **47** 346-51.

Hardwiger, L.A. and Schwochau, M.E. (1971b) UV light-induced formation of pisatin and PAL. *Plant Physiol.*, **47** 588-90.

Hasan, A. and Ahmad, I. (1996) Antibacterial activity of flavonoid glycosides from the leaves of *Rumex chalepensis*. *Fitoterapia*, **LXVII** 182-83.

Hayashi, K. and Hayashi, T. (1996) Scopadulciol is an inhibitor of herpes simplex virus type 1 and a potentiator of acyclovir. *Antiviral Chem. Chemother.*, **7** 79-85.

Hayashi, K., Hayashi, T. and Morita, N. (1992) Mechanism of action of the anti-herpesvirus biflavone, ginkgetin. *Antimicrob. Agents Chemother.*, **36** 1890-93.

Hayashi, K., Hayashi, H., Hiraoka, N. and Ikeshiro, Y. (1997) Inhibitory activity of soyasaponin II on virus replication *in vitro*. *Planta Med.*, **63** 102-105.

Heath-Pagliuso, S., Matlin, S.A., Fang, N., Thompson, R.H. and Rappaport, L. (1992) Stimulation of furanocoumarin accumulation in celery and celeriac tissues by *Fusarium oxysporum* f. sp. *apii*. *Phytochemistry*, **31** 2683-88.

Hili, P., Evans, C.S. and Veness, R.G. (1997) Antimicrobial action of essential oils: the effect of dimethylsulphoxide on the activity of cinnamon oil. *Lett. Appl. Microbiol.*, **24** 269-75.

Hill, A.M., Cane, D.E., Mau, C.J.D. and West, C.A. (1996) High level expression of *Ricinus communis* casbene synthase in *Escherichia coli* and characterization of the recombinant enzyme. *Arch. Biochem. Biophys.*, **15** 283-89.

Hinou, J.B., Harvala, C.E. and Hinou, E.B. (1989) Antimicrobial activity screening of 32 common constituents of essential oils. *Pharmazie*, **44** 302.

Hirai, N., Ishida, H. and Koshimizu, K. (1994) A phenalenone-type phytoalexin from *Musa acuminata*. *Phytochemistry*, **37** 383-85.

Houghton, P.J., Woldermariam, T.Z., Khan, A.I., Burke, A. and Mahmood, N. (1994) Antiviral activity of natural and semisynthetic chromone alkaloids. *Antiviral Res.*, **25** 235-44.

Hu, C-Q, Chen, K. and Shi, Q. (1994) Anti-AIDS agents. 10. Acacetin-7-*O*-β-D-galactopyranoside, an anti-HIV principle from *Chrysanthemum morifolium* and a structure-activity correlation with some related flavonoids. *J. Nat. Prod.*, **57** 42-51.

Iinuma, M., Tsuchiya, H., Sato, M., Yokoyama, J., Ohyama, M., Ohkawa, Y., Tanaka, T., Fujiwara, S. and Fujii, T. (1994) Flavanones with potent antibacterial activity against methicillin-resistant *Staphylococcus aureus*. *J. Pharm. Pharmacol.*, **46** 892-95.

Iinuma, M., Tosa, H., Tanaka, T., Kanamaru, S., Asai, F., Kobayashi, Y., Miyauchi, K. and Shimano, R. (1996) Antibacterial activity of some *Garcinia* benzophenone derivatives against methicillin-resistant *Staphylococcus aureus*. *Biol. Pharmaceut. Bull.*, **19** 311-14.

Ito, M., Nakashima, H., Baba, M., Pauwels, R., De Clercq, E., Shigeta, S. and Yamamoto, N. (1987) Inhibitory effects of glycyrrhizin and cytopathic activity of the human immunodefiency virus (HIV [HTLV-III/LAV]). *Antiviral Res.*, **7** 127-37.

Jalsenjak, V., Peljnjak, S. and Kustrak, D. (1987) Microcapsules of sage oil: essential oils content and antimicrobial activity. *Pharmazie*, **42** 419-20.

Janssen, A.M., Scheffer, J.J.C. and Svendsen, A.B. (1988) Antimicrobial activities of essential oils: a 1976–1988 literature review on possible applications. *Pharmaceutisch Weekblab*, **9** 193-97.

Kalodera, Z., Pepeljnjak, S., Vladimir, S. and Blazevic, N. (1994) Antimicrobial activity of essential oil from *Micromeria thymifolia (Scop.) Fritsch*. *Pharmazie*, **49** 376-77.

Kartnig, T., Still, F. and Reinthaler, F. (1991) Antimicrobial activity of the essential oil of young pine shoots (*Picea abies*). *J. Ethnopharmacol.*, **35** 155-57.

Kato, H., Kodama, O. and Akatsuka, T. (1993) Oryzalexin E, a diterpene phytoalexin from UV-irradiated rice leaves. *Phytochemistry*, **33** 79-81.

Kato, H., Kodama, O. and Akatsuka, T. (1994) Oryzalexin F, a diterpene phytoalexin from UV-irradiated rice leaves. *Phytochemistry*, **36** 299-301.

Kaul, T.N., Middelton, J.E. and Orga, P.L. (1985) Antiviral effects of flavonoids on human viruses. *J. Med. Virol.*, **15** 71-79.

Kauss, H. (1987) Callose-Synthese. *Naturwissenschaften*, **74** 275-81.

Keen, N.T. (1986) Phytoalexins and their involvement in plant disease resistance. *Iowa State J. Res.*, **60** 477-99.

Keen, N.T. and Yoshikawa, M. (1983) β-1,3-Endoglucanase from soybean releases elicitor-active carbohydrates from fungus cell walls. *Plant Physiol.*, **71** 460-65.

Kernan, M.R., Sendl, A., Chen, J.L., Jolad, S.D., Blanc, P., Murphy, J.T., Stoddart, C.A., Nanakorn, W., Balick, M.J. and Rozhon, E.J. (1997) Two new lignans with activity against influenza virus from the medicinal plant, *Rhinacanthus nasutus*. *J. Nat. Prod.*, **60** 635-37.

Kessmann, H., Daniel, S. and Barz, W. (1988) Elicitation of pterocarpan phytoalexins in cell suspension cultures of different chickpea (*Cicer arietinum*) cultivars by an elicitor from the fungus, *Ascochyta rabiei*. *Z. Naturforsch.*, **43c** 529-35.

Kilibarda, V., Nanusevic, N., Dogovic, N., Ivanic, R. and Savin, K. (1996) Content of the essential oil of carrot and its antibacterial activity. *Pharmazie*, **51** 777-78.

Kim, H.J., Woo, E.R., Shin, C.G. and Park, H. (1998) A new flavonol glycoside gallate ester from *Acer okamotoanum* and its inhibitory activity against human immunodeficiency virus-1 (HIV-1) integrase. *J. Nat. Prod.*, **61** 145-48.

Kobayashi, M. and Otha, Y. (1983) Induction of stress metabolite formation in suspension cultures of *Vigna angularis*. *Phytochemistry*, **22** 1257-61.

Kobayashi, A., Akiyama, K. and Kawazu, K. (1993) A pterocarpan, (+)-2-hydroxypisatin from *Pisum sativum*. *Phytochemistry*, **32** 77-78.

Kobayashi, A., Kim, M.J. and Kawazu, K. (1996) Uptake and exudation of phenolic compounds by wheat and antimicrobial components of the root exudate. *Z. Naturforsch. C*, **51** 527-33.

Kodoma, M., Wada, H., Otani, H., Kohmoto, K. and Kimura, Y. (1998) 3,5-di-*O*-caffeoylquinic acid, an infection-inhibiting factor from *Pyrus pyrifolia* induced by infection with *Alternaria alternata*. *Phytochemistry*, **47** 371-73.

Kokubun, T., Harborne, J.B. and Eagles, J. (1994) 2′,6′-dihydroxy-4′-methoxyacetophenone, a phytoalexin from the roots of *Sanguisorba minor*. *Phytochemistry*, **35** 331-33.

Kokubun, T., Harborne, J.B., Eagles, J. and Waterman, P.G. (1995a) Antifungal biphenyl compounds are the phytoalexins of the sapwood of *Sorbus aucuparia*. *Phytochemistry*, **40** 57-59.

Kokubun, T., Harborne, J.B., Eagles, J. and Waterman, P.G. (1995b) Dibenzofuran phytoalexins from the sapwood tissue of *Photinia*, *Pyracantha* and *Crataegus* species. *Phytochemistry*, **39** 1033-37.

Kokubun, T., Harborne, J.B., Eagles, J. and Waterman, P.G. (1995c) Four dibenzofuran phytoalexins from sapwood of *Mespilus germanica*. *Phytochemistry*, **39** 1039-42.

Kombrink, E. and Hahlbrock, K. (1985) Dependence of the level of phytoalexin and enzyme induction by fungal elicitor on the growth stage of *Petroselinum crispum* cell cultures. *Plant Cell Rep.*, **4** 277-80.

Konoshima, T., Yasuda, I., Kashiwada, Y., Cosentino, L.M. and Lee, K.H. (1995) Anti-AIDS agents. 21. Triterpenoids saponins as anti-HIV principles from fruits of *Gleditsia japonica* and *Gymnocladus chinensis*, and a structure-activity correlation. *J. Nat. Prod.*, **58** 1372-77.

Kubo, I., Muroi, H. and Himjima, M. (1992) Antibacterial activity of totarol and its potentiation. *J. Nat. Prod.*, **55** 1436-40.

Kuc, J. and Rush, J.S. (1985) Phytoalexins. *Arch. Biochem. Biophys.*, **236** 455-72.

Kuo, Y.-H. and Kuo, L.-M.Y. (1997) Antitumor and anti-AIDS triterpenes from *Celastrus hindsii*. *Phytochemistry*, **44** 1275-81.

Lamb, C.J., Ryals, J.A., Ward, E.R. and Dixon, R.A. (1992) Emerging strategies for enhancing crop resistance to microbial pathogens. *Biotechnology NY*, **10** 1436-45.

Lee, S.C. and West, C.A. (1981) Polygalacturonase from *Rhizopus stolonifer*, an elicitor of casbene synthetase activity in castor bean (*Ricinus communis* L.) seedlings. *Plant Physiol.*, **67** 633-39.

Li, B.-Q., Fu, T., Yan, Y.-D., Baylor, N.W., Ruscetti, F.W. and Kung, H.-F. (1993) Inhibition of HIV infection by baicalin, a flavonoid compound purified from Chinese herbal medicine. *Cell. Mol. Biol. Res.*, **39** 119-24.

Li, H.Y., Sun, N.J., Kashiwada, Y. and Sun, L. (1993) Anti-AIDS agents. 9. Suberol, a new C_{31} lanostane-type triterpene and anti-HIV-principle from *Polyalthia suberosa*. *J. Nat. Prod.*, **56** 1130-33.

Li, D., Chung, K.R., Smith, D.A., Schardl, C.L. (1995) The *Fusarium solani* gene encoding kievitone hydratase, a secreted enzyme that catalyzes detoxification of a bean phytoalexin. *Mol. Plant. Microbe Interact.*, **8** 388-97.

Li, E., Clark, A.M. and Hufford, C.D. (1995) Antifungal evaluation of pseudolaric acid B, a major constituent of *Pseudolarix kaempferi*. *J. Nat. Prod.*, **58** 57-67.

Li, X.G., Cai, L. and Wu, C.D. (1997) Antimicrobial compounds from *Ceanothus americanus* against oral pathogens. *Phytochemistry*, **46** 97-102.

Lin, Y.M., Anderson, H., Flavin, M.T. and Pai, Y. (1997) *In vitro* anti-HIV activity of biflavonoids isolated from *Rhus succedanea* and *Garcinia multiflora*. *J. Nat. Prod.*, **60** 884-88.

Liu, S., Oguntimein, B.O., Hufford, C.D. and Clark, A.M. (1990) 3-methoxysampangine, a novel antifungal copyrine alkaloid from *Cleistopholis patens*. *Antimicrob. Agents Chemother.*, **34** 529-33.

Loya, S., Bakhanashvili, M., Kashman, Y. and Hizi, A. (1995) Peyssonols A and B, two novel inhibitors of the reverse transcriptases of human immunodeficiency virus type 1 and 2. *Arch. Biochem. Biophys.*, **316** 789-96.

Lucchini, J.J., Corre, J. and Cremieux, A. (1990) Antibacterial activity of phenolic compounds and aromatic alcohols. *Res. Microbiol.*, **141** 499-510.

Lyons, P.C., Wood, K.V. and Nicholson, R.L. (1990) Caffeoyl ester accumulation in corn leaves inoculated with fungal pathogens. *Phytochemistry*, **29** 97-101.

Maatooq, G.T., Stumpf, D.K., Hoffmann, J.J., Hutter, L.K. and Timmermann, B.N. (1996) Antifungal eudesmanoids from *Parthenium argentatum* x *P. tomentosa*. *Phytochemistry*, **41** 519-24.

MacRae, W.D., Hudson, J.B. and Towers, G.H.N. (1989) The antiviral action of lignans. *Planta Med.*, **55** 531-35.

Madar, Z., Gottlieb, H.E., Cojocaru, M., Riov, J., Solel, Z. and Sztejnberg, A. (1995) Antifungal terpenoids produced by Cypress after infection by *Diplodia pinea* f. sp. *cupressi*. *Phytochemistry*, **38** 351-54.

Mahmood, N., Piacente, S., Pizza, C., Burke, A., Khan, A.I. and Hay, A.J. (1996) The anti-HIV activity and mechanisms of action of pure compounds isolated from *Rosa damascena*. *Biochem. Biophys. Res. Commun.*, **229** 73-79.

Mahmud, Z., Musa, M., Ismail, N. and Lajis, N.H. (1993) Cytotoxic and bactericidal activities of *Psychotria rostrata*. *Int. J. Pharmacol.*, **31** 142-46.

Maloney, A.P. and van Etten, H.D. (1994) A gene from the fungal plant pathogen, *Nectria haematococca*, that encodes the phytoalexin-detoxifying enzyme, pisatin demethylase, defines a new cytochrome P_{450} family. *Mol. Gen. Genet.*, **243** 506-14.

Mariee, N.K., Khalil, A.A., Nasser, A.A., al-Hiti, M.M., Ali, W.M. (1988) Isolation of the antimicrobial alkaloid, stemmadenine, from Iraqi *Rhazya stricta*. *J. Nat. Prod.*, **51** 186-87.

Marquina, G., Laguna, A., Franco, P., Fernandez, L., Perez, R. and Valiente, O. (1989) Antimicrobial activity of pyrrolizidine alkaloids from *Heliotropium bursiferum*. *Pharmazie*, **44** 870-71.

Marston, A., Hostettmann, K. and Msonthi, J.D. (1995) Isolation of antifungal and larvicidal constituents of *Diplolophium buchanani* by centrifugal partition chromatography. *J. Nat. Prod.*, **58** 128-30.

Martinez-Vazquez, M., Martinez, R., Diaz, M. and Sanchez, M.H. (1994) Antimicrobial properties of argentatine A, isolated from *Parthenium argentatum*. *Fitoterapia*, **LXV** 371-72.

Maruta, Y., Fukushi, Y., Ohkawa, K., Nakanishi, Y., Tahara, S. and Mizutani, J. (1995) Antimicrobial stress compounds from *Hypochoeris radicata*. *Phytochemistry*, **38** 1169-73.

Marquina, G., Laguna, A., Franco, P., Fernandez, L., Perez, R. and Valiente, O. (1989) Antimicrobial activity of pyrrolizidine alkaloids from *Heliotropium bursiferum*. *Pharmazie*, **44** 870-71.

Masuda, T., Inazumi, A., Yamada, Y., Padolina, W.G., Kikuzaki, H. and Nakatani, N. (1991) Antimicrobial phenylpropanoids from *Piper sarmentosum*. *Phytochemistry*, **30** 3227-28.

Masuda, T., Takasugi, M. and Anetai, M. (1998) Psoralen and other linear furanocoumarins as phytoalexins in *Glehnia littoralis*. *Phytochemistry*, **47** 13-16.

Matthews, D.E. and van Etten, H.D. (1983) Detoxification of the phytoalexin, pisatin, by fungal cytochrome P_{450}. *Arch. Biochem. Biophys.*, **224** 494-505.

Mayer, A.M. (1989) Plant-fungal interaction: a plant physiologist's viewpoint. *Phytochemistry*, **28** 311-17.

Mazumder, A., Raghavan, K., Weistein, J., Kohn, K.W. and Pommier, Y. (1995) Inhibition of human immunodeficiency virus type-1 integrase by curcumin. *Biochem. Pharmacol.*, **49** 1165-70.

McChesney, J.D. and Clark, A.M. (1991) Antimicrobial diterpenes of *Croton sonderianus*, hardwickic and 3,4-secotrachylobanoic acids. *J. Nat. Prod.*, **54** 1625-33.

McCormick, J.L., McKee, T.C., Cardellina, J.H. and Boyd, M.R. (1996) HIV inhibitory natural products. 26. Quinoline alkaloids from *Euodia roxburghiana*. *J. Nat. Prod.*, **59** 469-71.

McKee, T.C., Bokesch, H.R., McCormick, J.L., Rashid, M.A., Spielvogel, D., Gustafson, K.R., Alavanja, M.M., Cardellina, J.H. and Boyd, M. (1997) Isolation and characterization of new anti-HIV and cytotoxic leads from plants, marine and microbial organisms. *J. Nat. Prod.*, **60** 431-38.

Mehrotra, S., Rawat, A.K. and Shome, U. (1993) Antimicrobial activity of the essential oils of some Indian *Artemisia* species. *Fitoterapia*, **LXIV** 65-68.

Messana, I., Ferrari, F., Cavalcanti, M. and Morace, G. (1991) An anthraquinone and three naphthopyrone derivatives from *Cassia pudibunda*. *Phytochemistry*, **30** 708-10.

Meyer, J.J.M., Afolayan, A.J., Taylor, M.B. and Erasmus, D. (1997) Antiviral activity of galangin isolated from the aerial parts of *Helichrysum aureonitens*. *J. Ethnopharmacol.*, **56** 165-69.

Miao, V.P., Matthews, D.E. and van Etten, H.D. (1991) Identification and chromosomal location of a family of cytochrome P_{450} genes for pisatin detoxification in the fungus *Nectria haematococca*. *Mol. Gen. Genet.*, **226** 214-23.

Miles, D.H., de Medeiros, J.M., Chittawong, V., Hedin, P.A., Swithenbank, C. and Lidert, Z. (1991) 3′-Formyl-2′,4′,6′-trihydroxydihydrochalcone from *Psidium acutangulum*. *Phytochemistry*, **30** 1131-31.

Miles, D.H., Chittawong, V., Hedin, P.A. and Kokpol, U. (1993) Potential agrochemicals from leaves of *Wedelia biflora*. *Phytochemistry*, **32** 1427-29.

Mitrokotsa, D., Mitaku, S., Demetzos, C., Harvala, C., Mentis, A., Perez, S. and Kokkinopoulos, D. (1993) Bioactive compounds from the buds of *Plantanus orientalis* and isolation of a new kaempferol glycoside. *Planta Med.*, **59** 517-20.

Miyagawa, H., Ishihara, A., Kuwahara, Y., Ueno, T. and Mayama, S. (1996) A stress compound in oats induced by victorin, a host-specific toxin from *Helminthosporium victoriae*. *Phytochemistry*, **41** 1473-75.

Modafar, C., Clerivet, A., Fleuriet, A. and Machaix, J.J. (1993) Inoculation of *Platanus acerifolia* with *Ceratocystis fimbriata* f. sp. *platani* induces scopoletin and umbelliferone accumulation. *Phytochemistry*, **34** 1271-76.

Mohrig, A. (1996) Melissenextrakt bei Herpes simplex: die Alternative zu Nucleosid-Analoga. *Deutsche Apotheker Zeitung*, **136** 109-14.

Monde, K., Sasaki, K., Shirata, A. and Takasugi, M. (1990a) 4-methoxybrassinin, a sulphur-containing phytoalexin from *Brassica oleracea*. *Phytochemistry*, **29** 1499-500.

Monde, K., Oya, T., Shirata, A. and Takasugi, M. (1990b) A guaianolide phytoalexin, cichoralexin, from *Cichorium intybus*. *Phytochemistry*, **29** 3449-51.

Monde, K., Sasaki, K., Shirata, A. and Takasugi, M. (1991) Methoxybrassenins A and B, sulphur-containing stress metabolites from *Brassica oleracea* var. *capitata*. *Phytochemistry*, **30** 3921-22.

Montanaro, S., Bardon, A. and Catalan, C. (1996) Antibacterial activity of various sesquiterpene lactones. *Fitoterapia*, **LXVII** 185-87.

Mori, A., Nishino, C., Enoki, N. and Tawata, S. (1987) Antibacterial activity and mode of action of plant flavonoids against *Proteus vulgaris* and *Staphylococcus aureus*. *Phytochemistry*, **26** 2231-34.

Müller, K.O. and Boerger, H. (1940) Experimentelle Untersuchung über die Phytophthora-Resistenz der Kartoffel. *Arb. Biol. Reichsanstalt Land- und Forstwirtschaft Berlin*, **23** 189-231.

Munoz, V., Moretti, C., Sauvain, M., Caron, C., Porzel, A., Massiot, G., Richard, B. and Le-Men-Olivier, L. (1994) Isolation of bis-indol alkaloids with antileishmanial and antibacterial activities from *Peschiera van heurkii. Planta Med.*, **60** 455-59.
Nagafuji, T., Matsumoto, T., Takahashi, K., Kubo, S., Haraoka, M., Tanaka, M., Ogata, N. and Kumazawa, J. (1993) Enhancement of superoxide production of polymorphonuclear neutrophil by oflocacin and the effect of the inhibitors of protein kinase C. *Chemotherapy*, **39** 70-76.
Nagai, T., Miyaichi, Y. (1992) *In vivo* anti-influenza virus activity of plant flavonoids possessing inhibitory activity for influenza virus sialidase. *Antiviral Res.*, **19** 207-17.
Nagai, T., Moriguchi, R., Suzuki, Y., Tomimori, T. and Yamada, H. (1995) Mode of action of the anti-influenza virus activity of plant flavonoid, 5,7,4′-trihydroxy-8-methoxyflavone, from the roots of *Scutellaria baicalensis. Antiviral Res.*, **26** 11-25.
Naigre, R., Kalck, P., Roques, C., Roux, I. and Michel, G. (1996) Comparison of antimicrobial properties of monoterpenes and their carbonylated products. *Planta Med.*, **62** 275-77.
Nahrstedt, A. (1979) Chemische Waffen bei höheren Pflanzen. *Pharmazie in unserer Zeit*, **8** 129-38.
Nandy, A.K., Chakraborty, A. and Podder, G. (1997) Antimicrobial activity of *Terminalia bellerica. Fitoterapia*, **LXVIII** 178-80.
Ndounga, M. and Ouamba, J.M. (1997) Antibacterial and antifungal activities of essential oils of *Ocimum gratissimum* and *O. basilicum* from Congo. *Fitoterapia*, **LXVIII** 190-91.
Nenoff, P., Haustein, U.F. and Brandt, W. (1996) Antifungal activity of the essential oil of *Melaleuca alternifolia* (tea tree oil) against pathogenic fungi *in vitro*. *Skin Pharmacol.*, **9** 388-94.
Nick, A., Wright, A.D. and Sticher, O. (1994) Antibacterial triterpenoid acids from *Dillenia papuana. J. Nat. Prod.*, **57** 1245-50.
Niemann, D.J. (1993) The anthranilamide phytoalexins of the Caryophyllaceae and related compounds. *Phytochemistry*, **34** 319-28.
Niemann, G.J., Liem, J., Pureveen, J. and Boon, J.J. (1991) The amide-type phytoalexin activity of carnation extracts is partly due to an artifact. *Phytochemistry*, **30** 3923-27.
Nishizawa, M., Yamagishi, T., Dutschmann, G.E., Parker, W.B., Border, A.J., Kilkuski, R.E., Cheng, Y.C. and Lee, K.H. (1989) Anti-AIDS agents. 1. Isolation and characterization of four new tetragalloylquinic acids as a new class of HIV reverse transcriptase inhibitors from tannic acid. *J. Nat. Prod.*, **52** 762-68.
Nonaka, G.I., Nishioka, I., Nishizawa, M., Yamagishi, T., Kashiwada, Y., Dutschman, G.E., Border, A.J., Kilkuskie, R.E., Cheng, Y.C. and Lee, K.H. (1990) Anti-AIDS agents. 2. Inhibitory effects of tannins on HIV reverse transcriptase and HIV replication in H9 lymphocyte cells. *J. Nat. Prod.*, **53** 587-95.
Okuda, T., Yoshida, T. and Hatano, T. (1992) Pharmacologically active tannins isolated from medicinal plants. *Basic Life Sci. (Plant Phenols)*, **59** 539-69.
Okunade, A.L., Hufford, C.D., Richardson, M.D., Peterson, I.R. and Clark, A.M. (1994) Antimicrobial properties of alkaloids from *Xanthorhiza simplicissima. J. Pharmaceut. Sci.*, **83** 404-406.
Orabi, K.Y., Mossa, J.S. and El-Feraly, S. (1991) Isolation and characterization of two antimicrobial agents from mace (*Myristica fragrans*). *J. Nat. Prod.*, **54** 856-59.
Orjala, J., Wright, A.D., Behrends, H., Folkers, G., Sticher, O., Rüegger, H. and Rali, T. (1994) Cytotoxic and antibacterial dihydrochalcones from *Piper aduncum. J. Nat. Prod.*, **57** 18-26.
Osawa, K., Yasuda, H., Maruyama, T., Morita, H., Takeya, K. and Itokawa, H. (1992) Isoflavanones from the heartwood of *Swartzia polyphylla* and their antibacterial activity against cariogenic bacteria. *Chem. Pharmaceut. Bull.*, **40** 2970-74.
Osawa, K., Yasuda, H., Maruyama, T., Morita, H., Takeya, K. and Itokawa, H. (1994) Antibacterial trichorabdal diterpenes from *Rabdosia trichocarpa. Phytochemistry*, **36** 1287-91.

Pabuccuoglu, V., Rozwadowska, M.R., Brossi, A., Clark, A., Hufford, C.D., George, C. and Flippen-Anderson, J.L. (1991) Oxoaporphine alkaloids: conversion of lysicamine into liriodenine and its 2-*O*-methylether and antifungal activity. *Arch. Pharm.*, **324** 29-33.

Panizzi, L., Flamini, G., Cioni, P.L. and Morelli, I. (1993) Composition and antimicrobial properties of essential oils of four mediterranean Lamiaceae. *J. Ethnopharmacol.*, **39** 167-70.

Pare, P.W., Dmitrieva, N. and Maybry, T.J. (1991) Phytoalexin aurone induced in *Cephalocereus senilis* liquid suspension culture. *Phytochemistry*, **30** 1133-35.

Paris, A., Strukelj, B., Renko, M. and Turk, V. (1993) Inhibitory effect of carnosolic acid on HIV-1 protease in cell-fee assays. *J. Nat. Prod.*, **56** 1426-30.

Parniske, M., Ahlborn, B. and Werner, D. (1991) Isoflavonoid-inducible resistance to the phytoalexin, glyceollin, in soybean rhizobia. *J. Bacteriol.*, **173** 3432-39.

Paster, N., Juven, B.J. and Harshemesh, H. (1988) Antimicrobial activity and inhibition of aflatoxin B1 formation by olive plant tissue constituents. *J. Appl. Bacteriol.*, **64** 293-97.

Pattnaik, S., Subramanyam, V.R. and Rath, C.C. (1995a) Effect of essential oils on the viability and morphology of *Escherichia coli* (SP-11). *Microbios*, **84** 195-99.

Pattnaik, S., Subramanyam, V.R., Kole, C.R. and Sahoo, S. (1995b) Antibacterial activity of essential oils from *Cymbopogon*: inter- and intraspecific differences. *Microbios*, **84** 239-45.

Pattnaik, S., Subramanyam, V.R. and Kole, C.R. (1996) Antibacterial and antifungal activity of ten essential oils *in vitro*. *Microbios*, **86** 237-46.

Pattnaik, S., Subramanyam, V.R., Bapaji, M. and Kole, C.R. (1997) Antibacterial and antifungal activity of aromatic constituents of essential oils. *Microbios*, **89** 39-46.

Paulo, M., Barbosa-Filho, J.M., Lima, E.O., Maia, R.F., Barbosa, R. and Kaplan, M.A. (1992) Antimicrobial activity of benzylisoquinoline alkaloids from *Annona salzmanii* D.C. *J. Ethnopharmacol.*, **36** 39-41.

Paulo, A., Duarte, A. and Gomes, E.T. (1994) *In vitro* antibacterial screening of *Cryptolepis sanguinolenta* alkaloids. *J. Ethnopharmacol.*, **44** 127-30.

Paulo, A., Gomes, E., Duarte, A., Perrett, S. and Houghton, P.J. (1997) Chemical and antimicrobial studies on *Cryptolepis obtusa* leaves. *Fitoterapia*, **LXVIII** 558-559.

Pearce, G., Marchand, P.A., Griswold, J., Lewis, N.G. and Ryan, C.A. (1998) Accumulation of feruloyltyramine and *p*-coumaroyltyramine in tomato leaves in response to wounding. *Phytochemistry*, **47** 659-64.

Pengsuparp, T., Cai, L., Constant, H., Fong, H.H.S., Lin, F.L.-Z., Ingolfsdottir, K., Wagner, H. and Hughes, S. (1995) Mechanistic evaluation of new plant-derived compounds that inhibit HIV-1 reverse transcriptase. *J. Nat. Prod.*, **58** 1024-31.

Perese, M.T., Delle-Monache, E., Crūz, A.B., Pizzolatti, M.G. and Yūnes, R.A. (1997) Chemical composition and antimicrobial activity of *Croton urucurana* Baillon (Euphorbiaceae). *J. Ethnopharmacol.*, **56** 223-26.

Poehland, B.L., Carté, B.K., Francis, T.A., Hyland, L.J., Allaudeen, H.S. and Troupe, N. (1987) *In vitro* antiviral activity of dammar resin triterpenoids. *J. Nat. Prod.*, **50** 706-13.

Pomilio, A.B., Buschi, C.A., Tomes, C.N. and Viale, A.A. (1992) Antimicrobial constituents of *Gomphrena martiana* and *Gomphrena boliviana*. *J. Nat. Comp.*, **36** 155-61.

Preisig, C.L. and Kuc, J. (1985) Arachidonic acid-related elicitors of the hypersensitive response in potato and enhancement of their activities by glucans from *Phytophthora infestans*. *Arch. Biochem. Biophys.*, **236** 379-89.

Quetin-Leclercq, J., Favel, A., Balansard, G., Regli, P. and Angenot, L. (1995) Screening for *in vitro* antifungal activities of some indole alkaloids. *Planta Med.*, **61** 475-77.

Rahman, A.U., Ashraf, M., Choudhary, M.I., Rehman, H.U. and Kazmi, M.H. (1995) Antifungal aryltetralin lignans from leaves of *Podophyllum hexandrum*. *Phytochemistry*, **40** 427-31.

Rahman, A.U., Nasreen, A., Akhtar, F., Shekhani, S., Clardy, J., Parvez, M. and Choudhary, M.I. (1997) Antifungal diterpenoid alkaloids from *Delphinium denudatum*. *J. Nat. Prod.*, **60** 472-74.

Ratsimamanga-Urverg, S., Rasoanaivo, P., Rabemanantsoa, C., Ratsimamanga, A.R. and Frappier, F. (1994) Antimicrobial activity of flavonoids isolated from *Mundulea monantha* and *Tephrosia linearis*. *Fitoterapia*, **LXV** 551-53.

Renard-Nozaki, J., Kim, T., Imakura, Y., Kihara, M. and Kobayashi, S. (1989) Effect of alkaloids isolated from Amaryllidaceae on herpes simplex virus. *Res. Virol.*, **140** 115-28.

Rimando, A.M., Pezzuto, J.M. and Farnsworth, N.R. (1994) New lignans from *Anogeissus acuminata* with HIV-1 reverse transcriptase inhibitory activity. *J. Nat. Prod.*, **57** 896-904.

Roby, D., Toppan, A. and Esquerrétugayé (1986) Cell surfaces in plant-microorganism interactions. *Plant Physiol.*, **81** 228-33.

Rocha, L., Marston, A., Potterat, O., Kaplan, M.A., Stoeckli-Evans, H. and Hostettmann, K. (1995) Antibacterial phloroglucinols and flavonoids from *Hypericum brasiliense*. *Phytochemistry*, **40** 1447-52.

Saad, H.E., El-Sharkawy, S.H. and Shier, W.T. (1995) Biological activities of pyrrolidinoindoline alkaloids from *Calycodendron milnei*. *Planta Med.*, **61** 313-16.

San Feliciano, A., Gordaliza, M., del Corral, J.M.M., Castro, M.A., Garcia-Gravalos, M.D. and Ruiz-Lazaro, P. (1993) Antineoplastic and antiviral activities of some cyclolignans. *Planta Med.*, **59** 246-49.

Sartorelli, P., Young, M.C. and Kato, M.J. (1998) Antifungal lignans from the arils of *Virola oleifera*. *Phytochemistry*, **47** 1003-1006.

Sawer, I.K., Berry, M.I., Brown, M.W. and Ford, J.L. (1995) The effect of cryptolepine on the morphology and survival of *Escherichia coli*, *Candida albicans* and *Saccharomyces cerevisiae*. *J. Appl. Bacteriol.*, **79** 314-21.

Schales, C., Gerlach, H. and Köster, J. (1993) Investigation on the antibacterial effect of conifer needle oils on bacteria isolated from feces of captive capercaillies (*Tetrao urogallus*). *J. Vet. Med. B*, **40** 381-90.

Schloesser, E. (1983) Allgemeine Phytopathologie, Thieme, Stuttgart.

Schmitt, A., Telikepalli, H. and Mitscher, L.A. (1991) Plicatin B, the antimicrobial principle of *Psoralea juncea*. *Phytochemistry*, **30** 3569-70.

Schultz,T.P., Boldin, W.D., Fisher, T.H., Nicholas, D.D., Murtrey, K.D. and Pobanz, K. (1992) Structure-fungicidal properties of some 3- and 4-hydroxylated stilbenes and bibenzyl analogues. *Phytochemistry*, **31** 3801-06.

Serkedjieva, J. and Manolova, N. (1992) Plant polyphenolic complex inhibits the reproduction of influenza and herpes simplex viruses. *Basic Life Sci. (Plant Polyphenols)*, **59** 705-15.

Shukla, H.S., Dubey, P. and Chaturvedi, R.V. (1989) Antiviral properties of essential oils of *Foeniculum vulgare* and *Pimpinella anisum*. *Agronomie*, **9** 277-79.

Siddiqui, S., Faizi, S., Siddiqui, B.S. and Ghiasuddin, H.E. (1992) Constituents of *Azadirachta indica*: isolation and structure elucidation of a new antibacterial tetranortriterpenoid, mahmoodin, and a new protolimonoid, naheedin. *J. Nat. Prod.*, **55** 303-10.

Simeon, S., Rios, J.L. and Villar, A. (1990) Antimicrobial activity of *Annona cherimolia* stem bark alkaloids. *Pharmazie*, **45** 442-43.

Simeos, C.M., Amoros, M., Girre, L., Gleye, J. and Fauvel, M. (1990) Antiviral activity of ternatin and meliternatin, 3-methyloxyflavones from species of Rutaceae. *J. Nat. Prod.*, **53** 989-92.

Singh, S.P., Negi, S., Chand, L. and Singh, A.K. (1992) Antibacterial and antifungal activities of *Mentha arvensis*. *Fitoterapia*, **LXIII** 76-78.

Smirnov, V.V., Bondarenko, A.S. and Prikhodko, V.A. (1998) Antimicrobial activity of sesquiterpene phenol from *Bidens cernua*. *Fitoterapia*, **LXIX** 84-85.

Smith, D.W. and Banks, S.W. (1986) Biosynthesis, elicitation and biological activity of isoflavonoid phytoalexins. *Phytochemistry*, **25** 979-95.

Soby, S., Bates, R. and van Etten, H. (1997) Oxidation of the phytoalexin maackiain to 6,6-dihydroxy-maackiain by *Colletotrichum gloeosporioides*. *Phytochemistry*, **45** 925-29.

Sprecher, E. and Urbasch, J. (1983) Interaktionen zwischen höheren Pflanzen und phytopathogenen Pilzen. *Deutsche Apotheker Zeitung*, **123** 1961-71.

Srivastava, S.D. (1986) Limonoids from the seeds of *Melia azedarach*. *J. Nat. Prod.*, **49** 56-61.

Stevenson, N.R. and Lenard, J. (1993) Antiretroviral activities of hypericin and rose bengal: phytodynamic effects on Friend leukemia virus infection of mice. *Antiviral Res.*, **21** 119-27.

Suzuki, Y., Kono, Y., Inoue, T. and Sakurai, A. (1998) A potent antifungal benzoquinone in etiolated *Sorghum* seedlings and its metabolites. *Phytochemistry*, **47** 997-1001.

Takaisi-Kikuni, N.B., Krüger, D., Gnann, W. and Wecke, J. (1996) Microcalorimetric and electron microscopic investigation on the effects of essential oil from *Cymbopogon densiflorus* on *Staphylococcus aureus*. *Microbios*, **88** 55-62.

Takasugi, M. and Masuda, T. (1996) Three 4′-hydroxyacetophenone-related phytoalexins from *Polymnia sonchifolia*. *Phytochemistry*, **43** 1019-21.

Takechi, M., Tanaka, Y., Takehara, M., Nonaka, I., Nishioka, I. (1985) Structure and antiherpic activity among tannins. *Phytochemistry*, **24** 2245-50.

Tanaka, H. and Fujimori, T. (1985) Accumulation of phytuberin and phytuberol in tobacco callus inoculated with *Pseudomonas solanacereum* or *Pseudomonas syringae*. *Phytochemistry*, **24** 1193-95.

Tang, J., Colacino, J.M., Larsen, S.H. and Spitzer, W. (1990) Virucidal activity of hypericin against enveloped and non-enveloped DNA and RNA viruses. *Antiviral Res.*, **13** 313-26.

Tatematsu, H., Kilkuskie, R.E., Corrigan, A.J., Bodner, A.J. and Lee, K.-H. (1991) Anti-AIDS agents. 3. Inhibitory effects of colchicine derivatives on HIV replication in H9 lymphocyte cells. *J. Nat. Prod.*, **54** 632-37.

Tawara, J.N., Blokhi, A., Foderaro, T.A. and Stermitz, F.R. (1993) Toxic piperidine alkaloids from pine (*Pinus*) and spruce (*Picea*) trees: new structures and a biosynthetic hypothesis. *J. Org. Chem.*, **58** 4813-18.

Taylor, P.B., Culp, J.S., Debouck, C., Johnson, R.K., Patil, A.D., Woolf, D.J., Brooks, I. and Hertzberg, R.P. (1994) Kinetic and mutational analysis of human immunodeficiency virus type 1 reverse transcriptase inhibition by inophyllums, a novel class of non-nucleoside inhibitors. *J. Biol. Chem.*, **269** 6325-31.

Taylor, R.S. and Towers, G.H. (1998) Antibacterial constituents of the Nepalese medicinal herb, *Centipeda minima*. *Phytochemistry*, **47** 631-34.

Tereschuk, M.L., Riera, M.V., Castro, G.R. and Abdala, L.R. (1997) Antimicrobial activity of flavonoids from leaves of *Tagetes minuta*. *J. Ethnopharmacol.*, **56** 227-32.

Tirillini, B., Velaquez, E.R. and Pellegrino, R. (1996) Chemical composition and antimicrobial activity of essential oil of *Piper angustifolium*. *Planta Med.*, **62** 372-73.

Tomas-Barberan, F.A., Msonthi, J.D. and Hostettmann, K. (1988) Antifungal epicuticular methylated flavonoids from *Helichrysum nitens*. *Phytochemistry*, **27** 753-55.

Tomas-Lorente, F., Iniesta-Sanmartin, E., Tomas-Barberan, F.A., Trowitzsch-Kienast, W. and Wray, V. (1989) Antifungal phloroglucinol derivatives and lipophilic flavonoids from *Helichrysum decumbens*. *Phytochemistry*, **28** 1613-15.

Tomas-Lorente, F., Iniesta-Sanmartin, E. and Tomas-Barberan, F.A. (1991) Antimicrobial phenolics from *Helichrysum picardii*. *Fitoterapia*, **LXII** 521-23.

Tsai, I.L., Liou, Y.F. and Lu, S.T. (1989) Screening of isoquinoline alkaloids and their derivatives for antibacterial and antifungal activities. *Kaohsing J. Med. Sci.*, **5** 132-45.

Turbek, C.S., Smith, D.A., Schardl, C.L. (1992) An extracellular enzyme from *Fusarium solani* f. sp. *phaseoli* which catalyses hydration of the isoflavonoid phytoalexin, phaseollidin. *FEMS Microbiol. Lett.*, **73** 187-90.

Tverskoy, L., Dmitriev, A., Kozlovsky, A. and Grodzinsky, D. (1991) Two phytoalexins from *Allium cepa* bulbus. *Phytochemistry*, **30** 799-800.

Ulubelen, A., Evren, N., Tuziaci, E. and Johansson, C. (1988) Diterpenoids from the roots of *Salvia hypargeia. J. Nat. Prod.*, **51** 1178-83.

Ulubulen, A., Topcu, G., Eris, C., Sönmez, U., Kartal, M., Kurucu, S. and Bozok-Johansson, C. (1994) Terpenoids from *Salvia sclarea. Phytochemistry*, **36** 971-74.

Ulubelen, A., Sonmez, U., Topcu, G. and Bozok-Johansson, C. (1996) An abietane diterpene and two phenolics from *Salvia forskahlei. Phytochemistry*, **42** 145-47.

Van den Berghe, D.A., Ieven, M., Mertens, F., Vlietinck, A.J. and Lammens, E. (1978) Screening of higher plants for biological activities. II. Antiviral activities. *Lloydia*, **41** 463-71.

Van den Berghe, D.A., Vlietinck, A.J. and Van Hoof, L. (1986) Plant products as potential antiviral agents. *Bulletin de l'Institut Pasteur*, **84** 101-47

Van Hoof, L., Totté, J., Corthout, J., Pieters, L.A., Mertens, F., Vanden Berghe, D.A. and Vlietinck, A.J. (1989) Plant antiviral agents. VI. Isolation of antiviral phenolic glucosides from *Populus* cultivar Beaupre by droplet countercurrent chromatography. *J. Nat. Prod.*, **52** 875-78.

Van Puyvelde, L., de Kimpe, N., Costa, J., Munjabo, V., Nyirankuliza, S., Hakizamungu, E. and Schamp, N. (1989) Isolation of flavonoids and chalcone from *Helichrysum odoratissimum* and synthesis of helichrysetin. *J. Nat. Prod.*, **52** 629-33.

Vasanth, S., Hamsaveni Gopal, R. and Bhima Rao, R. (1997) Antibacterial activity of *Cryptolepsis buchanani. Fitoterapia*, **LXVIII** 463-64.

Verma, D.K., Tripathi, V.J., Rana, B.K. and Taneja, V. (1998) Antifungal activity of triterpenoids from *Latana indica* root. *Fitoterapia*, **LXIX** 188-89.

Verpoorte, R. (1998) Antimicrobially active alkaloids, in *Alkaloids, Biochemistry, Ecology and Medicinal Application* (eds. M.F. Roberts, M. Wink). Plenum Press, New York, pp. 397-425.

Vlietinck, A.J. and Vanden Berghe, D.A. (1991) Can ethnopharmacology contribute to the development of antiviral drugs? *J. Ethnopharmacol.*, **32** 141-53.

Vlietinck, A.J., De Bruyne, T., Apers, S. and Pieters, L.A. (1998) Plant-derived leading compounds for chemotherapy of human immunodeficiency virus (HIV) infection. *Planta Med.*, **64** 97-109.

Walker-Simmons, M., Jin, D., West, C.A., Hardwiger, L. and Ryan, C.A. (1984) Comparison of proteinase inhibitor-inducing activities and phytoalexin elicitor activities of a pure fungal endopolygalacturonase, pectic fragments and chitosans. *Plant Physiol.*, **76** 833-36.

Walton, T.J., Cooke, C.J., Newton, R.P. and Smith, C.J. (1993) Evidence that generation of inositol 1,4,5-trisphosphate and hydrolysis of phosphatidylinositol 4,5-bisphosphate are rapid response following addition of fungal elicitor which induces phytoalexin synthesis in lucerne (*Medicago sativa*) suspension culture cells. *Cell-Signal*, **5** 345-56.

Waterman, P.G. (1990) Searching for bioactive compounds: various strategies. *J. Nat. Prod.*, **53** 13-22.

Weber, N.D., Andersen, D.O., North, J.A., Murray, B.K., Lawson, L.D. and Hughes, B.G. (1992) *In vitro* virucidal effects of *Allium sativum* (garlic) extracts and compounds. *Planta Med.*, **58** 417-23.

Weidenbörner, M., Hindorf, H., Jha, H.C. and Tsotsonos, P. (1990a) Antifungal activity of flavonoids against storage fungi of the genus *Aspergillus. Phytochemistry*, **29** 1103-105.

Weidenbörner, M., Hindorf, H., Chandra, J., Tsotsonos, P. and Egge, H. (1990b) Antifungal activity of isoflavonoids in different reduced stages on *Rhizoctonia solani* and *Sclerotium rolfsii. Phytochemistry*, **29** 801-803.

West, C.A. (1981) Fungal elicitors of the phytoalexin response in higher plants. *Naturwissenschaften*, **68** 447-57.

Wink, M. (1993) Allelochemical properties or the raison d'être of alkaloids, in *The Alkaloids*, Vol. 43 (ed. G.A. Cordell) Academic Press, San Diego, pp. 1-118.

Wolters, B. and Eilert, U. (1983) Elicitoren: Auslöser der Akkumulation von Pflanzenstoffen. *Deutsche Apotheker Zeitung*, **123** 659-67.

Xu, H.X., Kadoto, S., Kurokawa, M., Shiraki, K., Matsumoto, T. and Namba, T. (1993) Isolation and structure of woodorien, a new glucoside having antiviral activity, from *Woodwardia orientalis*. *Chem. Pharmaceut. Bull.*, **41** 1803-806.

Xu, H.X., Zeng, F.Q., Wan, M. and Sim, K.Y. (1996) Anti-HIV triterpene acid from *Geum japonicum*. *J. Nat. Prod.*, **59** 643-45.

Yamamoto, N., Furukawa, H., Ito, Y., Yoshida, S., Maeno, K. and Nishiyama, Y. (1989) Anti-herpesvirus of citrusinine-I, a new acridone alkaloid, and related compounds. *Antiviral Res.*, **12** 21-36.

Zahir, A., Jossang, A. and Bodo, B. (1993) Knerachelins A and B, antibacterial phenylacyl-phenols from *Knema furfuracea*. *J. Nat. Prod.*, **56** 1634-37.

Zhao, W., Wolfender, J.L., Hostettmann, K., Xu, R. and Qin, G. (1998) Antifungal alkaloids and limonoid derivatives from *Dictamnus daysycarpus*. *Phytochemistry*, **47** 7-11.

Zeringue, H.J. (1990) Stress effects on cotton leaf phytoalexins elicited by cell-free mycelia extracts of *Aspergillus flavus*. *Phytochemistry*, **29** 1789-91.

6 New medical applications of plant secondary metabolites

Jörg Heilmann and Rudolf Bauer

6.1 Introduction

Plants continue to be a rich and valuable source of new compounds with potent pharmacological activity. Whereas only a few plant-derived secondary metabolites have been directly used as drugs, many pharmacologically active compounds have served as leading models for semisynthetic and synthetic drugs. Additionally, there is a growing interest in the application of standardized extracts, complex phytochemical mixtures with a well-defined content of the bioactive constituents. Emphasizing the therapeutic importance of natural compounds, various books and articles have been published in the past concerning their medical aspects (Kinghorn and Balandrin, 1993; Clark, 1996; Teuscher, 1997; Shu, 1998). The present chapter will provide an overview of new plant-derived compounds that have attracted medical and pharmacological interest in the last ten years.

6.2 Compounds with anticancer activity

In recent years, a strong interest has been maintained in natural products exhibiting antitumour activity. The development of drugs for the treatment of various human tumours has been focused mainly on three molecular targets: the eukaryotic deoxyribonucleic acid (DNA) topoisomerases; the microtubuli apparatus; and the different enzymes responsible for the accelerated cell cycle in cancer cells.

Two types of DNA topoisomerase have been described in the eukaryotic cell. They play an essential role in the transcription, replication and repair of DNA. Type II topoisomerase enables the supercoiling of DNA by catalysing double-strand breaks. It has been demonstrated that the antitumour activity of some well-known anticancer drugs, the epipodophyllotoxins, etoposide (Vepesid®) and teniposide (Vumon®) (Fig. 6.1), is due to the inhibition of topoisomerase II (Pommier and Kohn, 1989; Liu, 1989). Both compounds are semisynthetic derivatives of the lignan, podophyllotoxin, a constituent of the mayapple, *Podophyllum peltatum* (L.) (Berberidaceae), also known as American mandrake. Etoposide, in particular, is in extensive clinical use

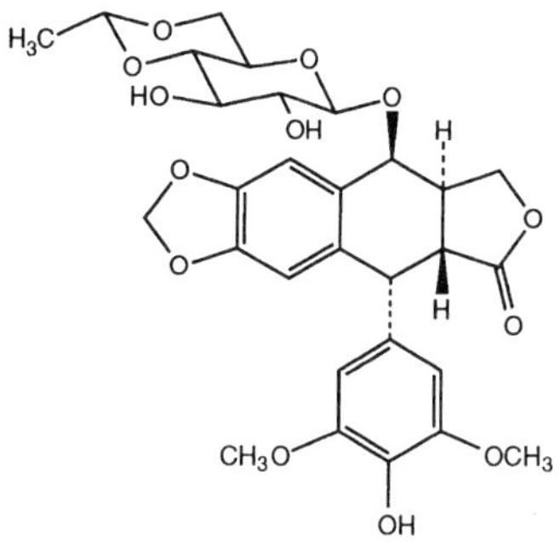

etoposide (**1**)

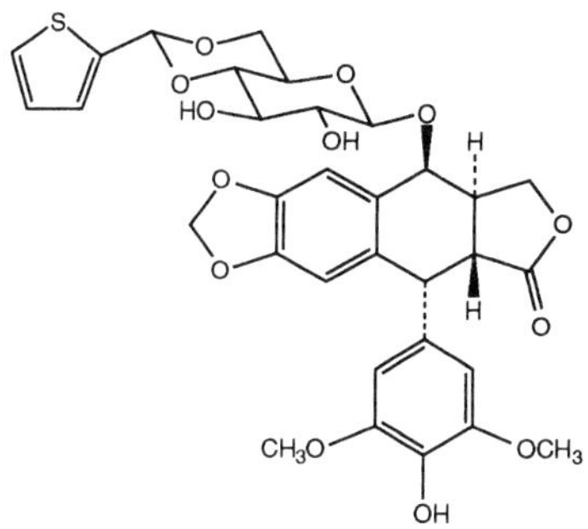

teniposide (**2**)

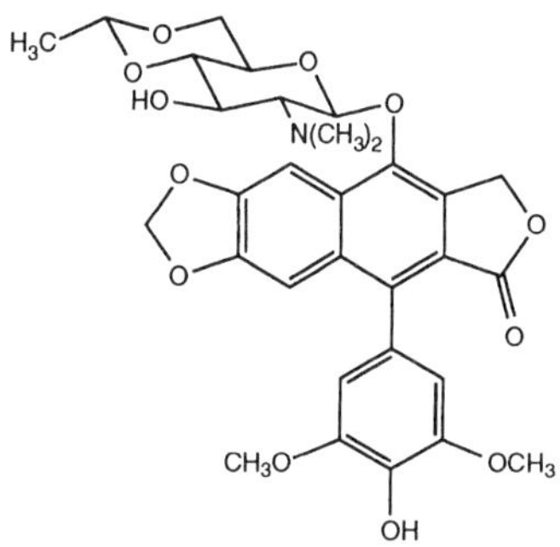

NK 611 (**3**)

Figure 6.1 Structures of the epipodophyllotoxins, etoposide (**1**) and teniposide (**2**), and NK 611 (**3**), a dimethylamino derivative of etoposide.

against various cancer types, e.g. small-cell lung cancer, non-small-cell lung cancer, ovarian cancer and breast cancer. Both are preferentially used in combination regimen with other anticancer drugs, such as cisplatin, carboplatin and cyclophosphamide (Gridelli *et al.*, 1997; Klumpp *et al.*, 1997; Rose *et al.*, 1998; Westeel *et al.*, 1998). Modification in the sugar ring resulted in the development of a dimethylamino derivative of etoposide (NK 611) (Fig. 6.1), which is at present in clinical trials (Pagani *et al.*, 1996; Rassmann *et al.*, 1996; for literature see Damayanthi and Lown, 1998).

Another group of plant-derived cancer drugs with inhibitory activity on topoisomerase II are the ellipticines. The alkaloids, ellipticine, 9-methoxyellipticine and olivacin (Figs. 6.2 and 6.3) have been isolated from *Ochrosia* and *Aspidosperma* species (Apocynaceae). They have exhibited marked *in vitro* antitumour activity in different cancer cell lines, such as L1210, P388 and P1534 leukaemias (for literature see Suffness and Cordell, 1985). 9-Hydroxyellipticine (Fig. 6.2), one of the 9-methoxyellipticine (Fig. 6.2) metabolites, was significantly more potent

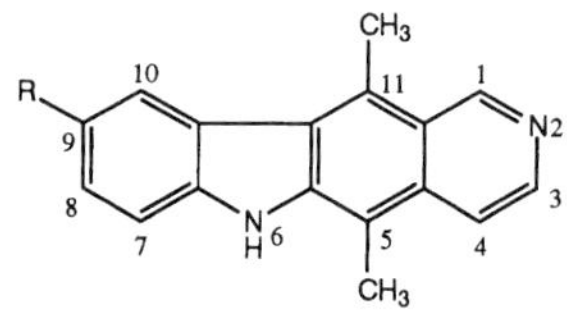

R = H	ellipticine (**4**)
R = OCH_3	9-methoxyellipticine (**5**)
R = OH	9-hydroxyellipticine (**7**)

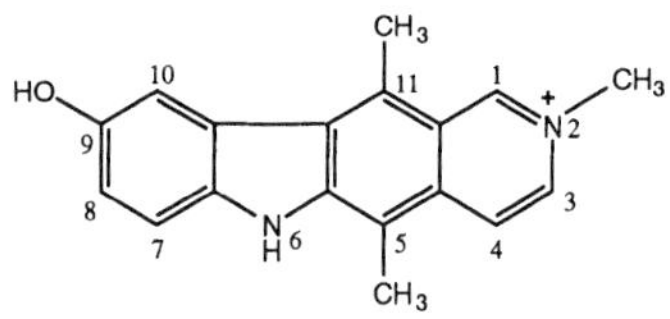

elliptinium (9-hydroxy-2-methylellipticine) (**8**)

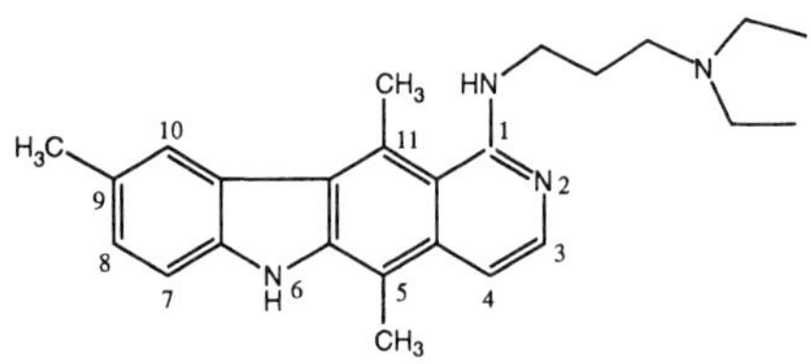

retelliptine (1-diethyl-aminopropylamino-9-methylecllipticine) (**9**)

Figure 6.2 Structures of the alkaloids, ellipticine (**4**) and 9-methoxyellipticine (**5**), of the 9-methoxyellipticine metabolite, 9-hydroxyellipticine (**7**), and of the semisynthetic water soluble derivatives, elliptinium (**8**) and retelliptine (**9**).

than its parent compound but its low solubility in water has hampered progression to clinical trials and has led to the development of the semisynthetic water soluble derivatives, elliptinium (2-methyl-9-hydroxyellipticine) (Fig. 6.2), retelliptine (1-diethyl-aminopropylamino-9-methylellipticine) (Fig. 6.2), pazelliptine (Fig. 6.3) and S 16020-2 (1-diethylaminoethylolivacine) (Fig. 6.3). These compounds have shown good *in vitro* cytotoxicity against various human and murine cell lines (Leonce *et al.*, 1996; Le Mée *et al.*, 1998) and *in vivo* antitumour activity in murine and human xenografts, e.g. P388 leukaemia and NCI-H460 (Langdon *et al.*, 1994; Guilbaud *et al.*, 1996; Krausberthier *et al.*, 1997). Retelliptine (Fig. 6.2) has already been tested in a dose-finding phase-I clinical trial (Kattan *et al.*, 1994). In some of these examinations, S 16020-2 (Fig. 6.3) was significantly more efficient than the other ellipticines (Guilbaud *et al.*, 1996; Leonce *et al.*, 1996) and has therefore been considered to be the most promising derivative for extensive clinical trials.

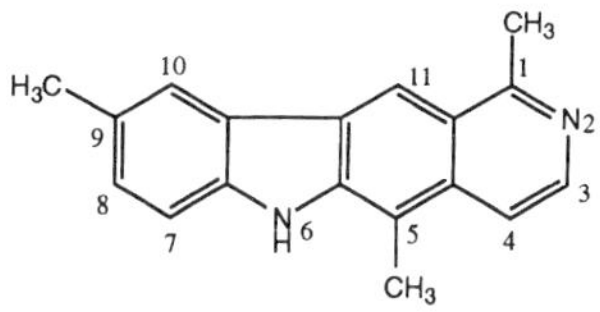

olivacine (**6**)

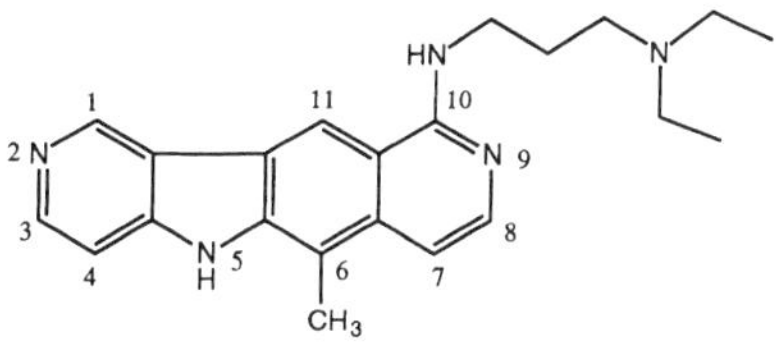

pazelliptine (**10**)

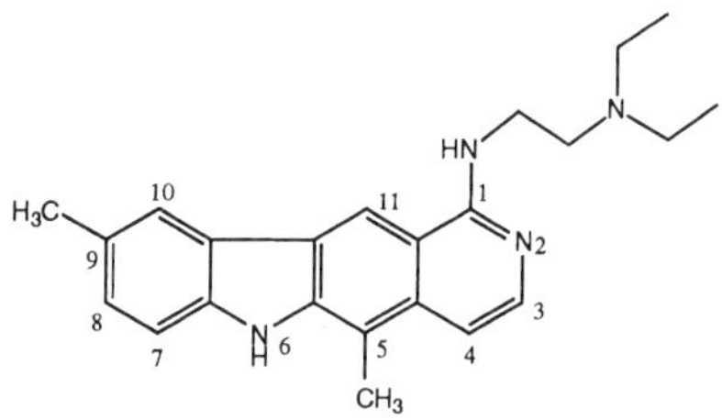

1-diethylaminoethylolivacine (**11**)

Figure 6.3 Structures of the cytotoxic compounds, olivacine (**6**), pazelliptine (**10**) and 1-diethylaminoethylolivacine (S 16020-2) (**11**).

It is noteworthy that the principal mechanism of the antitumour activity of ellipticines is not only related to inhibition of topoisomerase II but also to their planar structure and to their ability to intercalate with DNA. Moreover, it was shown that ellipticine and 9-hydroxyellipticine (Fig. 6.2) caused selective inhibition of p53 protein phosphorylation in Lewis lung carcinoma and SW480, a human colon cancer cell line. 9-Hydroxyellipticin suppressed cdk2 kinase activity in a concentration-dependent manner. The resulting higher concentrations of unphosphorylated proteins led to the expression of an apoptosis-inducing gene and cell death (Ohashi *et al.*, 1995). The ellipticines are also important antiviral compounds (see Section 6.3).

Another target of anticancer drugs is topoisomerase I. This enzyme allows the relaxation of supercoiled DNA by catalysing breaks of one DNA strand. The intermediates generated, enzyme-linked DNA breaks called the 'cleavable complex', form gates that can be passed by one DNA strand. At the end of the strain passage reaction, topoisomerases religate the DNA without loss of bases or change in the DNA sequence. It has been shown that topoisomerase I is the main cellular target of camptothecin (Fig. 6.4) and its semisynthetic derivatives, e.g. topotecan (hycamptamine = 9-dimethylaminomethyl-10-hydroxycamptothecin), 9-aminocamptothecin, 10,11-methylenedioxycamptothecin, GI 147211C (7N-methylpiperazinomethylene-10,11-ethylenedioxycamptothecin), irinotecan (CPT-11 = 7-ethyl-10-[4-(1-piperidino)-1-piperidino]carbonyloxycamptothecin) and 9-nitrocamptothecin (Fig. 6.4).

The antileukaemic activity of camptothecin, first isolated from *Camptotheca acuminata* Decne. (Nyssaceae), was demonstrated in various *in vitro* models in the late 1960s (Wall *et al.*, 1966; for overview

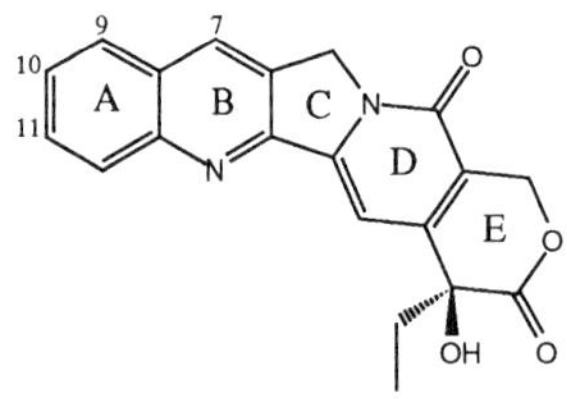

camptothecin (**12**)

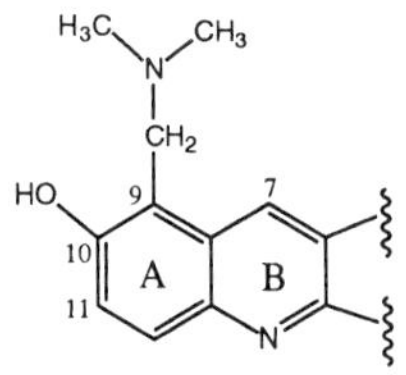

topotecan (hycamptamine) (**13**)

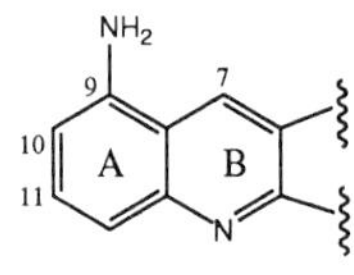

9-aminocamptothecin (9-AC) (**14**)

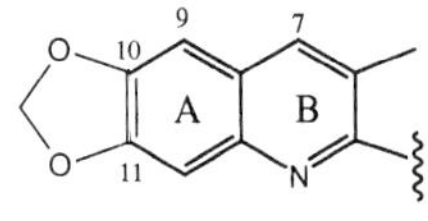

10,11-methylenedioxycamptothecin (**15**)

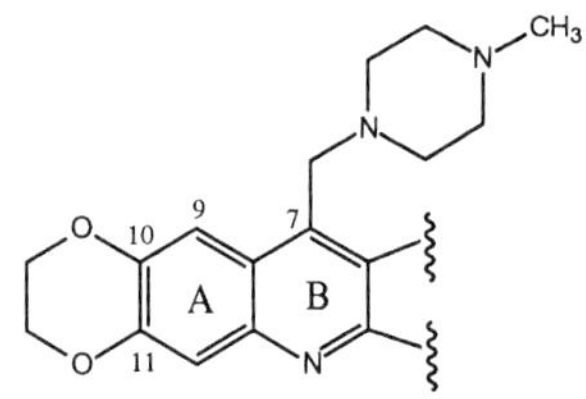

7N-methylpiperazinomethylene-10,11-ethylene-dioxycamptothecin (**16**)

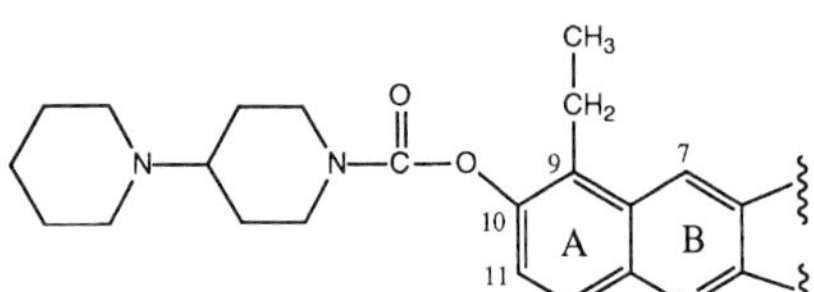

irinotecan (CPT 11)
7-ethyl-10-[4-(1-piperidino)-1-piperidino]
carbonyloxycamptothecin (**17**)

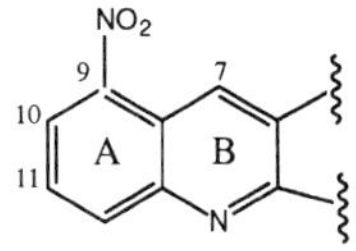

9-nitrocamptothecin (18)

Figure 6.4 Structures of camptothecin (**12**) and its semisynthetic derivatives.

see Wall and Wani, 1995). Because of its extreme insolubility in water, clinical trials were performed with the readily soluble sodium salt (Gottlieb and Luce, 1972; Muggia *et al.*, 1972), not knowing that this compound was only one-tenth as active as camptothecin (Wani *et al.*, 1980). The trials were halted because of disappointing response rates and severe side-effects (Moertel *et al.*, 1972). Since it was found that camptothecins possess topoisomerase I inhibitory properties, several more water-soluble semisynthetic analogues (Fig. 6.4) have been synthesized and evaluated for their anticancer activity. Hsiang and co-workers (1985) showed that camptothecin-induced DNA breaks are mediated by blocking topoisomerase I-cleavable complexes and inhibiting the religation of the topoisomerase I reaction (Svejstrup *et al.*, 1991). The correlation between inhibition of purified topoisomerase I by camptothecin and antitumour activity has been clearly demonstrated (Hsiang *et al.*, 1989). Racemic camptothecins were separated into the corresponding 20-(*S*) and 20-(*R*) analogues (Wani *et al.*, 1987) and it was shown that only the 20-(*S*)-camptothecins are active in topoisomerase I inhibition and in *in vivo* assays (Jaxel *et al.*, 1989; Giovanella *et al.*, 1991).

In aqueous solution, camptothecins occur in an equilibrium of two forms, the α-hydroxylactone and the α-hydroxycarboxyl form (see Fig. 6.5). The α-hydroxylactone moiety is considered to be essential for topoisomerase I activity, whereas α-hydroxycarboxyl forms are less active

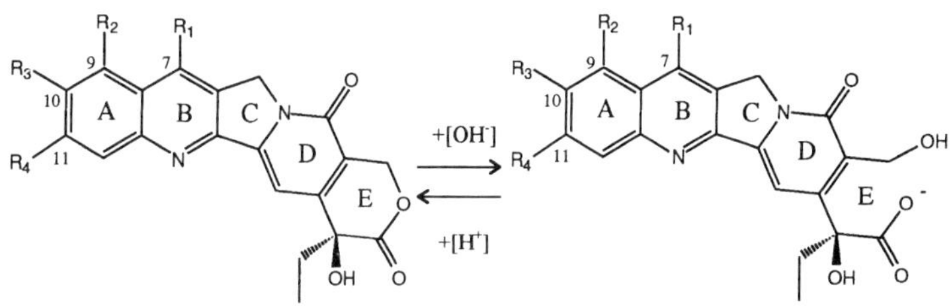

Figure 6.5 Equilibrium between the hydroxylactone and hydroxycarboxyl of camptothecins.

antitumour agents (Giovanella *et al.*, 1991). The carboxylate form can reverse to the lactone at acidic pH and it has been concluded that the cellular pharmacokinetics and the antitumour activity of the camptothecins are influenced by the pH of the target tissue (Teicher *et al.*, 1993; Pommier, 1996; Gabr *et al.*, 1997). It has been hypothesized that the limited recycling of the sodium salt to camptothecin under physiological conditions was the reason for its disappointing results in the original trials. The mechanism by which camptothecins achieve selective toxicity in cancer cells is not fully understood. Some published data (Perego *et al.*, 1994) contradict the assumption that cancer cells contain more topoisomerase I than normal cells (Giovanella *et al.*, 1989). From other results, it can be assumed that deficiencies of cancer cells in DNA repair or the inability to repress apoptopsis may contribute to the selective cytotoxity (Pommier *et al.*, 1994a,b).

A wide range of clinical and scientific data has been published on the subject of camptothecins. Numerous recent studies have shown the *in vitro* activity of camptothecin derivatives in different cell systems or xenograft models (Emerson *et al.*, 1995; Potmesil *et al.*, 1995) and antitumour activity against colorectal-, gastric-, cervical-, small-cell lung cancer and various other malignancies in clinical trials (for literature see Potmesil, 1994; Bonneterre, 1995). Combination with other antitumour agents, e.g. 5-fluorouracil or cisplatin, has produced promising results. As with many antitumour agents, the main toxic side-effects of camptothecins were evident in tissues with high cell turnover, and led to haematological toxicity. Further adverse events are nausea, vomiting and diarrhoea (Rougier and Bugat, 1996). In view of their broad spectrum of antitumour activity, the camptothecins will be valuable components for future cancer treatment. Up to now, topotecan (Hycamtin®) and irinotecan (Campto®) (Fig. 6.4) are indicated for the second-line treatment of adult patients with metastatic ovarian carcinoma or with metastatic colorectal cancer, respectively. Expansion of the therapeutic indications is in progress. Clinical resistance against the anticancer effects of the camptothecins is based, for instance, on: a reduced level of topoisomerase I protein (Madeleine *et al.*, 1993); mutation of the topoisomerase I gene, resulting in the formation of enzymes with altered structure (Fujimori *et al.*, 1995); and lack of apoptosis (Solary *et al.*, 1994).

Taxane derived compounds were isolated for the first time in the 1960s. However, the approval for marketing of paclitaxel (Taxol®) (Fig. 6.6) occurred in December 1992. Excellent reports on the discovery, development, plant sources, synthesis and semisynthesis, preclinical and clinical studies of paclitaxel (Fig. 6.6) until 1994 have been summarized by Suffness (1995). The present chapter, therefore, will discuss this period very briefly and will focus on the clinical results of the last six years and on the new semisynthetic derivative, docetaxel (Taxotere®) (Fig. 6.6).

Paclitaxel is a diterpenoid with an alkaloidal side chain. It was first isolated from *Taxus brevifolia* NUTT. (Taxaceae) following discovery of the cytotoxicity of the compound to human epidermoid carcinoma cells (KB cells) and its toxicity in leukaemia cell systems (Wani *et al.*, 1971). Prior to the isolation of paclitaxel (Fig. 6.6), there was no report on the antitumour activity of natural taxane derivatives. Subsequent studies with various other tumour models have shown only low activity. But since it was noted that paclitaxel possessed very high activity in the B-16

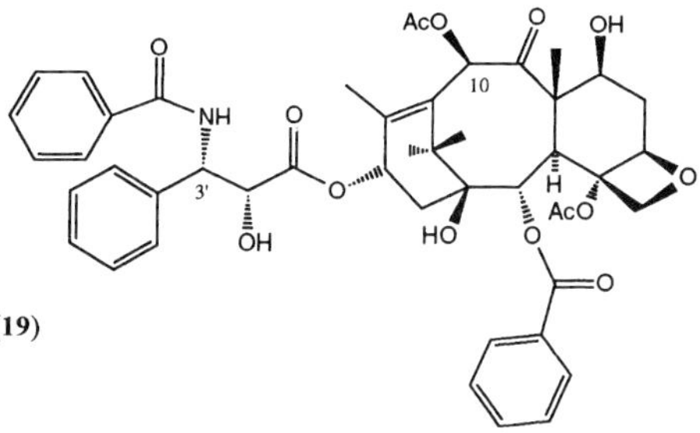

paclitaxel (**19**)

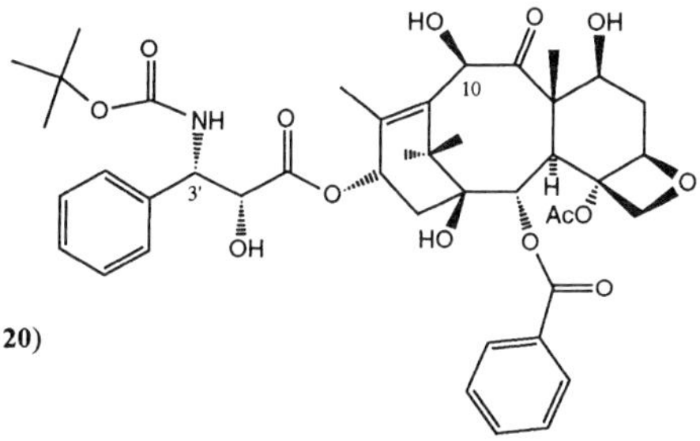

docetaxel (**20**)

Figure 6.6 Structure of paclitaxel (Taxol®) (**19**) and the new semisynthetic derivative, docetaxel (Taxotere®) (**20**).

melanoma assay (Suffness and Wall, 1995), the compound fulfilled the United States National Cancer Institute (US NCI) criteria to become a candidate for development. Clinical Phase I and Phase II trials were conducted over the period from 1983 to 1986. In the later stage of the development of Taxol®, its use was limited by the lack of sufficient supplies of the therapeutic agent. The restricted number of *Taxus brevifolia* trees, the small yield of bark per tree, their slow growth rate and the low content of paclitaxel (Fig. 6.6) in the bark (approximately 0.02%) raised concerns about the reliability of this source (Croom, 1995).

Nowadays, paclitaxel and docetaxel (Fig. 6.6) can be obtained on a large scale by semisynthesis from 10-deacetylbaccatin III, a precursor that is extracted from the needles of *Taxus baccata* (L.), the european yew, and other *Taxus* species. Total synthesis and production in plant cell cultures is not, so far, economically feasible (Suffness, 1995). However, clinical interest received a significant boost by the announcement of its remarkable efficacy against ovarian cancer (McGuire *et al.*, 1989; Einzig *et al.*, 1992). Subsequently, paclitaxel has been studied for activity against other solid tumours. Since the first report on the activity of paclitaxel in breast cancer was published in 1991 (Holmes *et al.*, 1991), many papers have reported on encouraging response rates (22–62%) of patients with breast cancer after the application of paclitaxel (e.g. Reichman *et al.*, 1993). Clinical activity has also been observed in lung, head and neck cancer (Forastiere *et al.*, 1993; Kirschling *et al.*, 1994). Recent investigations have emphasized the increase of response rates, the extension of therapy on other tumour types and reduction of severe side-effects by combining paclitaxel with other anticancer drugs, such as epirubicin (Conte *et al.*, 1998), topotecan (Takimoto and Arbuck, 1997), etoposide (Rosell *et al.*, 1997), and others (for details see Goldspiel, 1997).

The development of docetaxel (Taxotere®) (Fig. 6.6) was enforced because of its better solubility and higher potency compared to paclitaxel (in *in vitro* assays inhibiting cold- or calcium-induced depolymerization of tubulin, it is twice as potent as paclitaxel; Guéritte-Voegelein *et al.*, 1991). Phase I clinical trials were started in 1990 (e.g. Extra *et al.*, 1993). In subsequent Phase I and II trials, it was shown, that docetaxel possesses activity in ovarian, breast, lung, head and neck cancer (Dreyfuss *et al.*, 1996; Kavanagh *et al.*, 1996; Marty *et al.*, 1997; Millward *et al.*, 1997; for literature see Goldspiel, 1997). Interestingly, there is no complete cross-resistance between paclitaxel and docetaxel (Fig. 6.6) in some *ex-vivo* assays, so that some difference in their mode of action can be assumed (Hill *et al.*, 1994). Side-effects during treatment with paclitaxel and docetaxel are, for example, neutropenia, myelosuppression, hypersensitivity reactions, peripheral neuropathy and alopecia (Fumoleau, 1997). Paclitaxel is already registered in several countries for the first- or second-line treatment of ovarian and breast cancer, and docetaxel for the second-line treatment of metastatic breast cancer.

The importance of paclitaxel and docetaxel (Fig. 6.6) is due not only to their broad anticancer activity but also to their special mode of action. Whereas other classic spindle poisons, such as colchicine, bind to soluble tubulin and inhibit its polymerization (Stryer, 1995), both of these compounds stabilize microtubules and inhibit depolymerization back to tubulin (Schiff *et al.*, 1979); therefore, a new candidate for the treatment of cancer has become available. In the meantime, more natural compounds with a similar mode of action, such as diterpenoids, eleutherobin or sarcodictyin, isolated from corals, have reached clinical or preclinical trials and are promising drug candidates for the future (Ciomei *et al.*, 1997; Long *et al.*, 1998).

In addition to alkaloids, flavonoids display a remarkable array of biochemical and pharmacological activity (for review see Harborne, 1994). With regard to their anticancer activity, the growth inhibitory effects of quercetin and genistein (Fig. 6.7) on several malignant tumour cell lines, e.g. gastric, colon and breast cancer cells (for review see Harborne, 1994) and their chemopreventive effects on, e.g. colonic

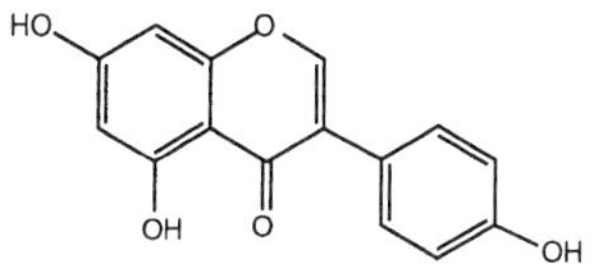

genistein (**21**)

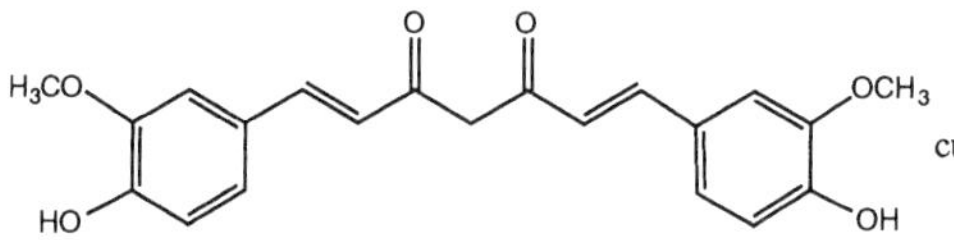

curcumin (**22**)

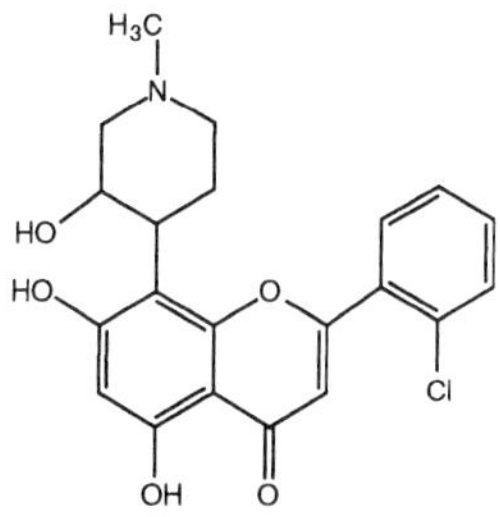

flavopiridol (**23**)

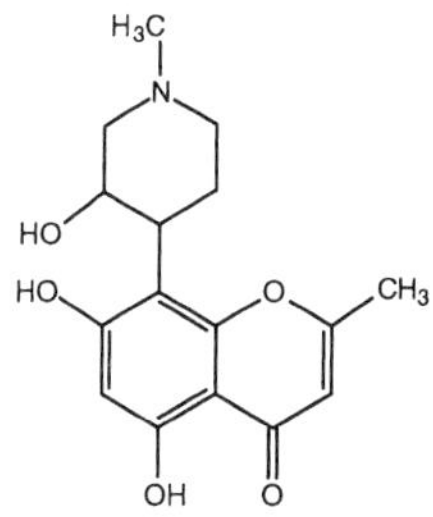

(+)-*cis*-5,7-dihydroxy-2-methyl-8-[4-(3-hydroxy-1-methyl)-piperidinyl]-4H-1-benzopyran-4-one (**24**)

Figure 6.7 Structures of the flavonoids, genistein (**21**) and curcumin (**22**), and of the new semisynthetic flavone, flavopiridol, which is closely related to (+)-*cis*-5,7-dihydroxy-2-methyl-8-[4-(3-hydroxyl-1-methyl)-piperidinyl]-4H-1-benzopyran-4-one.

cancer, breast adenocarcinoma and bladder carcinoma cells, have already been described (Kuo, 1996; Lu *et al.*, 1996). There is growing interest in compounds with chemopreventive properties, since they may reduce the incidence of cancer in human populations. The daily intake in the human diet of plant-derived compounds that have such activities contributes markedly to human health. The pharmacological significance and nutrition benefit of polyphenols, prominent components in various fruits, vegetables and spices, as well as in tea and wine, is the subject of intensive discussion (Aruoma, 1994; Gescher *et al.*, 1998).

Genistein (Fig. 6.7), e.g. from *Glycine max* Sieb. et Zucc. emend. Benth. (Fabaceae) and curcumin (Fig. 6.7) from *Curcuma xanthorrhiza* Roxb. or *Curcuma longa* (L.) (Zingiberaceae) appear to be the most interesting compounds. Both underwent intensive chemoprevention trials by the US NCI (Kelloff *et al.*, 1996a und b) after showing strong *in vitro* inhibition of enzymes related to the genesis of cancer, e.g. protein kinase C (Lin J.K. *et al.*, 1997) or ornithine decarboxylase (White *et al.*, 1998), and preventive activity in breast and colonic cancer cell lines, as well as in animal models (Adlercreutz and Mazur, 1997). In addition to its oestrogenic and antioestrogenic effects (for literature see Herman *et al.*, 1995), genistein is a specific inhibitor of tyrosine protein kinase and interferes with epidermal growth factor (EGF), $pp^{60v\text{-}src}$ and $pp110^{gag\text{-}fes}$ (Akiyama *et al.*, 1987; Akiyama and Ogawara, 1991), which may be the basis of its antitumour potency. Conjugation of genistein (Fig. 6.7) with recombinant EGF led to binding on the EGF-receptor of human breast cancer cells and a rapid cell death at nanomolar concentrations. The EGF-receptor is associated with protein tyrosine kinases at the cell membrane and so the conjugation of genistein with EGF resulted in a selective drug targeting (Uckun *et al.*, 1998b).

Curcumin (Fig. 6.7) exhibited strong inhibitory activity against 12-*O*-tetradecanoylphorbol-13-acetate (TPA)-induced generation of superoxide radical anions via inhibition of protein kinase C (Nakamura *et al.*, 1998) and radical scavenger activity against NO radicals (Sreejayan and Rao, 1997). It is also a strong inhibitor of phosphorylase kinase (Reddy and Aggarwal, 1994). Other observations include: suppression of the activation of transcription factors, such as nuclear factor (NF)-kappa B and activating protein 1 (AP-1) (Pendurthi *et al.*, 1997); enhancement of antibody response (South *et al.*, 1997); and as inhibition of cyclooxygenase and lipoxygenase (Ammon *et al.*, 1993). The chemopreventive effects of curcumin after topical or oral application in mice models have also been reported (Huang *et al.*, 1997).

Flavopiridol (Fig. 6.7), a new semisynthetic flavone closely related to (+)-cis-5,7-dihydroxy-2-methyl-8-[4-(3-hydroxy-1-methyl)-piperidinyl]-4H-1-benzopyran-4-one (Fig. 6.7), which was originally isolated from the stem bark of *Dysoxylum binectariferum* (Meliaceae) (Naik *at al.*, 1988), is currently undergoing Phase I clinical trials as a potential antineoplastic agent (Senderowicz *et al.*, 1996). Flavopiridol is a potent inhibitor of members of the cyclin-dependent kinase (CDK) family, e.g. CDK1, CDK2 and CDK4, which are important enzymes in the regulation of the cell cycle (Worland *et al.*, 1993; Losiewicz *et al.*, 1994; Carlson *et al.*, 1996). At low concentrations (0.1–0.4 μmol) flavopiridol binds to the adenosine triphosphate (ATP)-binding site of CDK1 and CDK2, resulting in competitive CDK inhibition with respect to ATP and, therefore, blocking cell cycle progression in the G1 and G2 phase (Carlson *et al.*, 1996). Further inhibitory effects on other enzymes, such as protein kinase C, have also been observed but only at higher concentrations (concentration required to produce half the maximum inhibition $[IC_{50}] = 6\ \mu M$) (Losiewicz *et al.*, 1994). The antitumour activity of flavopiridol has been demonstrated *in vitro* in different cycling and noncycling cancer cell lines, including prostate, breast, lung and leukaemia cells (Kaur *et al.*, 1992; Bible and Kaufmann, 1996). In addition, it was active in a wide range of human tumour xenografts *in vivo*

(Drees *et al.*, 1997; Arguello *et al.* 1998), and the Phase I study in patients with refractory neoplasms produced encouraging results for planning Phase II trials (Senderowicz *et al.*, 1996). Flavopiridol appears to be an interesting synergistic or additive partner of other agents, such as 5-fluorouracil and cisplatin but, on the other hand, it was shown that combination with cell cycle phase-specific agents, such as paclitaxel (Fig. 6.6) or topotecan (Fig. 6.4) led to a sequence-dependent cytotoxicity *in vitro*. When cells were treated with paclitaxel and flavopiridol simultaneously or in the flavopiridol-paclitaxel sequence, strong antagonism was observed. When paclitaxel preceded flavopiridol, the cytotoxicity was greater than additive. The same effects were observed between topotecan and flavopiridol (Bible and Kaufmann, 1997).

6.3 Antiviral compounds

Due to the alarming increase of viral diseases, especially of acquired immunodeficiency syndrome (AIDS) and of different types of hepatitis, there is strong interest in new antiviral drugs. Many compounds of plant origin with inhibitory effects against human immunodeficiency virus (HIV) have been identified. However, in most cases, the activity observed was not sufficient for therapeutic application and, therefore, only a few compounds have entered clinical trials. Compounds often exhibited potent activity both against cancer cell lines and viruses. In a clinical trial, 2-methyl-9-hydroxyellipticine (Fig. 6.2) was successfully combined with conventional virustatics attacking classical virus targets, such as reverse-transcriptase or protease, simultaneously with topoisomerase II (Mathe *et al.*, 1997).

Another example is the usage of cytotoxic ribosome-inactivating proteins (RIPs), a large class of proteins widely distributed in higher plants, against different types of cancer and viruses. These proteins are potent inhibitors of eukaryotic protein synthesis and inactivate 60S ribosomal subunits by breaking specific bonds in ribosomal 28S ribosomal ribonucleic acid (rRNA) (RNAse activity) (Endo *et al.*, 1987 and 1988b; Zhang and Liu, 1992). Various RIPs, such as trichosanthin, pokeweed antiviral protein (PAP), pokeweed antiviral protein from seeds (PAP-S), saporins, gelonin, ricin, mistletoe lectin I (viscumin) and many others, have exhibited irreversible damage to ribosomes with consequent arrest of protein synthesis, antiviral activity, immunosuppression or cytotoxicity in different *in vitro*, *ex-vivo* and animal models (e.g. Bolognesi *et al.*, 1990; Battelli *et al.*, 1992; Benigni *et al.*, 1995; for review see Barbieri *et al.*, 1993; Girbes *et al.*, 1996).

Several clinical trials have been conducted on patients with B- and T-cell leukaemias and lymphomas, malignant melanoma, breast cancer or HIV infections (see for example, Weiner *et al.*, 1989; Grossbard *et al.*, 1992; Amlot *et al.*, 1993; Byers *et al.*, 1994; for review see Kreitman and Pastan, 1998). Linking these nonspecific toxins to different monoclonal antibodies, fragments of a monoclonal antibody or growth factors has led to an immense increase in their pharmacological and clinical importance. Creation of these so-called immunotoxins or chimeric toxins, generally connected by disulfide bond chemistry, allows the selective targeting of RIPs to a particular cell type, such as cancer cells and virus-infected cells. Among the RIPs, trichosanthin, PAP, PAP-S and the saporins will be discussed in more detail.

Trichosanthin is a protein with a molecular weight of 27 kDa, extracted from a Chinese herbal drug, Tian-hua-fen, the root tuber of *Trichosanthes kirilowii* Maxim. (Cucurbitaceae) (e.g. Maraganore *et al.*, 1987; Collins *et al.*, 1990). Tian-hua-fen has been used in China to induce midterm abortion and trichosanthin was isolated as the abortificient active constituent. It belongs to the family of single chain ribosome inactivating proteins (type 1 RIPs) with RNA *N*-glycosidase activity, hydrolysing a single N-C glycosidic bond of an adenosine residue (Zhang and Liu, 1992). Further studies have reported on the inhibitory activity of trichosanthin on *in vitro* translation in cell-free systems and its abortificient activity (Nie *et al.*, 1998), as well as on its cytotoxicity against choriocarcinoma cells (Tsao *et al.*, 1990). In addition, it was found that trichosanthin inhibited the replication of HIV-1 in T-lymphoblastoid cells, primary monocytes and macrophages *in vitro* (McGrath *et al.*, 1989) and it was, therefore, considered as a therapeutic agent for AIDS. A Phase II study showed that addition of trichosanthin to zidovudin treatment of HIV-infected patients led to a significant increase in the number of CD4+ T-lymphocytes (Byers *et al.*, 1994). Because of the specific cytoxicity of trichosanthin against HIV-infected macrophages and lymphocytes, it was applied without conjugation to a monoclonal antibody.

In contrast, PAP, PAP-S and saporin are typical immunotoxins that were linked to different types of monoclonal antibodies before pharmacological and clinical application. They are also type 1 RIPs, with a molecular weight of *ca.* 30 kDa, and have been isolated from leaves, seeds or roots of pokeweed, *Phytolacca americana* (L.) (Phytolaccaceae) (Barbieri *et al.*, 1982) or soapworth, *Saponaria officinalis* (L.) (Caryophyllaceae) (Ferreras *et al.*, 1993). Linkage to different types of specific antibodies has resulted in a great number of conjugates with high *in vitro* and *in vivo* activity against various cancer cell lines, e.g. lymphoma and leukaemia cells (Waddick *et al.*, 1995; Flavell *et al.*, 1997), virus-infected cells and in animal models, e.g. anti-CD7-PAP with anti-HIV activity (Uckun *et al.*, 1998a). A detailed enumeration is beyond the scope of this chapter (for literature see Barbieri *et al.*, 1993; Girbes *et al.*, 1996).

Membrane glycoproteins of infected or malignant cells, such as the HIV envelope proteins, gp120 and gp41, are generally good targets for these immunotoxins (VanOijen and Preijers, 1998). Various immunotoxins were administered to patients suffering from cancer, AIDS and graft-*versus*-host disease (GVHD) in a number of Phase I and II clinical trials. However, so far most of the clinical results, with the exception of those from the GVHD patients (e.g. Laurent *et al.*, 1989; for literature see Thrush *et al.*, 1996), have fallen short of expectations because of severe side-effects or marginal response rates (Barbieri *et al.*, 1993). Nonspecific binding to other cells, removal from the blood by the Kupffer cells in the liver, moderate potency or instability of the disulfide link between antibody and toxin against reducing enzymes may be some of the reasons for the disappointing results (Vitetta *et al.*, 1987; Winkler *et al.*, 1997). Nevertheless, immunotoxin therapy with plant-derived RIPs is still a field of research with great medicinal potency. On the basis of our increasing knowledge of physiological and pathophysiological cell-cell signalling and cell function, there are good possibilities for efficient clinical treatment of malignant or virus-infected cells with highly effective proteins in the future.

Another compound with great importance in the treatment of cancer and AIDS is the indolizidine alkaloid, swainsonine (Fig. 6.8), which was found in different species of locoweed, *Astragalus* spp. and *Oxytropis*

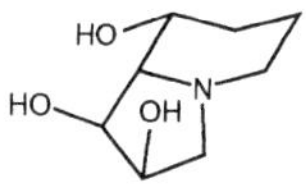

swainsonine (**25**)

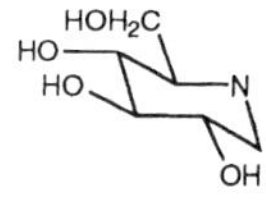

deoxynojirimycin (**26**)

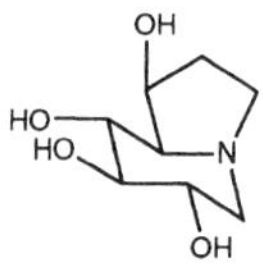

castanospermine (**27**)

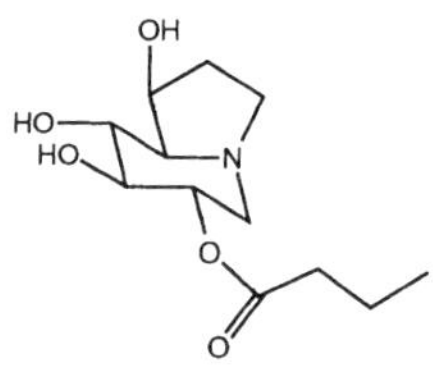

6-*O*-butyrylcastanospermine (**28**)

Figure 6.8 Structure of the indolizidine alkaloid, swainsonine (**25**), the α- and β-glucosidase I inhibitors, deoxynojirimycin (**26**) and castanospermine (**27**), and the chemically-prepared 6-*O*-acyl analogue of castanospermine, 6-*O*-butyrylcastanospermine (**28**).

spp., (Fabaceae), and became known for causing intoxication and behavioural abnormalities in domestic animals (Molyneux and James, 1981). Swainsonine is a potent inhibitor of Golgi α-mannosidase II and inhibits the synthesis and metabolism of glycoproteins (Tulsiani *et al.*, 1982). As a consequence, cell-cell signalling, transport of proteins and other cell functions are affected. Swainsonine has been shown to reduce the growth of tumour cells *in vitro* and *in vivo* (Dennis *et al.*, 1989) and of solid tumour growth in mice (Dennis *et al.*, 1990). Independent from the inhibition of Golgi α-mannosidase II, it also enhances the cellular immune response (Das *et al.*, 1995). Administration of swainsonine, in a phase IB clinical trial, to patients with advanced malignancies unsuitable for conventional therapy resulted in relatively low remission rates (Goss *et al.*,1997). Various tumour cells express other glycoproteins on their surface or have another glycosidase activity as nonmalignant cells. Therefore, the aim is to produce specific inhibitors of tumour glycosidases. Coating proteins of viruses are also good targets for glycosidase inhibitors.

The α- and β-glucosidase I inhibitors, desoxynojirimycin (*Morus* sp.) and castanospermine, from *Castanospermum australe* A. Cunn. & Fras. ex Hook. (Fabaceae) (Fig. 6.8) are potent inhibitors of the replication of HIV-1 viruses in tissue culture cells (Sunkara *et al.*, 1989; Gruters *et al.*, 1987). The chemically-prepared 6-*O*-acyl analogue of castanospermine, 6-*O*-butyrylcastanospermine, has been found to be even more potent and is a possible candidate for clinical development (Sunkara *et al.*, 1989). Desoxynojirimycin and castanospermine are not yet used in the therapy of infections with HIV but, like swainsonine, they are of great experimental importance. Modifications of HIV glycoproteins or of receptor glycoproteins on the cell surface provide valuable information on the infection pathways and on the mechanisms of the viruses (Montefiori *et al.*, 1993; Talbot *et al.*, 1995).

Dimeric naphthylisoquinoline alkaloids, called michellamines, represent another group of alkaloids with anti-HIV activity. They have recently been isolated from the tropical liana, *Ancistrocladus korupensis* D.W. Thomas & Gereau (Ancistrocladaceae) (Boyd *et al.*, 1994; Hallock *et al.*, 1997). Michellamine B (Fig. 6.9), the most abundant and potent member of this group, inhibited HIV-induced cell killing and viral replication in various human cell lines, as well as in cultures of human peripheral blood leukocytes and monocytes. In addition, it was active against a remarkable diversity of HIV-1 and HIV-2 strains *in vitro* (Hallock *et al.*, 1997). Studies on its mode of action demonstrated an inhibition of reverse transcriptase and, in later stages of the HIV life cycle, inhibition of cellular fusion and syncytium formation (McMahon *et al.*, 1995). Michellamine B has been committed to preclinical

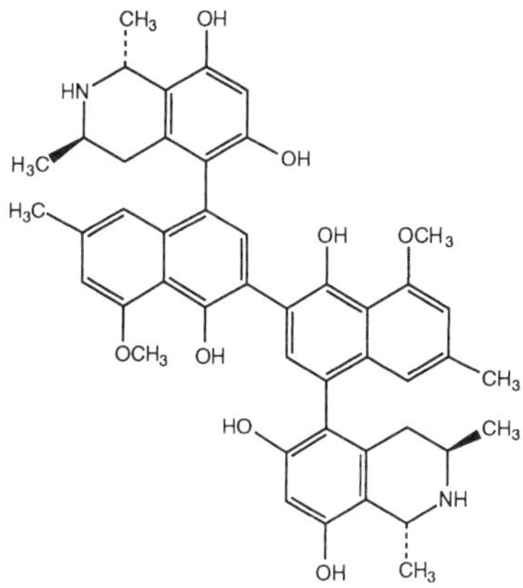

michellamine B (**29**)

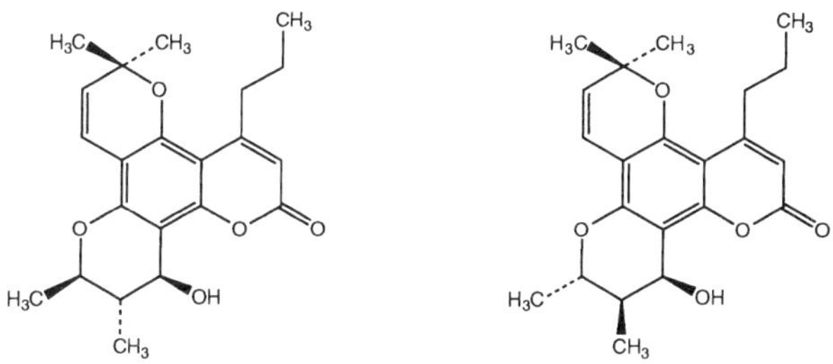

calanolide A (**30**) calanolide B (costatolide) (**31**)

Figure 6.9 Structure of the dimeric naphthylisoquinoline alkaloid, michellamine B (**29**), of the coumarin derivative, (+)-calanolide A (**30**) and the alternative, (-)- calanolide B (**31**).

development by the US NCI and pharmacokinetic data in mice and dogs have been published (Supko and Malspeis, 1995).

Another natural product that has been selected for preclinical development in the therapy of AIDS is the coumarin derivative, (+)-calanolide A (Fig. 6.9). This compound, isolated from the tropical rainforest tree, *Calophyllum lanigerum* MIQ. var. *austrocoriaceum* (T.C. WHITMORE) P.F. STEVENS (Guttiferae) showed *in vitro* inhibitory activity against various HIV-1 strains (concentration required to produce half the maximal effect [EC_{50}] values of 0.1–0.17 μM) but was inactive against HIV type 2 (Kashman *et al.*, 1992; Currens *et al.*, 1996b). (+)-Calanolide A has been found to be a non-nucleoside inhibitor of HIV reverse transcriptase but inhibition of non-nucleoside-resistant HIV strains and a synergistic activity with nevirapine indicated, that (+)-calanolide A comprizes a novel subclass of non-nucleoside RT inhibitors with a new mechanism of action (Currens *et al.*, 1996a,b). It can be assumed that (+)-calanolide binds near the active site of the enzyme and interferes with deoxynucleotide triphosphate (dNTP)-binding (Currens *et al.*, 1996b).

Since (+)-calanolide A is only a minor compound in the leaves of *Calophyllum lanigerum*, a synthesis has been established which has led to racemic (±)-calanolide A (Chenera *et al.*, 1993). Subsequently, it was shown that (-)-calanolide A is inactive against HIV and a synthesis with subsequent chromatographic separation of the enantiomers has been developed (Flavin *et al.*, 1996). Eventually, (-)-calanolide B (= costatolide) (Fig. 6.9) has been found to be an appropriate alternative because it is the main compound in latex extracts of *Calophyllum teysmanii* MIQ. var. *inophylloide* (KING.) P.F. STEVENS and only slightly less active (Cardellina *et al.*, 1995).

6.4 Antimalarial drugs

The potent antimalarial compound, artemisinin (quinghaosu) (Fig. 6.10) has been isolated from *Artemisia annua* (L.) (Asteraceae), a plant used in traditional Chinese medicine for malarial therapy (Trigg, 1989). *Artemisia*

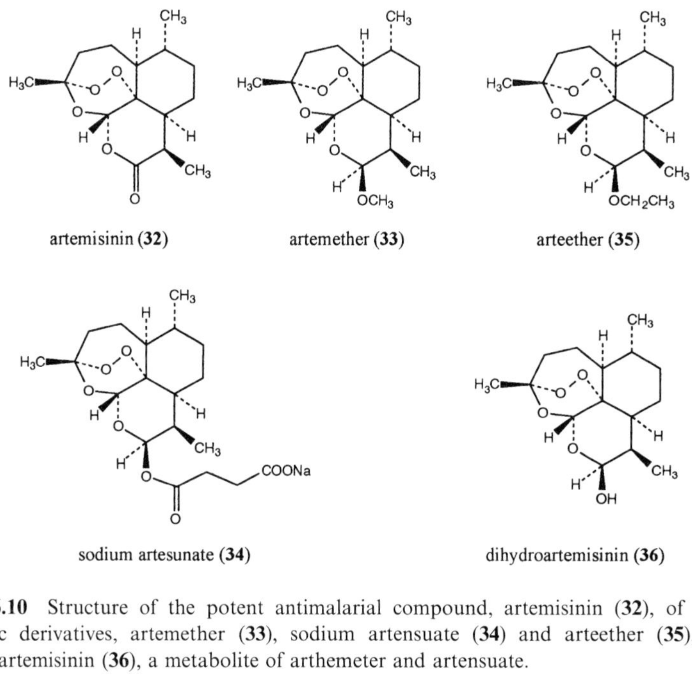

artemisinin (**32**) artemether (**33**) arteether (**35**)

sodium artesunate (**34**) dihydroartemisinin (**36**)

Figure 6.10 Structure of the potent antimalarial compound, artemisinin (**32**), of its semi-synthetic derivatives, artemether (**33**), sodium artensuate (**34**) and arteether (**35**), and of dihydroartemisinin (**36**), a metabolite of arthemeter and artensuate.

annua appears to be the only *Artemisia* species that contains appreciable amounts of artemisinin (approximately 0.1–0.2%). Artemisinin and its semisynthetic derivatives, artemether and sodium artesunate (Fig. 6.10), have been under intensive pharmacological, toxicological and clinical investigation since the mid 1970s. Most clinical studies of artemisinin, artemether and artesunate outside of China, especially those concerning the pharmacokinetics, metabolism and combination with other antimalarial drugs have been performed during the last decade. Previous investigations have been described in an excellent review by Trigg (1989).

Various clinical studies have shown, that application of artemisinin, artemether and artesunate to patients with severe or uncomplicated falciparum malaria are equally good or better than standard therapy with mefloquine (Jiang *et al.*, 1982) or quinine (e.g. Tran *et al.*, 1996). In addition, the combination of standard drugs, such as pyrimethamine (Na-Bangchang *et al.*, 1996), mefloquine (Price *et al.*, 1995), or quinine (Bich *et al.*, 1996) with artemisinin derivatives has led to a higher efficacy or tolerance in patients infected with (multiresistent) *Plasmodium falciparum* strains. Various studies were conducted to find the optimal regimen and optimal galenic formulations with high efficacy, low side-effects, low recrudescence rates and (important for development) with low costs (e.g. Cao *et al.*, 1997; Ha *et al.*, 1997; for further clinical data see deVries and Dien, 1996).

In spite of the extensive clinical data available for artemisinin and its derivatives (Fig. 6.10), their mode of action remains unclear. Older data indicated that the drugs damage parasite membranes and affect membrane-related processes (for literature see Trigg, 1989). More recent *in vitro* studies have suggested that the membrane damage is mediated by oxidative processes. Artemisinin enhanced the damage of haemin-induced ATPases in membranes via oxidation of thiol groups of the enzymes. Selective toxicity was claimed to be due to the high concentration of haemin in malarial parasites (Wie and Sadrzadeh, 1994) and a selective uptake of artemisinin by *Plasmodium falciparum*-infected red blood cells (Kamchonwongpaisan *et al.*, 1997). Other data have indicated an intraparasitic, iron-catalyzed cleavage of the endoperoxide bridge and the generation of free radicals. Subsequently, the authors postulated an alkylation of parasite proteins (Yang *et al.*, 1994; Kamchonwongpaisan and Meshnick, 1996).

Contradictory results have been published concerning the activity of artemisinin derivatives in combination with iron chelators. In one examination, iron chelators showed antagonism of the antimalarial activity of artemisinin and arteether (Fig. 6.10) *in vitro* (Meshnick *et al.*, 1993). However, covalent linking of artemisinin derivatives to iron chelators did not influence the antimalarial activity of the tested compounds *in vitro* (Kamchonwongpaisan *et al.*, 1995). Concerning the toxicity of artemisinin derivatives, an *in vivo* study on dogs revealed severe neurological effects after high-dose treatment with artemether and arteether (Brewer *et al.*, 1994). Investigations with undifferentiated neuroblastoma cells (NB2a) showed, after exposure to artemisinin and artemether and dihydroartemisin (Fig. 6.10), a metabolite of artemether and artesunate, in the presence of haemin, a dose-related decrease in the number of neurites formed (Smith *et al.*, 1997). From clinical data and recent results of an *in vitro* study, it can be assumed that severe neurotoxic effects are normally restricted to the application of high doses or prolonged exposure (Kamchonwongpaisan *et al.*, 1997). In the future, there is still great potential in these substances because they are leading compounds for the synthesis or semisynthesis of other even more potent antimalarial drugs (Lin *et al.*, 1997).

6.5 Anti-inflammatory drugs

Extracts of the gum resin of *Boswellia serrata* ROXB. (Burseraceae) have been used in traditional ayurvedic medicine as an antiphlogistic remedy. It has been shown that an alcoholic extract of the gum displayed marked anti-inflammatory activity in carrageenan- and dextran-induced oedema in mice and rats (Singh and Atal, 1986), and boswellic acids have been identified as the main anti-inflammatory active principle. Subsequently, anti-inflammatory activity was observed in papaya latex-induced rat paw inflammation (Gupta *et al.*, 1992) and anti-arthritic activity in bovine serum album-induced arthritis in rabbits (Sharma *et al.*, 1989). Investigation of the mechanism revealed that some of the boswellic acids are strong inhibitors of leukotriene B_4 synthesis via the 5-lipoxygenase pathway in neutrophils.

The highest inhibitory activity against 5-lipoxygenase was measured for acetyl-11-keto-β-boswellic acid (AKBA) (Fig. 6.11) (IC_{50} = 1.5 μM) (Safayhi *et al.*, 1992). It was shown that AKBA is a direct, non-competitive and nonredox type inhibitor of 5-lipoxygenase, acting at a selective site for pentacyclic triterpenes that is different from the arachidonate binding site (Safayhi *et al.*, 1992 and 1995). Studies on the structural requirements for binding and inhibitory activity led to the assumption that the pentacyclic triterpene ring is crucial for a selective binding, whereas the 11-keto group in combination with a hydrophilic group substituted at C-4 is essential for the 5-lipoxygenase inhibitory activity (Sailer *et al.*, 1996a, b). Further studies of its anti-inflammatory activity showed a lack of activity on cyclooxygenase-I and 12-lipoxygenase in human platelets as well as on peroxidation of arachidonic

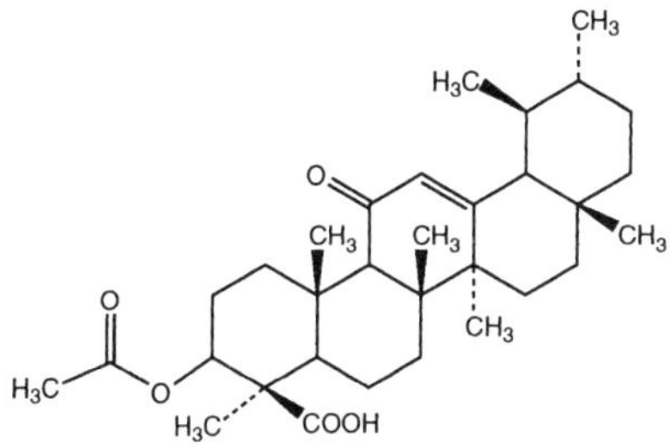

3-O-acetyl-11-keto-β-boswellic acid (**37**)

Figure 6.11 Structure of 3-O-acetyl-11-keto-β-boswellic acid (AKBA).

acid by Fe-ascorbate (Safayhi *et al.*, 1992). More recently, potent inhibition of human leukocyte elastase by AKBA was demonstrated (Safayhi *et al.*, 1997). Although studies with a commercial special extract of *Boswellia serrata* gum resin (H15®) have produced promising results in patients with chronic polyarthritis (Etzel, 1996) or ulcerative colitis (Gupta *et al.*, 1997), a final assessment of the therapeutic potential of boswellic acids remains open, until results of state-of-the-art clinical trials with a greater number of patients, representative data on its pharmacokinetics and the side-effects that occur are available.

6.6 Antidepressant drugs

In recent years, preparations with standardized extracts of St. John's wort, *Hypericum perforatum* (L.) (Guttifereae), have become of increasing importance. Various placebo-controlled clinical trials have confirmed its antidepressant potency (e.g. Schrader *et al.*, 1998), which is comparable to tricyclic antidepressants, such as imipramine (Vorbach *et al.*, 1997). Side-effects are usually less pronounced and less severe compared to standard antidepressants (Czekalla *et al.*, 1997; for overview see Linde *et al.*, 1996). Therefore, treatment of depressive disorders of mild-to-medium severity with *Hypericum* extracts has become a therapeutic alternative. Despite its widespread usage, the mechanism of the antidepressant action of St. John's wort remains unclear. The weak *in vitro* inhibition of MAO A and B (Cott, 1997; Müller *et al.*, 1997), the moderate inhibition of synaptosomal serotonin (5-HT), dopamine and norepinephrine uptake (Müller *et al.*, 1997), and the *in vitro* binding affinity for benzodiazepine receptors (Cott, 1997) are not sufficient to explain the antidepressant action *in vivo*.

Interestingly, recent studies have reported an upregulation of 5-HT_{1A} and 5-HT_{2A} receptors in the brain after treatment with *Hypericum* extract

(Teufel-Mayer and Gleitz, 1997); whereas, a downregulation of β-receptors has also been observed (Müller *et al.*, 1997). Furthermore, *Hypericum* extracts probably affect synaptic mechanisms of GABAergic and glutamatergic neurotransmission (Cott, 1997; Wonnemann *et al.*, 1997). Extracts of St. John's wort typically contain hypericin and pseudohypericin derivatives (Fig. 6.12), flavonoids and hyperforin (Fig. 6.12) as potential therapeutically-active components. Recent studies, using a forced swimming test with rats, indicated that hypericin and pseudohypericin contribute to the antidepressant activity of *Hypericum* extracts. Antagonism of the effect by sulpirid indicated that the dopaminergic system is involved (Butterweck *et al.*, 1998). Interesting but difficult to assess in their clinical relevance are new findings of a moderate *in vitro* affinity of hypericin to σ-receptors (Raffa, 1998).

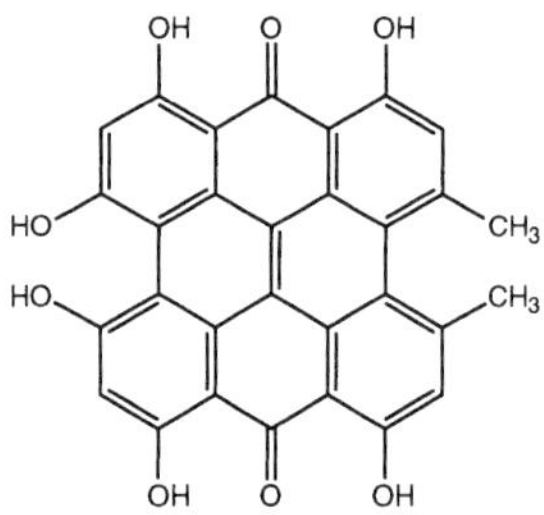

hypericin (**38**)

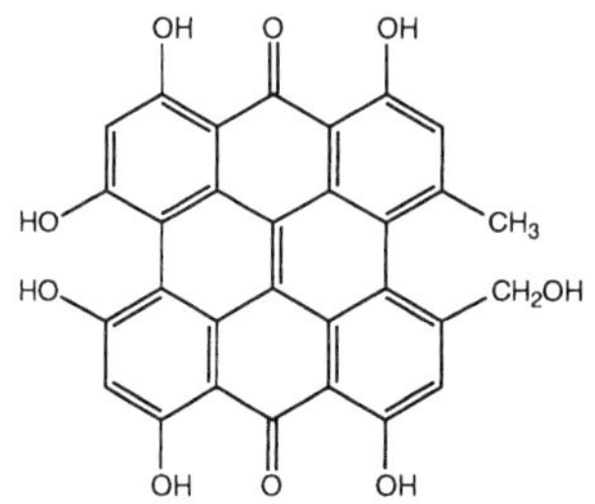

pseudohypericin (**39**)

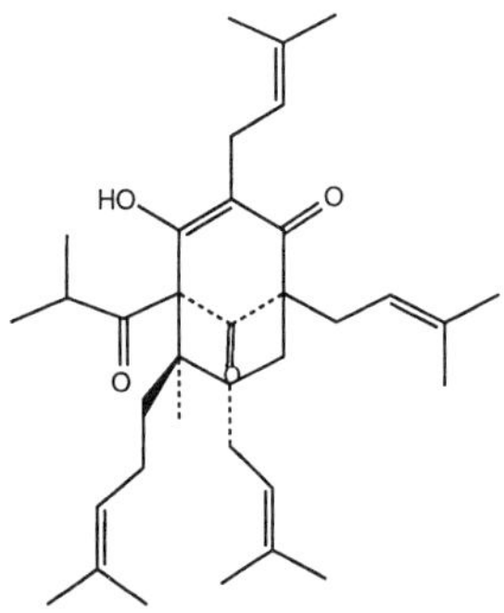

hyperforin (**40**)

Figure 6.12 Structure of the antidepressant compounds hypericin (**38**) and pseudohypericin (**39**), and of the phloroglucinol derivative, hyperforin (**40**).

The phloroglucinol derivative, hyperforin (Fig. 6.12) showed moderate but significant inhibition of nonspecific serotonin uptake in rat peritoneal cells (Chatterjee *et al.*, 1998) and in rat brain (Müller *et al.*, 1998). In addition, remarkable inhibition of norepinephrine and dopamine activity has been observed (Müller *et al.*, 1998). Another constituent of St. John's wort, amentoflavon (I3′,II8-biapigenin), showed an efficient inhibition of [^{3}H]flumazenil to rat brain benzodiazepine binding sites of $GABA_A$ receptors *in vitro* (Baureithel *et al.*, 1997). Therefore, it is likely that the antidepressant activity of *Hypericum* extracts *in vivo* is not due to a single group of constituents and to affinity to one receptor but to several pharmacologically active constituents that are present in the extract. This is substantiated by the above-mentioned data, showing a lack of prominent receptor affinity and the activity of both hypericin and hyperforin (Fig. 6.12) in behavioural models of depression (Butterweck *et al.*, 1998; Chatterjee *et al.*, 1998). Nevertheless, recent data have indicated that the antidepressant activity of *Hypericum* is closely related to the content of hyperforin (Fig. 6.12) (Laakmann *et al.*, 1998).

6.7 Anti-ischaemic drugs

Employing standardized extracts of *Ginkgo biloba* (L.) leaves with a defined content of Ginkgo-flavone glycosides and terpene lactones (24% Ginkgo-flavone glycosides and 6% terpene lactones; EGb 761) is well-accepted in the treatment of various peripheral or cerebral circulatory disorders. In a recent, randomized, double-blind, placebo-controlled, parallel-group, multicentre study, oral treatment with EGb 761 led to an objective improvement of cognitive performance and social functioning in patients suffering from Alzheimer's disease and multi-infarct dementia (Le Bars *et al.*, 1997). In the 1970s and 1980s, many pharmacological *in vitro* and *in vivo* as well as clinical studies were performed on *Ginkgo* extracts, mainly on EGb 761, and also on its single constituents: e.g. antioxidant and free radical scavenging activity; protection of hypoxia; effects on haemorheology and platelet aggregation; protection against hypoxia and ischaemia. These studies have been reviewed by Oberpichler-Schwenk and Krieglstein (1992) and Ahlemeyer and Krieglstein (1998).

Recent studies have emphasized the pharmacological effects of the terpene lactones, ginkgolide B and bilobalide (Fig. 6.13). Following the discovery of the potent platelet-activating factor (PAF) antagonism of ginkgolide B (BN 52021), it has been an invaluable tool in the investigation of the physiological and pathophysiological role of PAF (e.g. Lopes-Martins *et al.*, 1996). The cardioprotective and antioxidant effects of the ginkgolides and of bilobalide (Pietri *et al.*, 1997; Scholtyssek *et al.*, 1997) are also of importance, as well as the prevention of neurons from glutamate neurotoxicity by EGb 761 and ginkgolide B (Fig. 6.13) (Zhu *et al.*, 1997). EGb 761 and ginkgolide B have also been shown to

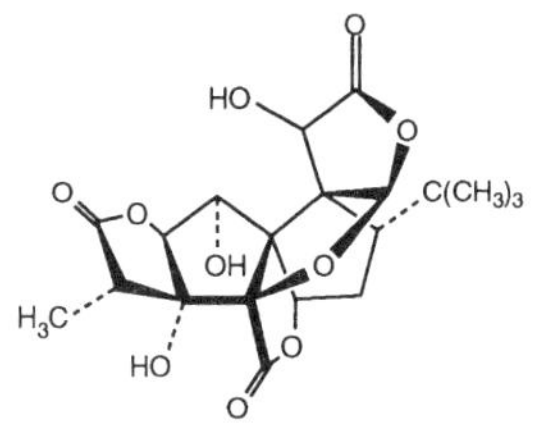

ginkgolide B (**41**)

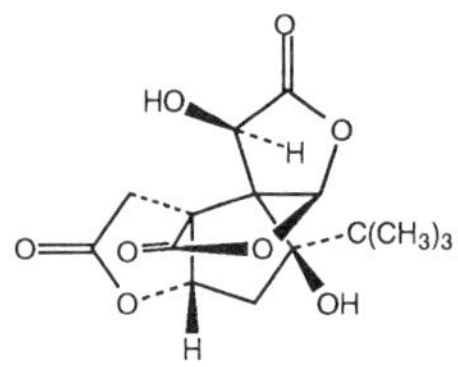

bilobalide (**42**)

Figure 6.13 Structure of the terpene lactones, ginkgolide B(**41**) and bilobalide (**42**).

reduce peripheral glucocorticoid levels in rats, affecting the adrenal peripheral benzodiazepine receptor (Amri *et al.*, 1996). *Ex-vivo* treatment resulted in a reduction of adrenocorticotrophic hormone (ACTH)-stimulated corticosterone production by adrenocortical rat cells (Amri *et al.*, 1997).

Bilobalide (Fig. 6.13) showed anticonvulsant activity in mice against convulsions induced by 4-*O*-methylpyridoxine. Together with an increase of hepatic 7-methoxycoumarin-*O*-demethylase activity, an accelerated hepatic metabolism of 4-*O*-methylpyridoxine was observed (Sasaki *et al.*, 1997). Ginkgo extract and bilobalide inhibited hypoxia-induced decreases of ATP levels in endothelial cells (Janssens *et al.*, 1995), phospholipid breakdown and choline release under hypoxic conditions *ex-vivo* (Klein *et al.*, 1997).

6.8 Immunostimulatory drugs

Extracts of mistletoe, *Viscum album* (L.), have frequently been applied in adjuvant cancer therapy (e.g. Lenartz *et al.*, 1996; Stein *et al.*, 1998). In recent years, major effects have been attributed to mistletoe lectin I, viscumin (ML I), which was demonstrated to be an immunomodulatory agent (Beuth *et al.*, 1995; Gabius and Gabius, 1998). Standardized aqueous mistletoe extracts with a constant lectin content are now on the market. However, their clinical efficacy is still a matter of controversy. Mistletoe lectin I is a double chain (type 2) RIP with a molecular weight of 60 kDa. The ribosome-inactivating properties are due to the A-chain, whereas the B-chain represents the lectin part and binds specifically to glycosylated cell surface proteins containing terminal galactose residues (Franz, 1986; Endo *et al.*, 1998a; Soler *et al.*, 1998).

ML I, mistletoe extracts exert cytotoxicity (e.g. Schumacher *et al.*,1995) and induce apoptosis (programmed cell death) in several cell types (Janssen *et al.*, 1993; Möckel *et al.*, 1997). In addition, inhibitory effects on tumour angiogenesis and metastasis of haematogenous and non-haematogenous tumour cells in mice have been observed (Yoon *et al.*, 1995). Stimulation of the cellular defence system has also been shown in several studies. ML I and mistletoe extracts enhanced the activity of natural killer cells and T-lymphocytes (Heiny *et al.*, 1998), as well as the levels and the activity of lymphocytes, peritoneal macrophages and monocytes *in vivo* and *in vitro* (Stein and Berg, 1996; Stein *et al.*, 1998). The production of cytokines, tumour necrosis factor-α (TNF-α) (Ribereau-Gayon *et al.*, 1996) and the activation of human polymorphonuclear leukocytes was observed *in vitro*, (Braun *et al.*, 1995). In a recent clinical trial, complementary administration of mistletoe extract to a standard oncological treatment led to significant immunostimulatory effects, which were correlated with an improved quality of life (Lenartz *et al.*, 1996).

The medical importance of lectins goes beyond their immunostimulatory effects. Specific binding properties of the lectins enable the characterization and location of glycoreceptors on the cell surface (Wu *et al.*, 1997). It can be assumed that the varying toxicity to different cell lines based on a selective binding to glycoproteins on the cell surface is a useful tool for drug targeting (Büssing *et al.*, 1998). Moreover, lectins activate specific types of ion channels and, thus, further information on signal transduction pathways can be obtained (Wenzel-Seifert *et al.*, 1996). The ability to express recombinant ML-I A-chain in *Escherichia coli* led to a molecule with *in vitro* RIP activity that was three times less than that of the plant ML-I A-chain (Langer *et al.*, 1996).

6.9 Conclusion

It is clear that plants continue to provide us with new drugs and leading structures. Low molecular weight compounds, peptides and proteins with influence on specific cell functions will play an important role in the development of new drugs in the future. Our knowledge of the molecular basis of diseases will dramatically increase in the near future, more specific bioassays based on receptors and enzymes will be developed and high-throughput screening will lead to even more interesting compounds from plant origins (Lin, 1995; Pezzuto, 1997; Shu, 1998). Rapid progress can also be expected in the field of combinatorial biosynthesis. Creation of novel gene combinations or hybrid genes may produce novel secondary metabolites, due to the effect of a new enzyme with new enzymatic properties on a metabolic pathway. Therefore, it may be possible to create new 'natural products', which have novel or more potent biological activities (Hutchinson, 1998; Khosla, 1998). However, the screening of plant extracts is hampered by the problem of nonspecific effects and false positive or negative results because of matrix compounds, such as lipids, tannins and chlorophylls (Beutler *et al.*, 1995). Therefore, automatic purification procedures are desirable. In this field, the opportunity for thorough academic research remains, contributing to the discovery of new drugs.

References

Adlercreutz, H. and Mazur, W. (1997) Phyto-oestrogens and Western diseases. *Ann. Med.*, **29** 95-120.

Ahlemeyer, B. and Krieglstein, J. (1998) Neuroprotective effects of *Ginkgo biloba* extract, in *Phytomedicines of Europe: Chemistry and Biological Activity* (eds. L.D. Lawson and R. Bauer), ACS Symposium Series 691, ACS Books, Washington, pp. 210-20.

Akiyama, T. and Ogawara, H. (1991) Use and specificity of genistein as inhibitor of protein tyrosine kinases. *Methods Enzymol.*, **201** 362-70.

Akiyama, T., Ishida, J., Nakagawa, S., Ogawara, H., Watanabe, S., Itoh, N., Shibuya, M. and Fukami, Y. (1987) Genistein, a specific inhibitor of tyrosine-specific protein kinases. *J. Biol. Chem.*, **262** 5592-95.

Amlot, P.L., Stone, M.J., Cunningham, D., Fay, J., Newman, J., Collins, R., May, R., McCarthy, M., Richardson, J., Ghetie, V., Ramilo, O., Thorpe, P.E., Uhr, J.W. and Vitetta, E.S. (1993) A Phase I study of an anti-CD22-deglycosylated ricin A-chain immunotoxin in the treatment of B-cell lymphomas resistant to conventional therapy. *Blood*, **82** 2624-33.

Ammon, H.P.T., Safayhi, H., Mack, T. and Sabieraj, J. (1993) Mechanism of anti-inflammatory actions of curcumine and boswellic acids. *J. Ethnopharmacol.*, **38** 113-19.

Amri, H., Ogwuegbu, S.O., Boujrad, N., Drieu, K. and Papadopoulos, V. (1996) *In vivo* regulation of the peripheral-type benzodiazepine receptor and glucocorticoid synthesis by the *Ginkgo biloba* extract, EGb 761, and isolated ginkgolides. *Endocrinology*, **130** 5707-18.

Amri, H., Drieu, K. and Papadopoulos, V. (1997) *Ex-vivo* regulation of adrenal cortical cell steroid and protein synthesis, in response to adrenocorticotropic hormone stimulation, by the *Ginkgo biloba* extract, EGb 761, and isolated ginkgolide B. *Endocrinology*, **138** 5415-26.

Arguello, F., Alexander, M., Sterry, J.A., Tudor, G., Smith, E.M., Kavalar, N.T., Greene, J.F. Jr, Koss, W., Morgan, C.D., Stinson, S.F., Siford, T.J., Alvord, W.G., Klabansky, R.L. and Sausville, A. (1998) Flavopiridol induces apoptosis of normal lymphoid cells, causes immunosuppression and has potent antitumor activity *in vivo* against human leukemia and lymphoma xenografts. *Blood*, **91** 2482-90.

Aruoma, O.I. (1994) Nutrition and health aspects of free radicals and antioxidants. *Food Chem. Toxicol.* **32** 671-83.

Barbieri, L., Aron, G.M., Irvin, J.D. and Stirpe, F. (1982) Purification and partial characterization of another form of the antiviral protein from the seeds of *Phytolacca americana* (L.) (pokeweed). *Biochem. J.*, **203** 55-59.

Barbieri, L., Battelli, M.G. and Stirpe, F. (1993) Ribosome-inactivating proteins from plants. *Biochim. Biophys. Acta*, **1154** 237-82.

Battelli, M.G., Montacuti, V. and Stirpe, F. (1992) High sensitivity of cultured human trophoblasts to ribosome-inactivating proteins. *Exp. Cell Res.*, **201** 109-12.

Baureithel, K.H., Büter, K.B., Engesser, A., Burkard, W. and Schaffner, W. (1997) Inhibition of benzodiazepine binding *in vitro* by amentoflavone, a constituent of various species of Hypericum. *Pharm. Acta Helv.*, **72** 153-57.

Benigni, F., Canevari, S., Gadina, M., Adobati, E., Ferreri, A.J.M., Di Celle, E.F., Comolli, R. and Colnaghi, M.I. (1995) Preclinical evaluation of the ribosome-inactivating proteins, PAP-1, PAP-S and RTA, in mice. *Int. J. Immunopharmacol.*, **17** 829-39.

Beuth, J., Stoffel, B., Ko, H.L., Jeljaszewicz, J. and Pulverer, G. (1995) Immunomodulating ability of galactoside-specific lectin standardized and depleted mistletoe extract. *Arzneim.-Forsch. Drug Res.*, **45** 1240-42.

Beutler, J.A., Cardelina II, J.H., McMahon, J.B., Shoemaker, R.H. and Boyd, M.R. (1995) Antiviral and antitumor metabolites, in *Phytochemistry of Medicinal Plants* (eds. J.T. Arnason, R. Mata and J.T. Romeo), Plenum Press, New York and London, pp. 47-64.

Bible, K.C. and Kaufmann, S.H. (1996) Flavopiridol: a cytotoxic flavone that induces cell death in noncycling A549 human lung carcinoma cells. *Cancer Res.*, **56** 4856-61.

Bible, K.C. and Kaufmann, S.H. (1997) Cytotoxic synergy between flavopiridol (NSC 649890, L86-8275) and various antineoplastic agents: the importance of sequence of administration. *Cancer Res.*, **57** 3375-80.

Bich, N.N., De-Vries, P.J., Van-Thien, H., Phong, T.H., Hung, L.N., Eggelte, T.A., Anh, T.K. and Kager, P.A. (1996) Efficacy and tolerance of artemisinin in short combination regimens for the treatment of uncomplicated falciparum malaria. *Am. J. Trop. Med. Hyg.*, **55** 438-43.

Bolognesi, A., Barbieri, L., Abbondanza, A., Falasca, A.I., Carnicelli, D., Battelli, M.G. and Stirpe, F. (1990) Purification and properties of new ribosome-inactivating proteins with RNA *N*-glycosidase activity. *Biochim. Biophys. Acta*, **1087** 293-302.

Bonneterre, J. (1995) Topoisomerase I inhibitors: review of clinical phase II trials of irinotecan (CPT-11) and topotecan (FRE). *Bull. Cancer*, **82** 623-28.

Boyd, M.R., Hallock, Y.F., Cardellina II, J.H., Manfredi, K.P., Blunt, J.W., McMahon, J.B., Buckheit, R.W. Jr, Bringmann, G., Schäffer, M., Cragg, G.M., Thomas, D.W. and Jato, J.G. (1994) Anti-HIV michellamines from *Ancistrocladus korupensis*. *J. Med. Chem.*, **37** 1740-45.

Braun, J.M., Gemmell, C.G., Beuth, J., Ko, H.L. and Pulverer, G. (1995) Respiratory burst of human polymorphonuclear leukocytes in response to the galactoside-specific mistletoe lectin. *Int. J. Med. Microbiol. Virol. Parasitol. Infect. Dis.*, **283** 90-94.

Brewer, T.G., Grate, S.J., Peggins, J.O., Weina, P.J., Petras, J.M., Levine, B.S., Heiffer, M.H. and Schuster, B.G. (1994) Fatal neurotoxicity of arteether and artemether. *Am. J. Trop. Med. Hyg.*, **51** 251-59.

Büssing, A., Stein, G.M. and Pfueller, U. (1998) Selective killing of CD8+ cells with a memory phenotype (CD62L(lo)) by the *N*-acetyl-D-galactosamine-specific lectin from *Viscum album* (L.) *Cell Death Differentiation*, **5** 231-40.

Butterweck, V., Petereit, F., Winterhoff, H. and Nahrstedt, A. (1998) Solubilized hypericin and pseudohypericin from *Hypericum perforatum* exert antidepressant activity in the forced swimming test. *Planta Med.*, **64** 291-94.

Byers, V.S., Levin, A.S., Malvino, A., Waites, L., Robins, R.A. and Baldwin, R.W. (1994) A phase II study of effect of addition of trichosanthin to zidovudine in patients with HIV disease and failing antiretroviral agents. *AIDS Res. Hum. Retroviruses*, **10** 413-20.

Cao, X.T., Bethell, D.B., Pham, T.P., Ta, T.T., Tran, T.N., Nguyen, T.T., Pham, T.T., Nguyen, T.T., Day, N.P. and White, N.J. (1997) Comparison of artemisinin suppositories, intramuscular artesunate and intravenous quinine for the treatment of severe childhood malaria. *Trans. R. Soc. Trop. Med. Hyg.*, **91** 335-42.

Cardellina II, J.H., Bokesch, H.R., McKee, T.C. and Boyd, M.R. (1995) Resolution and comparative anti-HIV evaluation of the enantiomers of calanolides A and B. *BioMed. Chem. Lett.*, **5** 1011-14.

Carlson, B.A., Dubay, M.M., Sausville, E.A., Brizuela, L. and Worland, P.J. (1996) Flavopiridol induces G_1 arrest with inhibition of cyclin-dependent kinase (CDK) 2 and CDK4 in human breast carcinoma cells. *Cancer Res.*, **56** 2973-76.

Chatterjee, S.S., Nöldner, M., Koch, E. and Erdelmeier, C. (1998) Antidepressant activity of *Hypericum perforatum* and hyperforin: the neglected possibility. *Pharmacopsychiat.*, **31** 7-15.

Chenera, B., West, M.L., Finkelstein, J.A. and Dreyer, G.B. (1993) Total synthesis of (±)-calanolide A, a non-nucleoside inhibitor of HIV-1 reverse transcriptase. *J. Org. Chem.*, **58** 5605-606.

Ciomei, M., Albanese, C., Pastori, W., Grandi, M., Pietra, F., D'Ambrosio, M., Guerriero, A. and Battistini, C. (1997) Sarcodictyins: a new class of marine derivatives with mode of action similar to taxol. *Proc. Am. Assoc. Cancer Res.*, **38** 5 (Abstract 30).

Clark, A.M. (1996) Natural products as a resource for new drugs. *Pharm. Res.*, **13** 1133-44.

Collins, E.J., Robertus, J.D., LoPresti, M., Stone, K.L., Williams, K.R., Wu, P., Hwang, K. and Piatak, M. (1990) Primary amino acid sequence of α-trichosanthin and molecular models for abrin A-chain and α-trichosanthin. *J. Biol. Chem.*, **265** 8665-69.

Conte, P.F., Gennari, A., Salvadori, B., Pazzagli, C. and Bengala, C. (1998) Paclitaxel plus epirubicin in advanced breast cancer. *Oncology Huntingt.*, **12** 40-44.

Cott, J.M. (1997) *In vitro* receptor-binding and enzyme inhibition by *Hypericum perforatum* extract. *Pharmacopsychiat.*, **30** 108-12.

Croom, E.M. Jr. (1995) *Taxus* for taxol and taxoids, in *Taxol® Science and Applications* (ed. Suffness, M.), CRC Press, Boca Raton, New York, London, Tokyo, pp. 37-70.

Currens, M.J., Gulakowski, R.J., Mariner, J.M., Moran, R.A., Buckheit, R.W. Jr, Gustafson, K.R., McMahon, J.B. and Boyd, M.R. (1996a) Antiviral activity and mechanism of action of calanolide A against the human immunodeficiency virus type-1. *J. Pharmacol. Exp. Ther.*, **279** 645-51.

Currens, M.J., Mariner, J.M., McMahon, J.B. and Boyd, M.R. (1996b) Kinetic analysis of inhibition of human immunodeficiency type-1 reverse transcriptase by calanolide A. *J. Pharmacol. Exp. Ther.*, **279** 652-61.

Czekalla, J., Gastpar, M., Hübner, W.D. and Jäger, D. (1997) The effect of Hypericum extract on cardiac conduction as seen in the electrocardiogram compared to that of imipramine. *Pharmacopsychiat.*, **30** 86-88.

Damayanthi, Y. and Lown, J.W. (1998) Podophyllotoxins: current status and recent developments. *Curr. Med. Chem.*, **5** 205-52.

Das, P.C., Roberts, J.D., White, S.L. and Olden, K. (1995) Activation of resident tissue-specific macrophages by swainsonine. *Oncol. Res.*, **7** 425-33.

Dennis, J.W., Koch, K. and Beckner, D. (1989) Inhibition of human HT29 colon carcinoma growth *in vitro* and *in vivo* by swainsonine and human interferon-alpha 2. *J. Natl. Cancer Inst.*, **81** 1028-33.

Dennis, J.W., Koch, K., Yousefi, S. and Vanderelst, I. (1990) Growth inhibition of human melanoma tumor xenografts in athymic nude mice by swainsonine. *Cancer Res.*, **50** 1867-72.

deVries, P.J. and Dien, T.K. (1996) Clinical pharmacology and therapeutic potential of artemisinin and its derivatives in the treatment of malaria. *Drugs*, **52** 818-36.

Drees, M., Dengler, W.A., Roth, T., Labonte, H., Mayo, J., Malspeis, L., Grever, M., Sausville, E.A. and Fiebig, H.H. (1997) Flavopiridol (L86-8275): selective antitumor activity *in vitro* and activity *in vivo* for prostate carcinoma cells. *Clin. Cancer Res.*, **3** 273-79.

Dreyfuss, A.I., Clerk, J.R., Norris, C.M., Rossi, R.M., Lucarini, J.W., Busse, P.M., Poulin, M.D., Thornhill, L., Costello, R. and Posner, M.R. (1996) Docetaxel: an active drug for squamous cell carcinoma of the head and neck. *J. Clin. Oncol.*, **14** 1672-78.

Einzig, A.I., Wiernik, P.H., Sasloff, J., Runowicz, C.D. and Goldberg, G.L. (1992) Phase II study and long-term follow-up of patients treated with taxol for advanced ovarian cancer. *J. Clin. Oncol.*, **10** 1748-53.

Emerson, D.L., Besterman, J.M., Brown, H.R., Evans, M.G., Leitner, P.P., Luzzio, M.J., Shaffer, J.E., Sternbach, D.D., Uehling, D. and Vuong, A. (1995) *In vivo* antitumor activity of two new seven-substituted water-soluble camptothecin analogs. *Cancer Res.*, **55** 603-609.

Endo,Y., Mitsui, K., Motizuki, M. and Tsurugi, K. (1987) The mechanism of action of ricin and related toxic lectins on eukaryotic ribosomes: the site and the characteristics of the modification in 28 S ribosomal RNA caused by the toxins. *J. Biol. Chem.*, **262** 5908-12.

Endo, Y., Tsurugi, K. and Franz, H. (1988a) The site of action of the A-chain of mistletoe lectin I on eukaryotic ribosomes: the RNA *N*-glycosidase activity of the protein. *FEBS Lett.*, **231** 378-80.

Endo, Y., Tsurugi, K. and Lambert, J.M. (1988b) The site of action of six different ribosome-inactivating proteins from plants on eukaryotic ribosomes: the RNA *N*-glycosidase activity of the proteins. *Biochem. Biophys. Res. Commun.*, **150** 1032-36.

Etzel, R. (1996) Special extract of *Boswellia serrata* (H 15) in the treatment of rheumatoid arthritis. *Phytomedicine*, **3** 91-94.

Extra, J.-M., Rousseau, F., Bruno, R., Clavel, M., Le Bail, N. and Marty, M. (1993) Phase I and pharmacokinetic study of taxotere (RP 56976; NSC 628503) given as short intravenous infusion. *Cancer Res.*, **53** 1037-42.

Ferreras, J.M., Barbieri, L., Girbés, T., Batelli, M.G., Rojo, M.A., Arias, F.J., Rocher, M.A., Soriano, F., Mendéz, E. and Stirpe, F. (1993) Distribution and properties of major ribosome-inactivating proteins (28S rRNA *N*-glycosidases) of the plant *Saponaria officinalis* (L.) (Caryophyllaceae). *Biochim. Biophys. Acta*, **1216** 31-42.

Flavell, D.J., Noss, A., Pulford, K.A., Ling, N. and Flavell, S.U. (1997) Systemic therapy with 3BIT, a triple combination cocktail of anti-CD19, -CD22, and -CD38-saporin immunotoxins, is curative of human B-cell lymphoma in severe combined immunodeficient mice. *Cancer Res.*, **57** 4824-29.

Flavin, M.T., Rizzo, J.D., Khilevich, A., Kucherenko, A., Sheinkman, A.K., Vilaychack, V., Lin, L., Chen, W., Greenwood, E.M., Pengsuparp, T., Pezzuto, J.M., Hughes, S.H., Flavin, T.M., Cibulski, M., Boulanger, W.A., Shone, R.L. and Xu, Z.-Q. (1996) Synthesis, chromatographic resolution and anti-human immunodeficiency virus activity of (±)-calanolide A and its enantiomers. *J. Med. Chem.*, **39** 1303-13.

Forastiere, A.A., Neuberg, D., Taylor, S.G., DeConti, R. and Adams, G. (1993) Phase II evaluation of taxol in advanced head and neck cancer: an Eastern Cooperative Oncology Group Trial. *Monogr. Natl. Cancer Inst.*, **15** 181-84.

Franz, H. (1986) Mistletoe lectins and their A and B chains. *Oncology*, **43** 23-34.

Fujimori, A., Harker, W.G., Kohlhagen, G., Hoki, Y. and Pommier, Y. (1995) Mutation at the catalytic side of topoisomerase I in CEM/C2, a human leukemia cell resistant to camptothecin. *Cancer Res.*, **55** 1339-46.

Fumoleau, P. (1997) Efficacy and safety of docetaxel in clinical trials. *Am. J. Health Syst. Pharm.*, **54** S19-S24.

Gabius, H.-J. and Gabius, S. (1998) Phytotherapeutic immunomodulation as a treatment modality in oncology: lessons from research with mistletoe, in *Phytomedicines of Europe: Chemistry and Biological Activity*, (eds. L.D. Lawson and R. Bauer), ACS Symposium Series 691, ACS Books, Washington, pp. 278-86.

Gabr, A., Kuin, A., Aalders, M., El-Gawly, H. and Smets, L.A. (1997) Cellular pharmacokinetics and cytotoxicity of camptothecin and topotecan at normal and acidic pH. *Cancer Res.*, **57** 4811-16.

Gescher, A., Pastorino, U., Plummer, S.M. and Manson, M.M. (1998) Suppression of tumour development by substances derived from the diet-mechanisms and clinical implications. *Br. J. Clin. Pharmacol.*, **45** 1-12.

Giovanella, B.C., Stehlin, J.S., Wall, M.E., Wani, M.C., Nicholas, A.W., Liu, L.F., Silber, R. and Potmesil, M. (1989) Highly effective DNA topoisomerase I-targeted chemotherapy of human colon cancer in xenografts. *Science*, **246** 1046-48.

Giovanella, B.C., Hinz, H.R., Kozielski, A.J., Stehlin, J.S., Silber, R. and Potmesil, M. (1991) Complete growth inhibition of human cancer xenografts in nude mice by treatment with 20-(*S*)-camptothecin. *Cancer Res.*, **51** 3052-55.

Girbes, T., Ferreras, J.M., Iglesias, R., Citores, L., DeTorre, C., Carbajales, M.L., Jimenez, P., DeBenito, F.M. and Munoz, R. (1996) Recent advances in the uses and applications of ribosome-inactivating proteins from plants. *Cell. Mol. Biol.*, **42** 461-71.

Goldspiel, B.R. (1997) Clinical overview of the taxanes. *Pharmacotherapy*, **17** 110S-125S.

Goss, E.P., Reid, C.L., Bailey, D. and Dennis, J.W. (1997) Phase IB clinical trial of the oligosaccharide processing inhibitor, swainsonine, in patients with advanced malignancies. *Clin. Cancer Res.*, **3** 1077-86.

Gottlieb, J.A. and Luce, J.K. (1972) Treatment of malignant melanoma with camptothecin (NSC-100880). *Cancer Chemother. Rep.*, **56** 103-105.

Gridelli, C., Rossi, A., Scognamiglio, F., Guida, C., Fiore, F., Gatani, T., Scoppa, G. and Pergola, M. (1997) Carboplatin plus oral etoposide in elderly patients with advanced non-small-cell lung cancer: a phase II study. *Anticancer Res.*, **17** 4755-58.

Grossbard, M.L., Freedman, A.S., Ritz, J., Coral, F., Goldmacher, V.S., Eliseo, L., Spector, N., Dear, K., Lambert, J.M., Blättler, W.A., Taylor, J.A. and Nadler, L.M. (1992) Serotherapy of B-cell neoplasms with anti-B4-blocked ricin: a phase I trial of daily bolus infusion. *Blood*, **79** 576-85.

Gruters, R.A., Neefjes, J.J., Tersmette, M., de Goede, R.E., Tulp, A., Huisman, H.G., Miedema, F. and Ploegh, H.L. (1987) Interference with HIV-induced syncytium formation and viral infectivity by inhibitors of trimming glucosidase. *Nature*, **330** 74-77.

Guéritte-Voegelein, F., Guénard, D., Lavelle, F., LeGoff, M.-T., Mangatal, L. and Potier, P. (1991) Relationships between the structure of taxol analogues and their antimitotic activity. *J. Med. Chem.*, **34** 992-98.

Guilbaud, N., KrausBerthier, L., SaintDizier, D., Rouillon, M.H., Jan, M., Burbridge, M., Visalli, M., Bisagni, E., Pierre, A. and Atassi, G. (1996) *In vivo* antitumor activity of S 16020-2, a new olivacine derivative. *Cancer Chemother. Pharmacol.*, **38** 513-21.

Gupta, I., Parihar, A., Malhotra, P., Singh, G.B., Ludtke, R., Safayhi, H. and Ammon, H.P.T. (1997) Effects of *Boswellia serrata* gum resin in patients with ulcerative colitis. *Eur. J. Med. Res.*, **2** 37-43.

Gupta, O.P., Sharma, N. and Chand, D. (1992) A sensitive and relevant model for evaluating anti-inflammatory activity, papaya latex-induced rat paw inflammation. *J. Pharmacol. Toxicol. Methods*, **28** 15-19.

Ha, V., Nguyen, N.H., Tran, T.B., Bui, M.C., Nguyen, H.P., Tran, T.H., Phan, T.Q., Arnold, K. and Tran, T.H. (1997) Severe and complicated malaria treated with artemisinin, artesunate or artemether in Vietnam. *Trans. R. Soc. Trop. Med. Hyg.*, **91** 465-67.

Hallock, Y.F., Manfredi, K.P., Dai, J.-R., Cardelina II, J.H., Gulakowski, R.J., McMahon, J.B., Schäffer, M., Stahl, M., Gulden, K.-P., Bringmann, G., Francois, G. and Boyd, M.R. (1997) Michellamines D–F, new HIV-inhibitory dimeric naphthylisoquinoline alkaloids, and korupensamine E, a new antimalarial monomer, from *Ancistrocladus korupensis*. *J. Nat. Prod.*, **60** 677-83.

Harborne, J.B. (ed) (1994) *The Flavonoids: Advances in Research Since 1986*, Chapman & Hall, London.

Heiny, B.M., Albrecht, V. and Beuth, J. (1998) Correlation of immune cell activities and beta-endorphin release in breast carcinoma patients treated with galactose-specific lectin standardized mistletoe extract. *Anticancer Res.*, **18** 583-86.

Herman, C., Adlercreutz, T., Goldin, B.R., Gorbach, S.L., Höckerstedt, K.A.V., Watanabe, S., Hämäläinen, E.K., Markannen, M.H., Mäkela, T.H., Wähälä, K.T., Hase, T.A. and Fotsis, T. (1995) Soybean phytoestrogen intake and cancer risk. *J. Nutr.*, **125** 757S-770S.

Hill, B.T., Whelan, R.D., Shellard, S.A., McClean, S. and Hosking, L.K. (1994) Differential cytotoxic effects of docetaxel in a range of mammalian tumor cell lines and certain drug resistant sublines *in vitro*. *Invest. New Drugs*, **12** 169-82.

Holmes, F.A., Walters, R.S., Theriault, R.L., Forman, A.D., Newton, L.K., Raber, M.N., Buzdar, A.U., Frye, D.K. and Hortobagyi, G.N. (1991) Phase II trial of taxol: an active drug in the treatment of metastatic breast cancer. *J. Natl. Cancer Inst.*, **83** 1797-805.

Hsiang, Y.H., Liu, L.F., Wall, M.E., Wani, M.C., Kirschenbaum, S., Silber, R. and Potmesil, M. (1989) DNA topoisomerase I-mediated DNA cleavage and cytotoxicity of camptothecin analogs. *Cancer Res.*, **49** 4385-89.

Hsiang, Y.H., Hertzberg, R., Hecht, S. and Liu, L. (1985) Camptothecin induces protein-linked DNA breaks via mammalian DNA topoisomerase I. *J. Biol. Chem.*, **260** 14873-78.

Huang, M.T., Newmark, H.L. and Frenkel, K. (1997) Inhibitory effects of curcumin on tumorgenesis in mice. *J. Cell Biochem.*, **27** 26-34.

Hutchinson, C.R. (1998) Combinatorial biosynthesis for new drug discovery. *Curr. Opin. Microbiol.*, **1** 319-29.

Janssen, O., Scheffler, A. and Kabelitz, D. (1993) *In vitro* effects of mistletoe extracts and mistletoe lectins: cytotoxicity towards tumor cells due to the induction of programmed cell death (apoptosis). *Arzneim. Forsch. Drug Res.*, **43** 1221-27.

Janssens, D., Michiels, C., Delaive, F., Eliaers, K., Drieu, K. and Remacle, J. (1995) Protection of hypoxia-induced ATP decrease in endothelial cells by *Ginkgo biloba* extract and bilobalide. *Biochem. Pharmacol.*, **50** 991-99.

Jaxel, C., Kohn, K.W., Wani, M.C., Wall, M.E. and Pommier, Y. (1989) Structure-activity study of the actions of camptothecin derivatives on mammalian topoisomerase I: evidence for a specific receptor site and for a relation to antitumor activity. *Cancer Res.*, **49** 1465-69.

Jiang, J.B., Li, G.Q., Guo, X.B., Kong, Y.C. and Arnold, K. (1982) Antimalarial activity of mefloquine and qinghaosu. *Lancet*, **ii** 285-88.

Kamchonwongpaisan, S., Paitayatat, S., Thebtaranonth, Y., Wilairat, P. and Yuthavong, Y. (1995) Mechanism-based development of new antimalarials: synthesis of derivatives of artemisinin attached to iron chelators. *J. Med. Chem.*, **38** 2311-16.

Kamchonwongpaisan, S. and Meshnick, S.R. (1996) The mode of action of the antimalarial, artemisinin, and its derivatives. *Gen Pharmacol.*, **27** 587-92.

Kamchonwongpaisan, S., McKeever, P., Hossler, P., Ziffer, H. and Meshnick, S.R. (1997) Artemisinin neurotoxicity: neuropathology in rats and mechanistic studies *in vitro*. *Am. J. Trop. Med. Hyg.*, **56** 7-12.

Kashman, Y., Gustafson, K.R., Fuller, R.W., Cardellina II, J.H., McMahon, J.B., Currens, M.J., Buckheit, R.W. Jr, Hughes, S.H., Cragg, G.M. and Boyd, M.R. (1992) The calanolides, a novel HIV-inhibitory class of coumarin derivatives from the tropical rainforest tree, *Calophyllum lanigerum*. *J. Med. Chem.*, **35** 2735-43.

Kattan, J., Durand, M., Droz, J.P., Mahjoubi, M., Marino, J.P. and Azab, M. (1994) Phase I study of retelliptine dihydrochloride (SR 95325 B) using a single two hour intravenous infusion schedule. *Am. J. Clin. Oncol.*, **17** 242-45.

Kaur, G., Stettler-Stevenson, M., Sebers, S., Worland, P., Sedlacek, H., Myers, C., Czech, J., Naik, R. and Sausville, E. (1992) Growth inhibition with reversible cell cycle arrest of carcinoma cells by flavone, L86-8275. *J. Natl. Cancer Inst.*, **84** 1736-40.

Kavanagh, J.J., Kudelka, A.P., deLeon, C.G., Tresukosol, D., Hord, M., Finnegan, M.B., Kim, E.E., Varma, D., Forman, A., Cohen, P., Edwards, C.L., Freedman, R.S. and Verschraegen, C.F. (1996) Phase II study of docetaxel in patients with epithelial ovarian carcinoma refractory to platinum. *Clin. Cancer Res.*, **2** 837-42.

Kelloff, G.J., Boone, C.W., Crowell, J.A., Steele, V.E., Lubet, R.A., Doody, L.A., Malone, W.F., Hawk, E.T. and Sigman, C.C. (1996a) New agents for cancer chemoprevention. *J. Cell Biochem.*, **26** 1-28.

Kelloff, G.J., Crowell, J.A., Hawk, E.T., Steele, V.E., Lubet, R.A., Boone, C.W., Covey, J.M., Doody, L.A., Omenn, G.S., Greenwald, P., Hong, W.K., Parkinson, D.R., Bargheri, D., Baxter, G.T., Blunden, M., Doeltz, M.K., Eisenhauer, K.M., Johnson, K., Knapp, G.G., Longfellow, D.G., Malone, W.F., Nayfield, S.G., Seifried, H.E., Swall, L.M. and Sigman, C.C. (1996b) Strategy and planning for chemopreventive drug development: clinical Development Plans II. *J. Cell. Biochem.*, **26**, 54-71.

Khosla, C. (1998) Combinatorial biosynthesis: new tools for the medicinal chemist. *Chemtracts*, **11** 1-15.

Kinghorn, A.D. and Balandrin, M.F. (eds) (1993) *Human Medicinal Agents from Plants*, ACS Symposium Series 534, American Chemical Society, Washington, DC.

Kirschling, R.J., Jung, S.H. and Jett, J.R. (1994) A phase II trial of taxol and GCSF in previously untreated patients with extensive stage small-cell lung cancer (SCLC). *Proc. Am. Soc. Clin. Oncol.*, **13** (Abstract) 326.

Klein, J., Chatterjee, S.S. and Löffelholz, K. (1997) Phospholipid breakdown and choline release under hypoxic conditions: inhibition by bilobalide, a constituent of *Ginkgo biloba*. *Brain Res.*, **755** 347-50.

Klumpp, T.R., Goldberg, S.L., Magdalinski, A.J. and Mangan, K.F. (1997) Phase II study of high-dose cyclophosphamide, etoposide, and carboplatin (CEC) followed by autologous hematopoietic stem cell rescue in women with metastatic or high-risk non-metastatic breast cancer: multivariate analysis of factors affecting survival and engraftment. *Bone Marrow Transplant.*, **20** 273-81.

Krausberthier, L., Guilbaud, N., Jan, M., SaintDizier, D., Rouillon, M.H., Burbridge, M.F., Pierre, A. and Atassi, G. (1997) Experimental antitumour activity of S 16020-2 in a panel of human tumours. *Eur. J. Cancer*, **33** 1881-87.

Kreitman, R.J. and Pastan, I. (1998) Immunotoxins for targeted cancer therapy. *Adv. Drug Delivery Rev.*, **31** 53-88.

Kuo, S.M. (1996) Antiproliferative potency of structurally distinct dietary flavonoids on human colon cancer cells. *Cancer Lett.*, **110** 41-48.

Laakmann, G., Schüle, C., Baghai, T. and Kieser, M. (1998) St. John's wort in mild-to-moderate depression: the relevance of hyperforin for the clinical efficacy. *Pharmacopsychiat.*, **31** 54-59.

Langdon, S.P., Hendriks, H.R., Braakhuis, B.J., Pratesi, G., Berger, D.P., Fodstad, O., Fiebig, H.H. and Boven, E. (1994) Preclinical phase II studies in human tumor xenografts: a European multicenter follow-up study. *Ann. Oncol.*, **5** 415-22.

Langer, M., Rothe, M., Eck, J., Möckel, B. and Zinke, H. (1996) A nonradioactive assay for ribosome-inactivating proteins. *Anal. Biochem.*, **243** 150-53.

Laurent, G., Maraninchi, D., Gluckman, E., Vernant, J.P., Derocq, J.M., Gaspard, M.H., Rio, B., Michalet, M., Reiffers, J., Dreyfus, F., Casellas, P., Schneider, P., Blythman, H.E., Bouloux, C. and Jansen, F.K. (1989) Donor bone marrow treatment with T101 Fab fragment-ricin A-chain immunotoxin prevents graft-*versus*-host disease. *Bone Marrow Transplant.*, **4** 367-71.

Le Bars, P.L., Katz, M.M., Berman, N., Itil, T.M., Freedman, A.M. and Schatzberg A.F. (1997) A placebo-controlled, double-blind, randomized trial of an extract of *Ginkgo biloba* for dementia: North American EGb Study Group. *J. Am. Med. Assoc.*, **278** 1327-32.

Le Mée, S., Pierré, A., Markovits, J., Atassi, G., Jacquemin-Sablon, A. and Saucier, J.-M. (1998) S16020-2, a new highly cytotoxic antitumor olivacine derivative: DNA interaction and DNA topoisomerase II inhibition. *Mol. Pharmacol.*, **53** 213-20.

Lenartz, D., Stoffel, B., Menzel, J. and Beuth, J. (1996) Immunoprotective activity of the galactoside-specific lectin from mistletoe after tumor destructive therapy in glioma patients. *Anticancer Res.*, **16** 3799-802.

Leonce, S., Perez, V., Casabianca-Piegnede, M.R., Anstett, M., Bisagni, E., Pierre, A. and Atassi, G. (1996) *In vitro* cytotoxicity of S16020-2, a new olivacine derivative. *Invest. New Drugs*, **14** 169-80.

Lin, A.J., Zikry, A.B. and Kyle, D.E. (1997) Antimalarial activity of new dihydroartemisinin derivatives. 7. 4-(*p*-substituted phenyl)-4(*R* or *S*)-[10(alpha or beta)-dihydroartemisininoxy] butyric acids. *J. Med. Chem.*, **40** 1396-400.

Lin, B.B. (1995) High throughput screening for new drug discovery. *Yaowu Shipin Fenxi*, **3** 233-41.

Lin, J.K., Chen, Y.C., Huang, Y.T. and LinShiau, S.Y. (1997) Suppression of protein kinase C and nuclear oncogene expression as possible molecular mechanisms of cancer chemoprevention by apigenin and curcumin. *J. Cell. Biochem.*, **28-29** 39-48.

Linde, K., Ramirez, G., Mulrow, C.D., Pauls, A., Weidenhammer, W. and Melchart, D. (1996) St John's wort for depression: an overview and meta-analysis of randomised clinical trials. *Br. Med. J.*, **313** 253-58.

Liu, L.F. (1989) DNA topoisomerase poisons as antitumor drugs. *Annu. Rev. Biochem.*, **58** 351-75.

Long, B.H., Carboni, J.M., Wasserman, A.J., Cornell, L.A., Casazza, A.M., Jensen, P.R., Lindel, T., Fenical, W. and Fairchild, C.R. (1998) Eleutherobin, a novel cytotoxic agent that induces tubulin polymerisation similar to paclitaxel (Taxol®). *Cancer Res.*, **58** 1111-15.

Lopes-Martins, R., Catelli, M., Araujo, C., Estato, V., Cordeiro, R. and Tibirica, E. (1996) Pharmacological evidence of a role for platelet-activating factor as a modulator of vasomotor tone and blood pressure. *Eur. J. Pharmacol.*, **308** 287-94.

Losiewicz, M.D., Carlson, B.A., Kaur, G., Sausville, E.A. and Worland, P.J. (1994) Potent inhibition of cdc2 kinase activity by the flavonoid, L86-8275. *Biochem. Biophys. Res. Commun.*, **201** 589-95.

Lu, H.Q., Niggemann, B. and Zanker, K.S. (1996) Suppression of the proliferation and migration of oncogenic *ras*-dependent cell lines, cultured in a three dimensional collagen matrix, by flavonoid-structured molecules. *J. Cancer Res. Clin. Oncol.*, **122** 335-42.

Madeleine, I., Prost, S., Naudin, A., Riou, G., Lavelle, F. and Riou, J.F. (1993) Sequential modifications of topoisomerase I activity in a camptothecin-resistant cell line established by progressive adaptation. *Biochem. Pharmacol.*, **45** 339-48.

Maraganore, J.M., Joseph, M. and Bailey, M.C. (1987) Purification and characterization of trichosanthin: homology to the ricin A-chain and implications as to mechanism of abortifacient activity. *J. Biol. Chem.*, **262** 11628-33.

Marty, M., Extra, J.M., Cottu, P.H. and Espie, M. (1997) Prospects with docetaxel in the treatment of patients with breast cancer. *Eur. J. Cancer*, **33** S26-S29.

Mathe, G., Morette, C., Hallard, M. and Blanquet, D. (1997) Combinations of three or four HIV-1 virostatics applied in short sequences which differ from each other by drug rotation: preliminary results of viral loads and CD4 numbers. *Biomed. Pharmacother.*, **51** 417-26.

McGrath, M.S., Hwang, K.M., Caldwell, S.E., Gaston, I., Luk, K.-C., Wu, P., Ng, V.L., Crowe, S., Daniels, J., Marsh, J., Deinhart, T., Lekas, P.V., Vennari, J.C., Yeung, H.-W. and Lifson, J.D. (1989) GLQ223: an inhibitor of human immunodeficiency virus replication in acutely and chronically infected cells of lymphocyte and mononuclear phagocyte lineage. *Proc. Natl. Acad. Sci. USA*, **86** 2844-48.

McGuire, W.P., Rowinsky, E.K., Rosenshein, N.B., Grumbine, F.C., Ettinger, D.S., Amstrong, D.K. and Donehower, R.C. (1989) Taxol: a unique antineoplastic agent with significant activity in advanced ovarian epithelial neoplasms. *Ann. Intern. Med.*, **111** 273-79.

McMahon, J.B., Currens, M.J., Gulakowski, R.J., Buckheit, R.W. Jr, Lackman-Smith, C., Hallock, Y.F. and Boyd, M.R. (1995) Michellamine B, a novel plant alkaloid, inhibits human immunodeficiency virus-induced cell killing by at least two distinct mechanisms. *Antimicrob. Agents Chemother.*, **39** 484-88.

Meshnick, S.R., Yang, Y.Z., Lima, V., Kuypers, F., Kamchonwongpaisan, S. and Yuthavong, Y. (1993) Iron-dependent free radical generation from the antimalarial agent, artemisinin (qinghaosu). *Antimicrob. Agents Chemother.*, **37** 1108-14.

Millward, M.J., Zalcberg, J., Bishop, J.F., Webster, L.K., Zimet, A., Rischin, D., Toner, G.C., Laird, J., Cosolo, W., Urch, M., Bruno, R., Loret, C., James, R. and Blanc, C. (1997) Phase

I trial of docetaxel and cisplatin in previously untreated patients with advanced non-small-cell lung cancer. *J. Clin. Oncol.*, **15** 750-58.

Möckel, B., Schwarz, T., Zinke, H., Eck, J., Langer, M. and Lentzen, H. (1997) Effects of mistletoe lectin I on human blood cell lines and peripheral blood cells: cytotoxicity, apoptosis and induction of cytokines *Arzneim. Forsch. Drug Res.*, **47** 1145-51.

Moertel, C.G., Schutt, A.J., Reitemerer, R.G. and Hahn, R.G. (1972) Phase II study of camptothecin (NSC-100880) in the treatment of advanced gastrointestinal cancer. *Cancer Chemother. Rep.*, **56** 95-101.

Molyneux, R.J. and James, L.F. (1981) Loco intoxication: indolizidine alkaloid of spotted locoweed. *Science* **216** 190-91.

Montefiori, D.C., Stewart, K., Ahearn, J.M., Zhou, J. and Zhou, J. (1993) Complement-mediated binding of naturally glycosylated and glycosylation-modified human immunodeficiency virus type 1 to human CR2 (CD21). *J. Virol.*, **67** 2699-706.

Muggia, F.M., Creaven, P.J., Hanson, H.H., Cohen, M.C. and Selawry, O.S. (1972) Phase I clinical trial of weekly and daily treatment with camptothecin (NSC-100880): correlation with preclinical studies. *Biochemistry*, **56** 515-21.

Müller, W.E., Rolli, M., Schäfer, C. and Hafner, U. (1997) Effects of Hypericum extract (LI 160) in biochemical models of antidepressant. *Pharmacopsychiat.*, **30** 102-107.

Müller, W.E., Singer, A., Wonnemann, M., Hafner, U., Rolli, M. and Schäfer, C. (1998) Hyperforin represents the neurotransmitter reuptake inhibiting constituent of Hypericum extract. *Pharmacopsychiat.*, **31** 16-21.

Na-Bangchang, K., Tipwangso, P., Thanavibul, A., Tan-ariya, P., Suprakob, K., Kanda, T. and Karbwang, J. (1996) Artemether-pyrimethamine in the treatment of pyrimethamine-resistant falciparum malaria. *Southeast Asian J. Trop. Med. Public. Health*, **27** 19-23.

Naik, R.G., Kattige, S.L., Bhat, S.V., Alreja, B., de Souza, N.J. and Rupp, R.H. (1988) An anti-inflammatory cum immunomodulatory piperidinylbenzopyranone from *Dysoxylum binectariferum*: isolation, structure and total synthesis. *Tetrahedron*, **44** 2081-86.

Nakamura, Y., Ohto, Y., Murakami, A., Osawa, T. and Ohigashi, H. (1998) Inhibitory effects of curcumin and tetrahydrocurcuminoids on the tumor promotor-induced reactive oxygen species generation in leukocytes *in vitro* and *in vivo*. *Jpn. J. Cancer Res.*, **89** 361-70.

Nie, H., Cai, X., He, X., Xu, L., Ke, X., Ke, Y. and Tam, S.-C. (1998) Position 120–123, a potential active site of trichosanthin. *Life Sci.*, **62** 491-500.

Oberpichler-Schwenk, H. and Krieglstein, J. (1992) Pharmakologische Wirkungen von Ginkgo biloba-Extrakt und -Inhaltsstoffen. *Pharm. unserer Zeit*, **21** 224-35.

Ohashi, M., Sugikawa, E. and Nakanishi, N. (1995) Inhibition of p53 protein phosphorylation by 9-hydroxyellipticine: a possible anticancer mechanism. *Jpn. J. Cancer Res.*, **86** 819-27.

Pagani, O., Zucchetti, M., Sessa, C., deJong, J., D'Incalci, M., DeFusco, M., KaeserFrohlich, A., Hanauske, A. and Cavalli, F. (1996) Clinical and pharmacokinetic study of oral NK611, a new podophyllotoxin derivative. *Cancer Chemother. Pharmacol.*, **38** 541-47.

Pendurthi, U.R., Williams, J.T. and Rao, L.V. (1997) Inhibition of tissue factor gene activation in cultured endothelial cells by curcumin: suppression of activation of transcription factors, Egr-1, AP-1 and NF-kappa B. *Arterioscler. Thromb. Vasc. Biol.*, **17** 3406-13.

Perego, P., Capranico, G., Supino, R. and Zunino, F. (1994) Topoisomerase I gene expression and cell sensitivity to camptothecin in human cell lines of different tumor types. *Anticancer Drugs*, **5** 645-49.

Pezzuto, J.M. (1997) Plant-derived anticancer agents. *Biochem. Pharmacol.*, **53** 121-33.

Pietri, S., Maurelli, E., Drieu, K. and Culcasi, M. (1997) Cardioprotective and antioxidant effects of the terpenoid constituents of *Ginkgo biloba* extract (EGb 761). *J. Mol. Cell. Cardiol.*, **29** 733-42.

Pommier, Y. (1996) Eukaryotic DNA topoisomerase I: genome gatekeeper and its intruders, camptothecins. *Semin. Oncol.*, **23** 3-10.

Pommier, Y. and Kohn, K.W. (1989) Topoisomerase II inhibition by antitumor intercalators and demethylepipodophyllotoxins, in *Developments in Cancer Chemotherapy* (eds. R.I. Glazer), CRC Press, Boca Raton, FL, pp. 175-96.

Pommier, Y., Bertrand, R. and Solary, E. (1994a) Apoptosis induced by DNA topoisomerase I and II inhibitors in human leukemia HL-60 cells. *Leuk. Lymph.*, **15** 21-31.

Pommier, Y., Leteurtre, F., Fesen, M.R., Fujimori, A., Bertrand, R., Solary, E., Kohlhagen, G. and Kohn, K.W. (1994b) Cellular determinants of sensitivity and resistance to DNA topoisomerase inhibitors. *Cancer Invest.*, **12** 530-42.

Potmesil, M. (1994) Camptothecins: from bench research to hospital wards. *Cancer Res.*, **54** 1431-39.

Potmesil, M., Vardeman, D., Kozielski, A.J., Mendoza, J., Stehlin, J.S. and Giovanella, B.C. (1995) Growth inhibition of human cancer metastases by camptothecins in newly developed xenograft models. *Cancer Res.*, **55** 5637-41.

Price, R.N., Nosten, F., Luxemburger, C., Kham, A., Brockman, A., Chongsuphajaisiddhi, T. and White, N.J. (1995) Artesunate *versus* artemether in combination with mefloquine for the treatment of multidrug-resistant falciparum malaria. *Trans. R. Soc. Trop. Med. Hyg.*, **89** 523-27.

Raffa, R.B. (1998) Screen of receptor and uptake-site activity of hypericin component of St. John's wort reveals σ receptor binding. *Life Sci.*, **62** 265-70.

Rassmann, I., Schrodel, H., Schilling, T., Zucchetti, M., KaeserFrohlich, A., Rastetter, J. and Hanauske, A.R. (1996) Clinical and pharmacokinetic phase I trial of oral dimethylaminoetoposide (NK611) administered for 21 days every 35 days. *Invest. New Drugs*, **14** 379-86.

Reddy, S. and Aggarwal, B.B. (1994) Curcumin is a noncompetitive and selective inhibitor of phosphorylase kinase. *FEBS Lett.*, **341** 19-22.

Reichman, B.S., Seidman, A.D., Crown, J.P.A., Heelan, R., Hakes, T.B., Lebwohl, D.E., Gilewski, T.A., Surbone, A., Currie, V., Hudis, C.A., Yao, T.J., Klecker, R., Jamis-Dow, C., Collins, J., Quinlivan, S., Berkery, R., Toomasi, F., Canetta, R., Fisherman, J., Arbuck, S. and Norton, T. (1993) Paclitaxel and recombinant human granulocyte colony-stimulating factor as initial chemotherapy for metastatic breast cancer. *J. Clin. Oncol.*, **11** 1943-51.

Ribereau-Gayon, G., Dumont, S., Müller, C., Jung, M.L., Poindron, P. and Anton, R. (1996) Mistletoe lectins I, II and III induce the production of cytokines by cultured human monocytes. *Cancer Lett.*, **109** 33-38.

Rose, P.G., Blessing, J.A., Mayer, A.R. and Homesley, H.D. (1998) Prolonged oral etoposide as second-line therapy for platinum-resistant and platinum-sensitive ovarian carcinoma: a gynecologic oncology group study. *J. Clin. Oncol.*, **16** 405-10.

Rosell, R., Felip, E., Massuti, B., Gonzalez-Larriba, J.L., Benito, D., Lopez-Cabrerizo, M.P., Salamanca, O., Camps, C. and Puerto-Pica, J. (1997) A sequence-dependent paclitaxel/etoposide phase II trial in patients with non-small-cell lung cancer. *Semin. Oncol.*, **24** S12-56-S12-60.

Rougier, P. and Bugat, R. (1996) CPT-11 in the treatment of colorectal cancer: clinical efficacy and safety profile. *Semin. Oncol.*, **23** 34-41.

Safayhi, H., Mack, T., Sabieraj, J., Anazodo, M.I., Subramanian, L.R. and Ammon, H.P.T. (1992) Boswellic acids: novel, specific, nonredox inhibitors of 5-lipoxygenase. *J. Pharmacol. Exp. Ther.*, **261** 1143-46.

Safayhi, H., Sailer, E.-R. and Ammon, H.P.T. (1995) Mechanism of 5-lipoxygenase inhibition by acetyl-11-keto-β-boswellic acid. *Mol. Pharmacol.*, **47** 1212-16.

Safayhi, H., Rall, B., Sailer, E.R. and Ammon, H.P.T. (1997) Inhibition by boswellic acids of human leukocyte elastase. *J. Pharmacol. Exp. Ther.*, **281** 460-63.

Sailer, E.-R., Hörnlein, R.F., Subramanian, L.R., Ammon, H.P.T. and Safayhi, H. (1996a) Preparation of novel analogues of the nonredox-type noncompetitive leukotriene biosynthesis inhibitor, AKBA. *Arch. Pharm.*, **329** 54-56.

Sailer, E.-R., Subramanian, L.R., Rall, B., Hörnlein, R.F., Ammon, H.P.T. and Safayhi, H. (1996b) Acetyl-11-keto-β-boswellic acid (AKBA): structure requirements for binding and 5-lipoxygenase inhibitory activity. *Br. J. Pharmacol.*, **117** 615-18.

Sasaki, K., Wada, K., Hatta, S., Ohshika, H. and Haga, M. (1997) Bilobalide, a constituent of *Ginkgo biloba* (L.) potentiates drug-metabolizing enzyme activities in mice: possible mechanism for anticonvulsant activity against 4-*O*-methylpyridoxine-induced convulsions. *Res. Commun. Mol. Pathol. Pharmacol.*, **96** 45-56.

Schiff, P.B., Fant, J. and Horwitz, S.B. (1979) Promotion of microtubule assembly *in vitro* by taxol. *Nature*, **22** 665-67.

Scholtyssek, H., Damerau, W., Wessel, R. and Schimke, I. (1997) Antioxidative activity of ginkgolides against superoxide in an aprotic environment. *Chem. Biol. Interact.*, **106** 183-90.

Schrader, E., Meier, B. and Brattstrom, A (1998) *Hypericum* treatment of mild-moderate depression in a placebo-controlled study: a prospective, double-blind, randomized, placebo-controlled, multicentre study. *Human Psychopharm.*, **13** 163-69.

Schumacher, U., Stamouli, A., Adam, E., Peddie, M. and Pfueller, U. (1995) Biochemical, histochemical and cell biological investigations on the actions of mistletoe lectins I, II and III with human breast cancer cell lines. *Glycoconj. J.*, **12** 250-57.

Senderowicz, A.M., Headlee, D., Stinson, S., Lush, R.M., Tompkins, A., Brawley, O., Bergan, R., Figg, W.D., Smith, A. and Sausville, E.A. (1996) Phase I trial of a novel cyclin-dependent kinase inhibitor, flavopiridol, in patients with refractory neoplasms. *Ann. Oncol.*, **7** 77.

Sharma, M.L., Bani, S. and Singh, G.B. (1989) Antiarthritic activity of boswellic acids in bovine serum albumin (BSA)-induced arthritis. *Int. J. Immunopharmacol.*, **11** 647-52.

Shu, Y.-Z. (1998) Recent natural products based drug development: a pharmaceutical industry perspective. *J. Nat. Prod.*, **61** 1053-71.

Singh, G.B. and Atal, C.K. (1986) Pharmacology of an extract of salai guggal ex-*Boswellia serrata*, a new nonsteroidal anti-inflammatory agent. *Agents Actions*, **18** 407-12.

Smith, S.L., Fishwick, J., McLean, W.G., Edwards, G. and Ward, S.A. (1997) Enhanced *in vitro* neurotoxicity of artemisinin derivatives in the presence of haemin. *Biochem. Pharmacol.*, **53** 5-10.

Solary, E., Bertrand, R. and Pommier, Y. (1994) Apoptosis induced by DNA topoisomerase I and II inhibitors in human leukemic HL-60 cells. *Leuk. Lymph.*, **15** 21-32.

Soler, M.H., Stoeva, S. and Voelter, W. (1998) Complete amino acid sequence of the B-chain of mistletoe lectin I. *Biochem. Biophys. Res. Commun.*, **246** 596-601.

South, E.H., Exon, J.H. and Hendrix, K. (1997) Dietary curcumin enhances antibody response in rats. *Immunopharmacol. Immunotoxicol.*, **19** 105-19.

Sreejayan and Rao, M.N. (1997) Nitric oxide scavenging by curcuminoids. *J. Pharm. Pharmacol.* **49** 105-107.

Stein, G.M. and Berg, P.A. (1996) Evaluation of the stimulatory activity of a fermented mistletoe lectin-I free mistletoe extract on T-helper cells and monocytes in healthy individuals *in vitro*. *Arzneim. Forsch. Drug Res.*, **46** 635-39.

Stein, G., Henn, W., von Laue, H. and Berg, P.I. (1998) Modulation of the cellular and humoral immune responses of tumor patients by mistletoe therapy. *Eur. J. Med. Res.*, **3** 194-202.

Stryer, L. (1995) *Biochemistry*, 4th edn, W.H. Freemann and Co., New York.

Suffness, M. (ed.) (1995) *Taxol® Science and Applications*, CRC Press, Boca Raton, New York, London, Tokyo.

Suffness, M. and Cordell, G.A. (1985) Antitumor alkaloids, in *The Alkaloids*, Vol. XXV (ed. A. Brossi), Academic Press, New York, pp. 1-369.

Suffness, M. and Wall, M.E. (1995) Discovery and development of taxol, in *Taxol® Science and Applications* (ed. Suffness, M.), CRC Press, Boca Raton, New York, London, Tokyo, pp. 3-25.

Sunkara, P.S., Bowlin, T.L., Kang, M.S., Liu, P.S., Tyms, A.S. and Sjoerdsma, A. (1989) Anti-HIV activity of castanospermine analogues. *Lancet*, **i** 1206.

Supko, J.G. and Malspeis, L. (1995) Pharmakokinetics of michellamine B, a naphthylisoquinoline alkaloid with *in vitro* activity against immunodeficiency virus types 1 and 2, in the mouse and dog. *Antimicrob. Agents Chemother.*, **39** 9-14.

Svejstrup, J.Q., Christiansen, K., Gromova, I.I., Andersen, A.H. and Westergaard, O. (1991) New techniques for uncoupling the cleavage and religation reactions of eukaryotic topoisomerase I: the mode of action of camptothecin at a specific recognition site. *J. Mol. Biol.*, **222** 669-78.

Takimoto, C.H. and Arbuck, S.G. (1997) Clinical status and optimal use of topotecan. *Oncology Huntingt.*, **11** 1635-46.

Talbot, S.J., Weiss, R.A. and Schulz, T.F. (1995) Reduced glycosylation of human cell lines increases susceptibility to CD4-independent infection by human immunodeficiency virus type 2 (LAV-2/B). *J. Virol.*, **69** 3399-406.

Teicher, B.A., Holden, S.A., Khandakar, V. and Herman, T.S. (1993) Addition of a topoisomerase I inhibitor to trimodality therapy (cis-diamminediochloroplatinum(II)/heat/radiation) in a murine tumor. *J. Cancer Res. Clin. Oncol.*, **119** 645-51.

Teufel-Mayer, R. and Gleitz, J. (1997) Effects of long-term administration of Hypericum extracts on the affinity and density of the central serotonergic, 5-HT1 A and 5-HT2 A, receptors. *Pharmacopsychiat.*, **30** 113-16.

Teuscher, E. (1997) *Biogene Arzneimittel*, 5th edn., Wissenschaftliche Verlagsgesellschaft mbH, Stuttgart.

Thrush, G.R., Lark, L.R., Clinchy, B.C. and Vitetta, E.S. (1996) Immunotoxins: an update. *Annu. Rev. Immunol.*, **14** 49-71.

Tran, T.H., Day, N.P., Nguyen, H.P., Nguyen, T.H., Tran, T.H., Pham, P.L., Dinh, X.S., Ly, V.C., Ha, V., Waller, D., Peto, T.E. and White, N.J. (1996) A controlled trial of artemether or quinine in Vietnamese adults with severe falciparum malaria. *N. Engl. J. Med.*, **335** 76-83.

Trigg, P.I. (1989) Quinghaosu (artemisinin) as an antimalarial drug, in *Economic and Medicinal Plant Research* (eds. H. Wagner, H. Hikino and N.R. Farnsworth), Academic Press, London, pp. 19-55.

Tsao, S.W., Ng, T.B. and Yeung, H.W. (1990) Toxicities of trichosanthin and alpha-momorcharin, abortifacient proteins from Chinese medicinal plants, on cultured tumor cell lines. *Toxicon.*, **28** 1183-92.

Tulsiani, D.R., Harris, T.M. and Touster, O. (1982) Swainsonine inhibits the biosynthesis of complex glycoproteins by inhibition of Golgi mannosidase II. *J. Biol. Chem.*, **257** 7936-39.

Uckun, F.M., Chelstrom, L.M., TuelAhlgren, L., Dibirdik, I., Irvin, J.D., Langlie, E.C. and Myers, D.E. (1998a) TXU (anti-CD7)-pokeweed antiviral protein as a potent inhibitor of human immunodeficiency virus. *Antimicob. Agents Chemother.*, **42** 383-88.

Uckun, F.M., Narla, R.K., Zeren, T., Yanishevski, Y., Myers, D.E., Waurzyniak, B., Ek, O., Schneider, E., Messinger, Y., Chelstrom, L.M., Gunther, R. and Evans, W. (1998b) *In vivo* toxicity, pharmacokinetics and anticancer activity of genistein linked to recombinant human epidermal growth factor. *Clin. Cancer Res.*, **4** 1125-43.

VanOijen, M.G.C.T. and Preijers, F.W.M.B. (1998) Rationale for the use of immunotoxins in the treatment of HIV-infected humans. *J. Drug Target.*, **5** 75-91.

Vitetta, E.S., Fulton, R.J., May, R.D., Till, M. and Uhr, J.W. (1987) Redesigning nature's poisons to create antitumor reagents. *Science*, **238** 1098-104.

Vorbach, E.U., Arnoldt, K.H. and Hübner, W.D. (1997) Efficacy and tolerability of St. John's wort extract, LI 160, *versus* imipramine in patients with severe depressive episodes according to ICD-10. *Pharmacopsychiat.*, **30** 81-85.

Waddick, K.G., Myers, D.E., Gunther, R., Chelstrom, L.M., Chandan-Langlie, M., Irvin, J.D., Tumer, N. and Uckun, F.M. (1995) *In vitro* and *in vivo* antileukemic activity of B43-

pokeweed antiviral protein against radiation resistant human B-cell precursor leukemia cells. *Blood*, **86** 4228-33.

Wall, M.E. and Wani, M.C. (1995) Camptothecin and taxol: discovery to clinic. *Cancer Res.*, **55** 753-60.

Wall, M.E., Wani, M.C., Cook, C.E., Palmer, K.H., McPhail, A.T. and Sim, G.A. (1966) Plant antitumor agents. I. The isolation and structure of camptothecin: a novel alkaloidal leukemia and tumor inhibitor from *Camptotheca acuminata*. *J. Am. Chem. Soc.* **88** 3888-90.

Wani, M.C., Nicholas, A.W. and Wall, M.E. (1987) Plant antitumor agents. 28. Resolution of a key tricyclic synthon, 5′ (RS)-1,5-dioxo-(5′-ethyl-5′-hydroxy-2′-H, 5′-H, 6′-H-6-oxopyrano) [3′,4′-f]-6,8-tetrahydroindolizine: total synthesis and antitumor activity of 20(*S*)- and 20(*R*)-camptothecin. *J. Med. Chem.*, **30** 2317-19.

Wani, M.C., Ronman, P.E., Lindley, J.T. and Wall, M.E. (1980) Plant tumor agents. 18. Synthesis and biological activity of camptothecin analogs. *J. Med. Chem.*, **23** 554-60.

Wani, M.C., Taylor, H.L., Wall, M.E., Coggon, P. and McPhail, A.T. (1971) Plant antitumor agents. VI. The isolation and structure of taxol, a novel antileukemic and antitumor agent from *Taxus brevifolia*. *J. Am. Chem. Soc.*, **93** 2325-27.

Weiner, L.M., O'Dwyer, J., Kitson, J., Comis, R.L., Frankel, A.E., Bauer, R.J., Konrad, M.S. and Groves, E.S. (1989) Phase I evaluation of an anti-breast-carcinoma monoclonal antibody 260F9-recombinant ricin A chain immunoconjugate. *Cancer Res.*, **49** 4062-67.

Wenzel-Seifert, K., Krautwurst, D., Lentzen, H. and Seifert, R. (1996) Concanavalin A and mistletoe lectin I differentially activate cation entry and exocytosis in human neutrophils: lectins may activate multiple subtypes of cation channels. *J. Leukoc. Biol.*, **60** 345-55.

Westeel, V., Murray, N., Gelmon, K., Shah, A., Sheehan, F., McKenzie, M., Wong, F., Morris, J., Grafton, C., Tsang, V., Goddard, K., Murphy, K., Parsons, C., Amy, R. and Page, R. (1998) New combination of the old drugs for elderly patients with small-cell lung cancer: a phase II study of the PAVE regimen. *J. Clin. Oncol.*, **16** 1940-47.

White, E.L., Ross, L.J., Schmid, S.M., Kelloff, G.J., Steele, V.E. and Hill, D.L. (1998) Screening of potential cancer-preventing chemicals for inhibition of induction of ornithine decarboxylase in epithelial cells from rat trachea. *Oncol. Rep.*, **5** 717-22.

Wie, N. and Sadrzadeh, S.M. (1994) Enhancement of hemin-induced membrane damage by artemisinin. *Biochem. Pharmacol.*, **48** 737-41.

Winkler, U., Barth, S., Schnell, R., Diehl, V. and Engert, A. (1997) The emerging role of immunotoxins in leukemia and lymphoma. *Ann. Oncol.*, **8** 139-46.

Wonnemann, M., Schäfer, C. and Müller, W.E. (1997) Effects of Hypericum extracts on glutamatergic and GABAergic receptor systems. *Pharmacopsychiat.*, **30** 237.

Worland, P.J., Kaur, G. and Stetler-Stevenson, M. (1993) Alteration of the phosporylation state of p34 cdc2 kinase by the flavone, l86-8275, in breast carcinoma cells. *Biochem. Pharmacol.*, **46** 1831-40.

Wu, A.M., Song, S.C., Sugii, S. and Herp, A. (1997) Differential binding properties of Gal/GalNAc specific lectins available for characterization of glycoreceptors. Indian *J. Biochem. Biophys.*, **34** 61-71.

Yang, Y.Z., Little, B. and Meshnick, S.R. (1994) Alkylation of proteins by artemisinin: effects of heme, pH and drug structure. *Biochem. Pharmacol.*, **48** 569-73.

Yoon, T.J., Yoo, Y.C., Choi, O.B., Do, M.S., Kang, T.B., Lee, S.W., Azuma, I. and Kim, J.B. (1995) Inhibitory effect of Korean mistletoe (*Viscum album* coloratum) extract on tumour angiogenesis and metastasis of haematogenous and non-haematogenous tumour cells in mice. *Cancer Lett.*, **97** 83-91.

Zhang, J.-S. and Liu, W.-Y. (1992) The mechanism of action of trichosanthin on eukaryotic ribosomes-RNA *N*-glycosidase activity of the cytotoxin. *Nucleic Acids Res.*, **20** 1271-75.

Zhu, L., Wu, J., Liao, H., Gao, J., Zhao, X.N. and Zhang, Z.X. (1997) Antagonistic effects of extract from leaves of *Ginkgo biloba* on glutamate neurotoxicity. *Acta Pharm. Sin.*, **18** 344-45.

7 Production of secondary metabolites in cell and differentiated organ cultures

N.J. Walton, A.W. Alfermann and M.J.C. Rhodes

7.1 Introduction

As a result of intensive work in many laboratories, beginning over a hundred years ago with the pioneering work of Gottlieb Haberlandt in 1898 (published in 1902), *in vitro* culture of plant cells (plant tissue culture) became a reality. In principle, *in vitro* cultures of all plant species can be initiated. Millions of plants are propagated every year by using tissue culture techniques. By removing the cell walls, protoplasts can be isolated. Protoplasts of different species can be fused to generate new hybrids, which cannot be obtained by sexual-crossing. Moreover, isolated foreign genes can be integrated into the genome of plant cells, leading to an altered genotype. Plant cells can be cultivated in suspension to produce natural products. Working volumes of up to 75,000 l have been achieved (Westphal, 1990). The aim of the present chapter is to present an overview of the possibilities and problems in the use of plant cell and organ cultures for natural product formation. For further information, the reader is referred to several recent reviews (Petersen and Alfermann, 1993; Berlin, 1997; Doran, 1997; Mühlbach, 1998) and to the original research papers cited therein. Section 7.2 deals with cell cultures; and Section 7.3 covers differentiated organ cultures and, in addition, addresses strategies to increase secondary-product formation, which in principle are applicable to both types of culture.

What are plant cell cultures? According to a widely-used definition by Street (1977), a callus culture is an unorganised plant tissue growing on a solidified medium. Cell cultures or cell suspension cultures are initated from callus cultures by transferring pieces of callus into liquid medium. Such cell suspension cultures ideally consist of only single cells but, in reality, they contain a range of cell aggregates, some containing up to several hundred cells. For detailed information on the techniques and media formulations used to initiate and subcultivate callus and cell cultures, the reader is referred to textbooks dealing with the theory and practice of tissue culture (e.g. Street, 1977; Bhojwani and Razdan, 1983; Seitz *et al.*, 1985; and Dixon and Gonzales, 1994). It is laborious to produce large amounts of callus cultures for biochemical investigations. It is much easier to use suspension cultures, since the growth rates in suspension are much higher. The highest growth rates were achieved with tobacco cells (Noguchi *et al.*, 1977), with doubling times as short as 15 h. To produce larger amounts of cell mass, cell suspensions are cultured in large-volume bioreactors (formerly called 'fermenters'). In the past, there was much debate as to the types of bioreactor that would be most suitable for the culture of plant cells. Plant cells show many differences from microorganisms (*cf.* e.g. Scragg, 1992; Petersen and Alfermann, 1993); a particular difference is their higher shear sensitivity. Due to their low shear forces, airlift reactors were initially preferred for plant cell cultures (Wagner and Vogelmann, 1977) but, more recently, it has become widely accepted that continuously-stirred tank reactors (CSTR) with appropriate stirrer configurations, e.g. a marine impeller (Westphal, 1990; and literature cited therein) or spiral stirrer (Ulbrich *et al.*, 1985; Spieler *et al.*, 1985), can be used equally well and may even be preferable (for more detailed discussions, see Scragg, 1992; Schlatmann *et al.*, 1996; Roberts and Shuler, 1997; and Mühlbach, 1998). Special types of bioreactor have been developed for the growth of organ cultures, especially hairy root cultures, as indicated in Section 7.3.4.4.

Why attempt to produce natural products with plant cell or organ cultures? Plants produce a large spectrum of 'natural products' or 'secondary plant products'. Nowadays, it is accepted that although these low molecular weight products are not important for the primary metabolism of the plant, they are in many cases of great importance for the plant to survive in its natural environment (Hartmann, 1985). Man uses many of these natural products, for example as fibres, food additives, cosmetics, dyestuffs or medicines. Farnsworth (1985) reported that one quarter of all prescription drugs used in the USA still contain plant-derived substances isolated from plant sources, since chemical synthesis is either not possible or uneconomic in such cases (Table 7.1).

Obtaining sufficient supplies of appropriate plant materials for drug isolation has become more difficult and expensive in recent years. Often these substances are still isolated from plants collected in the wild. *Podophyllum hexandrum*, the source for the isolation of podophyllotoxin used for the semisynthesis of anti-cancer drugs, and *Pilocarpus jaborandi*, formerly the source of pilocarpin used in glaucoma treatment, may be cited as examples. *Podophyllum hexandrum* is now an endangered species in the Himalayas (Gupta, 1991); and the alternative species, *Pilocarpus microphyllus*, is nowadays used for the isolation of pilocarpin, since *P. jaborandi* is no longer available due to severe overcollection (A. Basedow, personal communication). This explains why there is great interest in establishing alternative supplies of such drug material; plant cell or organ culture may provide such alternatives. Routien and Nickel (1952) claimed, in their patent on "Cultivation of Plant Tissue", that submerged, cultivated plant cells could be used for the production of useful products.

Table 7.1 Plant-derived drugs widely used in western medicine (after Farnsworth, 1985)

Acetyldigoxin	Ephedrine*	Pseudoephedrine*
Aescin	Hyoscyamine	Quinidine
Ajmalicine	Khellin	Quinine
Allantoin*	Lanatoside C	Rescinnamine
Atropine	Leurocristine	Reserpine
Bromelain	Lobeline	Scillarens
Caffeine*	Morphine	Scopolamine
Codeine	Narcotine	Sennosides
Colchicine	Ouabain	Sparteine
Danthron*	Papain	Strychnine
Deserpidine	Papaverine*	Tetrahydrocannabinol
Digitoxin	Physostigmine	Theobromine*
Digoxin	Picrotoxin	Theophylline*
L-Dopa*	Pilocarpine	Tubocurarine
Emetine	Protoveratrines	Vincaleukoblastine
		Xanthotoxin

*produced industrially by synthesis. Abbreviation: L-Dopa, l-dihydroxyphenylalanine.

At that time, the accumulation of anthocyanins in callus cultures demonstrated the possibility, in principle, of natural product formation by plant cell cultures.

As discussed further in Section 7.2.1, cell cultures of only a limited number of plant species, notably *Lithospermum*, *Coleus* and *Coptis*, have been successfully developed to give stable, high product yields on a process scale. Cell cultures of *Lithospermum* and *Coleus* can achieve accumulations of shikonins and rosmarinic acid, respectively, exceeding 20% of the dry matter of the culture (Fujita, 1988; Petersen *et al.*, 1994); however, such production rates remain uncommon. In many cases, high rates of production of commercially-important compounds have not been achieved. This lack of productivity is related, firstly, to the failure by empirical means to find conditions in the cell culture to stimulate production and, secondly, to the fact that production, even if established, is often unstable. Given this situation, alternative approaches to facilitate secondary product formation in culture have been sought. In particular, the use of differentiated organ cultures has offered at least a partial answer. In these cultures, in contrast to dispersed cell cultures, the endogenous control mechanisms that specify the production of secondary metabolites during the process of organ differentiation are intact and regulate both the expression of the biosynthetic pathways to the products and the stable generation of the product over many cycles of growth in culture.

The major successes with organ cultures have been achieved with root cultures, but some progress has also been made with shoot cultures. With both types of organ, transformed as well as untransformed cultures have been evaluated. Section 7.3 describes the properties of these different types of differentiated organ cultures and current progress in exploiting their biochemical potential. It then indicates the limitations of currently-available cultures and outlines how advances in molecular genetics might be applied to improve the productivity of these cultures (and also of cell cultures), and to extend organ-culture technology to a wider range of organ types.

7.2 Production of natural products by plant cell cultures

7.2.1 Production of known compounds

The principal objective of cell culture so far has been the production of compounds found in the normal differentiated plant; for example, as a means to overcome shortages of naturally-produced plant material for drug production, as discussed above. In order to increase product yields, Zenk and co-workers (1977) elaborated a strategy (Table 7.2) that proved to be very helpful and that was adopted by many researchers. Table 7.3 gives some examples of cell cultures that accumulate large amounts of secondary products. Shikonin was the first compound produced by cell cultures on a commercial scale. Cell cultures of *Coleus* accumulate more than 5 g/l of rosmarinic acid (Ulbrich *et al*., 1985; Hippolyte *et al*., 1992) and *Coptis* cells produce more than 7 g/l of the alkaloid, berberine (Matsubara and Fujita, 1991). Several other alkaloids have also been produced in high yields.

Table 7.2 Strategy to improve natural product formation in plant cell cultures

- Screen plants for high accumulation of the natural compound(s) desired
- Initiate callus cultures from selected high-producing parent plants
- Analyse these cultures for the desired product(s)
- Establish cell suspension cultures from producing-callus
- Analyse the suspension cultures
- Select high-producing cell lines via single cell cloning using random selection based on somaclonal variations or mutagenic treatment
- Ultimate objective: selection of stable high-producing cell lines.
- Further improvement of product yields by optimisation of the culture process (i.e. optimisation of the medium composition, using a two-stage or a fed-batch system as well as improving the physical parameters of the bioreactor process)

The problem remains, however, that one cannot predict that cell cultures of a given species will indeed be able to accumulate the compounds of interest; there is a large element of luck. Many species have turned out to be very recalcitrant. For instance, cell cultures of *Digitalis* or *Papaver* accumulate the interesting cardenolides and alkaloids characteristic of the parent plants in only tiny amounts, or not at all,

Table 7.3 Some examples of high-producing cell cultures

Metabolite	Species	Yield		Reference
		g/L	%dw	
Berberine	*Coptis japonica*	7.0	12	Fujita and Tabata, 1987
Jatrorrhizine	*Berberis wilsoniae*	3.0	12	Breuling *et al.*, 1985
Rosmarinic acid	*Coleus blumei*	5.6	20	Ulbrich *et al.*, 1985
	Salvia officinalis	6.4		Hippolyte *et al.*, 1992
Anthocyanins	*Perilla frutescens*	5.8		Zhong and Yoshida, 1995
Shikonin	*Lithospermum erythrorhizon*	3.5	12	Fujita *et al.*, 1982
Anthraquinones	*Morinda citrifolia*	2.5	18	Zenk *et al.*, 1975
Raucaffricine	*Rauwolfia serpentina*	1.6	3	Schübel *et al.*, 1989
Cinnamoylputrescine	*Nicotiana tabacum*	1.5	13	Schiel *et al.*, 1984
Arbutine[a]	*Datura innoxia*	7.1	43–50	Suzuki *et al.*, 1987
Paclitaxel	*Taxus* sp.	0.1–0.3		Yukimune *et al.*, 1996 Bringi *et al.*, 1995

[a]After biotransformation of hydroquinone added to the medium. Abbreviation: %dw, percentage dry weight.

and further attempts to improve product yields by traditional empirical approaches cannot be recommended (Berlin, 1997). On the other hand, however, surprises still occur. For many years, it seemed impossible to induce formation of azadirachtin in cell cultures of *Azadirachta indica* but, very recently, two laboratories have been successful (van der Esch, 1998; Wewetzer, 1998). The recent progress in the production of paclitaxel (Taxol®) has demonstrated, very impressively, that plant cell cultures can serve as an alternative system for the production of a desired plant product, if there is an urgent demand (Yukimune *et al.*, 1996; Bringi *et al.*, 1995; Venkat, 1996).

7.2.2 Production by cell cultures of new compounds not yet found in nature

Since the detection of the paniculides in callus cultures of *Andrographis paniculata* by Overton and co-workers (Allison *et al.*, 1968), it has been generally accepted that plant cell cultures may produce natural products that cannot otherwise be found in nature. Ruyter and Stöckigt (1989) counted more than 70 such compounds; the figure has since risen to 150 (Table 7.4 shows some examples). Of course, it sometimes transpires that so-called 'novel compounds' apparently found only in tissue cultures are present in differentiated plants as well. For example, 5-methoxypodophyllotoxin was first detected in tissue cultures of *Linum flavum* by Berlin and co-workers (1986), but Broomhead and Dewick (1990) subsequently showed that this lignan occurs quite commonly in the genus *Linum*.

Several companies have used, or are still using, plant cell cultures to search for new secondary products with interesting biological activities,

Table 7.4 Some examples of natural compounds found in cell cultures but not in the differentiated plant

Species	Compound	Reference
Andrographis paniculata	Paniculide	Allison *et al.*, 1968
Morinda citrifolia	5,6-Dihydrolucidin	Inoue *et al.*, 1981
Picralina nitida	Pericine	Arens *et al.*, 1982
Podophyllum versipelle	Podoverine	Arens *et al.*, 1986

since such new products can provide 'lead' structures and can be patented. In 1985, Kesselring (Nattermann Co., Köln) reported on the screening of tissue cultures for natural products with anti-inflammatory activity; 26 compounds with biological activity were isolated from different cell cultures and seven of these were new compounds. More recently, a similar strategy used by Phytera Ltd (UK) was reported (Stafford and Pazoles, 1997; Stafford, 1998). It is not known why these novel compounds are produced by plant cell cultures. It can be speculated that a general stress phenomenon activates 'silent' genes and induces the formation of these novel compounds, which may serve as phytoalexins.

7.2.3 Biotransformation of natural compounds by plant cell cultures

Plant cells can be used as chemists—and they are much better chemists than man. They can perform biotransformations, both regio- and stereospecifically, on substrates supplied in the culture medium. Only a very few examples will be mentioned in the present chapter; for more information, see recent reviews (e.g. Suga and Hirata, 1990; Pras, 1992; Yokoyama, 1996; and Franssen and Walton, 1999).

Digitalis lanata cell cultures are able to perform specific biotransformation reactions on cardiac glycosides, producing β-methyldigoxin from β-methyldigitoxin (Alfermann *et al.*, 1983). The enzyme involved has been identified, purified and immobilised (Petersen and Seitz, 1985, 1988; Petersen *et al.*, 1987, 1988). The biotransformation process was tested industrially on a 1 m^3-scale, but was found to be uneconomic (Wahl, 1985). A typical reaction of plant cells is to glucosylate substrates added to the medium. (This might be viewed as a generalised detoxification mechanism; see also Section 7.3.5.2). Several groups have shown that simple phenols, such as salicylic acid, salicyl alcohol or salicylaldehyde, can be glucosylated very effectively (Mizukami *et al.*, 1983, 1985, 1986, 1987; Pilgrim, 1970; Petersen *et al.*, 1992; Tabata *et al.*, 1976, 1988; Tanaka *et al.*, 1990). An especially interesting reaction in this context is the glucosylation of hydroquinone to arbutin, which is performed by cell lines of various plant species. Very high yields were achieved by Suzuki and co-workers (1987), but these were improved even further by Yokoyama and Yanagi (1991), of Shiseido Company, who achieved yields in excess of 9 g/l arbutin within 2–3 days of biotransformation.

Arbutin is used in traditional medicine in Europe as an antimicrobial agent in the treatment of urinary tract infections. However, it also inhibits melanin biosynthesis (Akiu *et al.*, 1988); therefore, Shiseido Company plans to use arbutin as an additive in cosmetics. More recently, Lutterbach and Stöckigt (1992) and Stöckigt (1993) have reported that cell cultures of *Rauwolfia serpentina* are able to transform even larger amounts of hydroquinone within a short period. The cells produce up to 18 g/l arbutin within 7 days. This is the highest value reported so far for natural product accumulation in plant cell biotechnology. A disadvantage of this system may be that the cells accumulate not only arbutin but also up to 6 g/l of *p*-hydroxyphenyl-*O*-β-D-primveroside as a byproduct, which is difficult to separate from arbutin. It is not yet known whether this latter compound has biological activities similar to those of arbutin itself.

7.2.4 Production of recombinant proteins by cell cultures

An important new development is the possibility of expressing recombinant proteins in plant cells. After regeneration of plants from transgenic cell cultures, these proteins may be harvested from transgenic plants grown in the field (Whitelam and Cockburn, 1996). An interesting alternative is to produce recombinant proteins by growing transgenic plant cells in bioreactors (see, e.g. Miele, 1997; Drossard *et al.*, 1998; Fischer *et al.*, 1998).

7.2.5 Use of plant cell cultures to study the biosynthesis of natural products at the biochemical and molecular-genetic levels

Plant cell suspension cultures are an ideal system to study various aspects of secondary product formation, including: the molecular biology and enzymology of biosynthesis; aspects of induction and regulation; compartmentation of biosynthesis and storage; and even certain aspects of degradation and transport. They are, in a very real sense, a 'pot of gold', as expressed by Zenk (1991). In Vol. 2 of Annual Plant Reviews, the reader is referred to chapters dealing with different classes of compounds. It should only be mentioned here that the use of cell cultures by Grisebach and Hahlbrock in the late 1960s enabled a breakthrough to be made in understanding the enzymology of natural product formation, especially in relation to flavonoids and other phenolic compounds. It was again the use of cell (and also latterly organ) culture systems that enabled the identification of genes involved in secondary product biosynthesis (see, for example, Hashimoto *et al.*, 1990; Hashimoto and Yamada, 1992; Kutchan, 1995; Kutchan and Zenk, 1993; Kutchan *et al.*, 1991; Matsuda *et al.*, 1991; and Scott, 1994).

7.2.6 Problems

Plant cell cultures have enabled very substantial progress to be made in our understanding of the biosynthesis of secondary plant products. However, their practical application in biotechnology is currently rather limited. Table 7.5 shows that only three systems have been or are

Table 7.5 Economical processes for the production of secondary compounds by plant cell cultures

Product	Species	Company	Reference
Shikonin	*Lithospermum erythrorhizon*	Mitsui Petro-chemical Ind. Ltd.	Fujita *et al.*, 1982
Ginsenosides	*Panax ginseng*	Nitto Denko Corp.	Ushiyama, 1991
Purpurin	*Rubia akane*	Mitsui Petro-chemical Ind. Ltd.	Personal communication

presently in use on a process scale; a fourth (paclitaxel production) is in the pipeline. The reason for this situation is that most of the compounds, e.g. those of medicinal importance (*cf.* Table 7.1), are not produced by cell cultures or, at least, not in yields necessary for a process to become economically feasible. A corresponding situation exists for the production of flavours and fragrances. Why are many secondary products, such as, for example, tropane alkaloids of the Solanaceae or cardenolides of *Digitalis*, still only accumulated in tiny amounts, or not at all, in plant cell suspension cultures? Both of these classes of compounds are produced only in organ cultures, such as hairy roots of *Hyoscyamus* or shoot organ cultures of *Digitalis*.

The fundamental questions to be addressed, therefore, relate to the essential differences in whole-pathway gene expression between disorganised and organised cultures. Although the appropriate pathway-structural genes can be shown to be present in disorganised cultures, they are frequently not expressed. This may be due to the fact that the appropriate transcription factors are not active. If the formation of an appropriate transcription factor can be switched on, then secondary product formation can be made to occur, as was recently shown by Grotewold and co-workers (1998), with the induction of formation of anthocyanins in formerly unproductive cells of maize (see also Section 7.3.5.2). When we understand more completely how molecular-genetic regulation of natural product formation occurs in whole organs and in organ cultures, it may become possible to trigger unorganised cell suspension cultures to produce natural compounds of interest. The biotechnological production of natural products by unorganised cell cultures may then become a reality. In the meantime, plant organ cultures offer valuable opportunities both for biochemical and molecular-genetic investigation and, perhaps, for biotechnological application. These aspects will now be discussed further (Section 7.3).

7.3 Differentiated organ cultures

Differentiated organ cultures fall into two categories: untransformed and transformed. The former are maintained by supplying plant growth substances exogenously in the medium, whereas transformed cultures, established following infection of plant material by *Agrobacterium* spp., are hormone-independent. These two classes of culture will be considered in turn.

7.3.1 Untransformed root cultures

The study of the culture of plant roots dates from the pioneering work of White (1934) in the 1930s. Cultures may be developed by inoculating sterile root tips into relatively small volumes of liquid culture medium, often containing low concentrations of auxins. Roberts and Street (1955) classified excised root cultures on the basis of their response to exogenously-supplied auxin. A range of responses were noted: in some, growth was entirely dependent upon the supply of auxin; in others, auxin was not essential for growth but stimulated it; and, in some species, addition of auxin actually inhibited root growth. Some excised root cultures can only be maintained in culture for limited periods before senescence of the culture occurs. However, in other species, normal roots may be maintained in culture for extended periods in medium lacking exogenous auxin. Good examples of this are the cultures of *Senecio* and *Hyoscyamus* spp.

Studies on secondary metabolism using normal roots in culture have been limited to a narrow range of species but two groups, Hartmann and co-workers (1989) using *Senecio* and Yamada and co-workers (Hashimoto *et al.*, 1986) using *Hyoscyamus*, have made extensive use of untransformed root cultures in their studies of pyrrolizidine alkaloid and tropane alkaloid biosynthesis, respectively. Both of these cultures show high growth rates and root morphology is maintained over extended periods in culture.

The root cultures of *Senecio* developed by Toppel and co-workers (1987) showed high rates of accumulation of pyrrolizidine alkaloids in the absence of auxin in the culture medium. In contrast, the root cultures of *Hyoscyamus alba* and *H. niger*, developed by Hashimoto and co-workers (1986) to study tropane alkaloid biosynthesis, required auxin to optimise growth but the levels of auxin required for maximal growth were unfortunately inhibitory to alkaloid accumulation. The procedure used was to maintain the root cultures in auxin-containing media and to transfer them to auxin-free medium to stimulate alkaloid production in a single cycle. In normal roots of *Hyoscyamus muticus*, auxins such as indole-3-acetic acid (IAA) or naphthaleneacetic acid (NAA) at low concentration (0.5–2.5 nM) stimulated the formation of abundant root branches and such branching may account for the auxin stimulation of growth (Biondi *et al.*, 1997). *Hyoscyamus* root cultures have been used for extensive studies of the biosynthesis of tropane alkaloids, leading, for example, to the cloning of the gene encoding hyoscyamine 6β-hydroxylase (Matsuda *et al.*, 1991).

7.3.2 Transformed root cultures

Transformed root cultures result from the ability of the soil bacterium, *Agrobacterium rhizogenes*, to infect plant tissues and to induce the formation of roots at the point of infection. This process involves the transfer of a section of plasmid deoxyribonucleic acid (DNA), transfer DNA (t-DNA), from the bacterium into the plant nucleus and its integration into the plant genome. Two sections of plasmid DNA may be involved, TR and TL. Of these, TL is essential for the induction of root formation. TL bears three *Rol* genes, A, B and C, all of which contribute to root initiation and maintenance. A detailed review of the involvement of the *Rol* genes in root induction and maintenance is given by Michael and Spena (1995). The role of TR is less clear and many stable transformed root lines lack TR. TR bears *tms1* and *tms2* genes, which encode enzymes of auxin biosynthesis; however, their significance in root induction is unclear. There are suggestions that root lines bearing TR and TL may be morphologically different from those having only TL integrated into their genomes (Amselem and Tepfer, 1992).

In practical terms, transformed roots may be developed by infecting sterile explants of the chosen plant species, typically a leaf or petiole, excising the roots from the point of inoculation after a suitable period and transferring single root tips (1–2 cm in length) into culture medium based on Gamborg's B5 or Murashige and Skoog (MS) medium lacking hormones but containing an antibiotic. After several cycles of growth, free-living *Agrobacteria* can be removed from the roots, which are thenceforth maintained on media such as Gamborg's B5 without hormones and in the absence of antibiotics.

These transformed roots (often referred to as 'hairy roots' because of the profusion of root hairs commonly associated with them) are valuable and versatile systems for the study of secondary metabolism. Hairy roots are robust in culture and may be maintained for extended periods without detectable changes in their growth and biosynthetic capacities. They can be maintained under sterile conditions in simple media containing salts and carbon and nitrogen sources without hormonal supplements or antibiotics. Transformed roots may be grown from small inocula and do not appear to be sensitive to the conditioning effects associated with untransformed roots. Typically, an inoculum of 200 mg in 50 ml of medium will lead to rapid growth without a significant lag phase. In contrast, untransformed roots may need up to tenfold more inoculum to achieve rapid initial growth (Rhodes, 1990).

The growth rate of transformed roots can be high by plant standards. In a fast-growing species, such as *Nicotiana rustica*, biomass accumulation rates of up to 1.5 g dry weight/l/day have been observed. An important feature of the growth of transformed roots is the formation of many new root meristems, to produce highly-branched root networks coupled with a high rate of linear extension. Hairy root cultures with high growth rates can be developed from plants of families such as Solanaceae and Asteraceae, but not all hairy root cultures are fast-growing and it is probable that intrinsic factors can determine the growth potential of transformed cultures of a particular species and may override those resulting from the transformation process. There are well-documented examples of transformed root cultures that grow only slowly in culture, for example woody species, such as *Cinchona* (Hamill *et al.*, 1989; Hallard *et al.*, 1997).

In general, the secondary metabolism of transformed roots mirrors that of the species from which they were developed. The main pattern of secondary metabolites accumulated reflects that of intact plant roots but there are a number of differences. For instance, detachment from the plant means that the normal transport of secondary metabolites from roots to the rest of the plant is disrupted and compounds normally exported may accumulate in the roots. Similarly, feedback signals associated with such transport may be modified. Metabolites not normally prominent in roots of the normal plant may occasionally be present at appreciable concentrations in hairy root cultures. For instance, acetyltropane is found in high concentrations in hairy root cultures of *Datura wrightii*, even though it is only a minor constituent of roots in the intact plant (Parr *et al.*, 1990). These features have made hairy roots attractive for studies on secondary metabolism (see review by Rhodes *et al.*, 1997). Table 7.6 lists typical studies carried out over the past two years and updates a similar table produced by Hamill and Lidgett (1997). It illustrates the wide range of species and products that are currently under study.

Table 7.6 Recent examples of studies of secondary product formation in hairy root cultures

Species	Product	Reference
Artemisia annua	Artemisinin	Liu *et al.* (1998)
Astragalus mongholicus	Cycloartane	Ionkova *et al.* (1997)
Catharanthus roseus	Tabersonine, serpentine	Rijhwani and Shanks (1998a,b)
Catharanthus roseus	19(*S*)-Epimisiline	PerazaSanchez *et al.* (1998)
Coleus forskohlii	Forskolin	Sasaki *et al.* (1998)
Datura stramonium	Polyamines, tropanes	Ford *et al.* (1998)
Datura candida x D. aurea	Hyoscyamine	Nussbaumer *et al.* (1998)
Digitalis lanata	Anthraquinones, flavones	Pradel *et al.* (1997)
Glycyrrhiza glabra	Isoprenylated flavonoids	Asada *et al.* (1998)
Hyssopus officinalis	Rosmarinic acid	Murakami *et al.* (1998)
Hyoscyamus muticus	Hysocyamine, lubimin	Sevon *et al.* (1998)
	Solvavetivone	Mehmetoglu and Curtis (1997)
Lawsonia inermis	Lawsone, tannins	Bakkali *et al.* (1997a,b)
		Bakkali *et al.* (1997a,b)
Lithospermum erythrorhizon	Hydroxyechinofuran	Fukui *et al.* (1998)
Lotus corniculatus	Tannins	Bavage *et al.* (1997)
Paulownia tomentosa	Verbascoside	Wysokinka and Rozga (1998)
Pimpinella anisum	Essential oils	Santos *et al.* (1998)
Scutellaria baicalensis	Flavonoid glycosides	Nushikawa and Ishimuru (1997)
Solanum aviculare	Solasidine	Kittipongpatan *et al.* (1998)
Trachelium caeruleum	Polyacetylenes	Murakami *et al.* (1998)
Trigonella foenum-graecum	Diosgenin	Merkli *et al.* (1997)
Valeriana wallichi	Valepotriates	Banerjee *et al.* (1998)
Wahlenbergia marginata	Polyacetylenes	Ando *et al.* (1997)

It has been shown in a number of studies that the ability of root cultures to produce their typical secondary products is dependent upon the maintenance of root morphology. For instance, mechanical wounding of *N. rustica* roots leads to loss of root morphology and of the ability of the culture to produce nicotine. Not surprisingly, in view of the hormone-independence of transformed root cultures, normal and transformed root cultures respond differently to exogenous hormones. This is illustrated in a study of *Hyoscyamus muticus*. In normal roots, auxins, such as IAA and NAA, at low concentrations stimulated the formation of abundant root branches, whereas these concentrations of hormones were generally inhibitory to growth of transformed roots (Biondi *et al.*, 1997). In other species, such as *Datura* and *Nicotiana*, auxins at high concentration

(NAA at 2 mg/l, in the presence of a low level of cytokinin) inhibited root growth and induced the formation of disorganised growth on the surfaces of the roots. In the case of *Nicotiana*, exposure to this medium for 28 days led to the development of a suspension culture of transformed cells, which could be maintained over an extended period of growth on subsequent subculture of this medium. During the first 28 days, there was a 93% reduction in the nicotine content of the culture and, associated with this, a rapid decrease in the activity of the enzymes of nicotine biosynthesis, putrescine *N*-methyltransferase (PMT) and *N*-methylputrescine oxidase (MPO). Similar results were obtained with *Datura stramonium* transformed roots, where auxin treatment led to loss of hyoscyamine formation and to a similar loss of activity of PMT and MPO. This system has been further exploited in a nuclear magnetic resonance (NMR) study by Ford and co-workers (1998), which showed that free putrescine appears to play a crucial role in mediating the auxin-induced dedifferentiation of the culture.

7.3.3 Normal and transformed shoot cultures

Shoot cultures may be developed either from axillary meristems, from shoot tips inoculated on solid or liquid media supplemented with a suitable combination of plant growth regulators, or from induction of organogenesis on callus by selection of suitable hormonal and/or environmental conditions. As an example, shoot cultures of feverfew (*Tanacetum parthenium*) were developed from nodal segments of seedlings on MS medium containing 1 mg/l 6-benzyladenine and 0.1 mg/l naphthaleneacetic acid (Stojakowska and Kisiel, 1996). The parthenolide content of these cultures was 60% of that of shoots on the plant. Shoot cultures of *Atropa belladonna* were also initiated from shoot tips on MS medium containing 6-benzylaminopurine (BAP) (Charlwood *et al.*, 1990). Shoot cultures have been extensively used to study rosmarinic acid production in *Rosmarinus* (Komali and Shetty, 1998), cardenolides in *Digitalis* (Hagimori *et al.*, 1983) and terpenoids in *Pelargonium* (Brown and Charlwood, 1986). Such cultures are normally grown heterotrophically, typically with sucrose as the carbon source. They are relatively slow-growing and need to be handled carefully to prevent damage to the developing shoots and to avoid microbial infection. Shoot cultures of lupinus spp., but not root cultures, produce quinolizidine alkaloids, as the site of synthesis is the leaf chloroplast (Wink, 1993).

With the objective of developing faster-growing, more stable shoot cultures, there have been attempts to develop transformed shoot cultures –in essence counterparts of hairy root cultures–which might be grown in the absence of hormones and in which intrinsic hormonal factors determine the maintenance of shoot morphology. It is well known from studies on tobacco callus that increasing the cytokinin level in the medium stimulates shoot formation and that transformation of the tissue with constructs bearing the *ipt* gene, which codes for an essential step in cytokinin biosynthesis, leads to the induction of shoot formation. This principle was applied to *Mentha* cultures and transformed shoot cultures were developed that expressed the *ipt* gene and that were maintained in culture for several years. However, it was found that the nopaline strain of *A. tumefaciens*, T37, would also induce shooty teratoma formation in *Mentha*. Transformed shoot cultures of *M. piperita* and *M. citrata* were subsequently developed in this way (Spencer *et al.*, 1993). Over 5 yrs of subculture, the *M. citrata* culture maintained production of an essential oil profile that resembled the parent plant from which it was derived, with linalool and linalool acetate as the major components. In contrast, the *M. piperita* shoot culture showed an initial divergence in composition from the parent plant oil and diverged still further during the 5 yr period (Spencer *et al.*, 1993; Hilton *et al.*, 1995).

Transformed shoot cultures were also developed by transformation of *Artemisia annua* with a nopaline strain of *A. tumefaciens*. Initally, crown galls were formed but shooty teratomas developed from them spontaneously (Ghosh *et al.*, 1997). These shooty teratomas contained 63 mg/100 g dry weight of the antimalarial substance, artemisinin. In another study, using the nopaline strain of *A. tumefaciens*, T37, it was shown that shooty teratomas of *Artemisia* could be developed and maintained in culture in hormone-free media for extended periods. Under these conditions, the culture produced artemisinin at up to 18 mg/100g dry weight. It was shown that addition of gibberellic acid (GA_3) to the medium stimulated artemisinin accumulation by three- to four fold (Paniego and Giulietti, 1996). This strain of *A. tumefaciens* was also used to develop transformed shoot cultures of *Pimpinella anisum* (anise) (Salem and Charlwood, 1995). These cultures produce an essential oil in which the relative amounts of the main components are similar to those of the parent plant but the overall yield of oil is, unfortunately, only 11% of that of the plant.

7.3.4 Current limitations on the production of secondary metabolites in differentiated organ cultures

7.3.4.1 Limitations on expression of pathways in organ cultures

The biosynthesis of many secondary products is closely associated with organ differentiation. Obvious examples include: the glands of the female flowers (cones) of hops, *Humulus lupulus*, responsible for the production of α-acids (Robins *et al.*, 1990); the laticifers associated with the production and translocation of long-chain isoprenoids, such as rubber (Fahn, 1979; John, 1992); and the glandular trichomes of leaves of the *Lamiaceae*, such as *Mentha* spp., which produce the monoterpenes and essential oils characteristic of these species (McGarvey and Croteau, 1995). As a general rule, organ cultures cannot be used to study the biosynthesis of classes of compounds that the normal organs of the corresponding plant species do not make. On the other hand, organs may synthesise secondary products that are then transported and accumulated elsewhere in the plant. Amongst the most well-known examples are nicotine in *Nicotiana* and the tropane alkaloids of *Datura* and *Atropa* and related genera, which are synthesised in the roots and transported and stored in other organs, such as the leaves (though high levels are maintained in the roots).

In these cases, root cultures have proved valuable tools. The site of biosynthesis of one of these alkaloids, scopolamine, the 6,7-epoxide of hyoscyamine, has been investigated in detail in root cultures of *Hyoscyamus*. Hyoscyamine 6β-hydroxylase, the 2-oxoglutarate-dependent dioxygenase responsible both for the 6β-hydroxylation and subsequent epoxidation of hyoscyamine to scopolamine, has been localised immunohistochemically to the cells of the pericycle, and ribonucleic acid (RNA) gel blot analysis has demonstrated that the transcript of the gene is similarly localised (Hashimoto *et al.*, 1991). Transcript was not found in the aerial parts of the plant or in cell cultures. It is very likely that long-distance transport of alkaloids is a widespread phenomenon, consistent with the role of these compounds as antifeedants and toxins. For example, there is evidence that the pyrrolizidine alkaloids of the *Asteraceae* are transported as their *N*-oxides (Hartmann *et al.*, 1989). As suggested above, the inability of organ cultures to export secondary products to aerial organs in the normal way may, in principle, have consequences for the productivity of the culture and for the spectrum of compounds produced but there is, at present, insufficient information available from which to draw generalised conclusions.

Where secondary products are produced and/or stored in specialised cellular structures, such as trichomes, glands etc., the failure of cell suspension cultures to produce such compounds has been attributed to the lack of these structures. This is because, in the first place, such structures are likely to be necessary for expression of the biosynthetic pathway and, secondly, the products are phytotoxic and failure to sequester them may inhibit growth or kill the culture. Good examples of this situation are found among the terpenoid essential oil components, which are normally stored in trichomes on the leaf surface and which have potential phytotoxicity. Whereas cell suspension cultures produce only low or negligible amounts of these compounds, shoot cultures which bear well-developed storage glands accumulate significant quantities of such terpenoids. Transformed shoot cultures of mint, which have well-formed oil glands, may accumulate levels of monoterpene flavour compounds at levels similar to those of the plant (Spencer *et al.*, 1993).

The two most important determinants of the overall rate of accumulation of a secondary metabolite are the growth rate of the culture and the specific production rate of the compound in question. These will now be considered in turn.

7.3.4.2 Limitations on growth rate

In root organ cultures, growth rate is determined by the linear rate of extension of the root meristem, the number of meristems generated in culture and, to a lesser extent, the increase in root girth. In some, but not all, untransformed cultures, growth is essentially linear and very few, if any, branches are generated. In the case of such untransformed cultures, the growth of the root is limited by the intrinsic controls that determine the rates of linear growth and branching of that species. It is possible to influence factors such as the degree of branching by exogenous hormonal treatment, and nutritional factors will influence the linear rate of growth within the limits set by the inherent controls.

In transformed roots, the expression of the *Rol* genes to some extent overrides, at least in part, this inherent control system and, in particular, increases the rate of branching of the culture and, thus, enhances the growth potential of the culture by increasing the number of growing points. There is little evidence to suggest that transformation and expression of the *Rol* genes enhances the linear rate of extension of the individual root branches or their increase in girth. But, as with untransformed roots, intrinsic factors limit the linear growth rate in individual species. As indicated previously, in some species, particularly from the Solanaceae and the Asteracae, high growth rates are common, whereas in other (particularly woody) species, such as *Cinchona*, relatively slow growth rates are the norm (Hamill *et al.*, 1989; Hallard *et al.*, 1997). The nature of these intrinsic control systems remains unknown.

7.3.4.3 Limitations on the site and rate of product formation

Limitations on the rate of formation of secondary products in organ cultures arise both from limitations on growth rate and on their inherent capacity to accumulate the product. Within an organ culture comprising a range of cell types and of cells in different physiological states, production may well be associated with particular cell types at defined stages in their development, and thus only a proportion of the cells within the differentiated organ culture will be productive at any specific stage. The ratio of mature to meristematic cells may well be a determinant of the overall production rate. This is well illustrated with transformed roots of *Beta vulgaris* (beetroot), in which the meristems are colourless, production of the red betalain pigments occurs in the elongation zone of the root and only the mature cells accumulate significant amounts of the pigment (Hamill *et al.*, 1986). Thus, only a few elongating cells are actually active in betalain biosynthesis at any particular time. This parallels the situation in many cell suspension cultures, where biosynthetic capacity is only expressed in a small subpopulation of cells.

Fundamental limitations on production are of course imposed by: the degree of expression of the biosynthetic pathway(s) in question; the presence or absence of effective sequestering mechanisms; the activity of degradative pathways leading to the concomitant loss of the desired secondary products; the activity of competing pathways which might lead to alternative products; and, finally, the action of possible feedback mechanisms leading to inhibition of biosynthesis. It has been possible to influence empirically this plethora of endogenous, largely-uncharacterised mechanisms by external factors, which may be nutritional or which may involve external inducers, such as fungal elicitors or abiotic factors (see Section 7.3.5.1). Increasingly, however, more fundamentally-based approaches are becoming feasible (see Section 7.3.5.2).

7.3.4.4 Problems of scale-up of production

Differentiated organ cultures pose special difficulties for scale-up. In particular, they are not amenable to the well-mixed stirred culture systems, which are used with microbial cells and which have been applied to plant cell suspension cultures. Stirred systems are not applicable to root cultures as mechanical disturbance and shear can lead to loss of root integrity and dedifferentiation, with serious impairment of root growth and of the capacity to form the required products.

For hairy root cultures, several groups have designed non-stirred fermenters. The largest such systems have a capacity of up to 500 l (Wilson, 1997), and in this case protocols for operation on this scale have been devised. Problems of sterile inoculation have been solved by dispersing, through a specially-designed transfer system, a roughly-chopped slurry of roots into a fermenter that incorporates a framework bearing an array of attachment points. Thus, roots are seeded from these attachment points, which act as centres for growth. In this way, relatively even mats of root networks have been achieved throughout the fermenter. Up to 40 kg of roots of *Datura stramonium* have been grown in 40 days in such fermenters, under conditions in which high levels of hyoscyamine have accumulated. Shoot cultures, both transformed and untransformed, have been grown in small fermenters (1–10 l) by a number of groups but there have been no reported studies to address the scale-up problems, as in the case of transformed root cultures. Similarly, there have been few studies with untransformed root cultures and no attempts to approach the scale-up problems.

7.3.5 Modifying and improving the production of secondary metabolites in differentiated organ cultures

7.3.5.1 Empirical approaches

Empirical approaches fall into two categories, based on: 1) selection or optimisation of the culture characteristics; and 2) manipulation of the culture medium or regimen.

Selection or optimisation of culture characteristics starts with the choice of suitable parent plant material. It is logical to select plants with high levels of the compound of interest, though within a population there is little evidence of an overriding relationship between levels of a secondary product in individual plants and the levels produced in cultures arising from them (Parr *et al.*, 1990). This may reflect translocation within the intact plant, as described previously. Individual cultures derived from the same parent plant, however, commonly vary in growth rate and morphology. Cultures often display growth characteristics (e.g. in roots, slow growth, lack of lateral branching, tendency to callus formation) that rule them out of consideration at an early stage; such cultures rarely, if ever, improve with successive subculturing. Occasionally, cultures harbour persistent *Agrobacterium* populations and require continual subculture into medium containing antibiotic to prevent overgrowth by the bacteria; these cultures are not ideal for scale-up or long-term secondary metabolite formation and such routine use of antibiotics is to be discouraged.

In the selection of cell-culture lines with potentially-enhanced secondary product biosynthetic capacity, toxic analogues of primary metabolites, particularly of amino acids such as phenylalanine or tryptophan, have historically been used as selection agents (Berlin *et al.*, 1982a,b; Gasse *et al.*, 1983). The rationale is that the secondary product pathway provides a detoxification mechanism which, if actively expressed, may afford resistance to the analogue. This approach has rarely been adopted with organised cultures, owing to the lack of genetic variation. However, it is, in principle, available if variation is deliberately introduced–most obviously by causing the temporary dedifferentiation of organ cultures by treatment with plant growth substances (Aird *et al.*, 1988).

Manipulation of the culture medium or regimen may, in principle, include: 1) variation of nutritional status; 2) use of abiotic factors; 3) application of elicitors; 4) exogenous provision of a metabolic precursor; or 5) provision of a metabolic sink. Some success has been achieved in increasing the production of secondary metabolites in organised cultures by the use of abiotic factors or yeast or fungal elicitors (see Furze *et al.*, 1991; Robbins *et al.*, 1991, 1995; Ballica *et al.*, 1993; Rhodes *et al.*, 1997). Attempting to increase secondary product accumulation by providing a metabolic precursor amounts, essentially, to a biotransformation (Franssen and Walton, 1999). Unless this precursor is cheap (relative to sucrose), is able to enter the cells of the culture and is non-toxic – all stringent requirements – this approach will not succeed. There is, in any case, little evidence that supplementing cultures with metabolic precursors overcomes a biosynthetic limitation. According to contemporary Metabolic Control Theory (Niederberger *et al.*, 1992; Fell, 1992; ap Rees and Hill, 1994), metabolic pathways are regulated by partial rate-limitation occurring at each of a number of enzymic steps and it is, therefore, likely that increasing the supply of an individual precursor compound will not substantially affect the formation of a distant end-product. It is important, however, to distinguish this situation from another in which analogues or homologues of normal metabolic precursors are fed to cultures and found to give rise to significant levels of novel end-products, as in the case of the formation of anabasine when cadaverine is fed to *Nicotiana* hairy root cultures (Walton *et al.*, 1988); in this case, the bioconversion occurs by competition with intermediates of the normal pathway.

Inherent in Metabolic Control Theory is the principle that biosynthesis is also affected by the reactions and processes by which a molecule is catabolised, derivatised or sequestered. Therefore, creating a new metabolic sink provides a strategy to increase the accumulation of a desired product. This seems rarely to have been attempted in the case of differentiated cultures, though Westcott and co-workers (1994) reported the use of charcoal to adsorb vanillin produced from ferulic acid by detached, cultured aerial roots of *Vanilla*. Some success has been reported with undifferentiated cultures; in a study of the stimulation of anthraquinone production in cell cultures of *Cinchona*, hydrophobic Amberlite ^{R} XAD resins were included in the culture medium (Robins and Rhodes, 1986). Anthraquinones appearing in the culture medium were adsorbed onto the resin; this had the effect of increasing the total production of anthraquinone by the culture. Whilst the product of interest must obviously be released to some extent into the medium, in principle, with a sufficiently high binding constant, it should be possible to sequester a product that is normally retained very largely within the cells. The problem may lie in devising adsorbents that do not sequester other metabolites or culture-medium components nonspecifically.

An interesting semi-empirical approach to manipulating culture yield is illustrated by recent co-culture experiments, in which the products of one organ culture were further metabolised by another (Subroto *et al.*, 1996, 1997). Co-culture of transformed shoot cultures and hairy roots of *Atropa belladonna* led to a scopolamine content in the shoot cultures of up to 0.84 mg/g dry weight, 3–11 times the average concentrations normally found in leaves of the whole plant and very much greater than the scopolamine content of *A. belladonna* hairy root cultures.

7.3.5.2 Fundamental approaches

Fundamental approaches make use of knowledge of the biochemical or molecular-genetic regulation of secondary-product pathways to alter the formation of pathway end-products. The objective may be to enhance the formation of an existing compound, but other potential goals include: introduction of a biosynthetic pathway leading to a novel compound or modification of the spectrum of end-products by controlling ancillary or competing pathways. Relatively little of our understanding in this area arises directly from studies of organ cultures; the large majority of the relevant work has been undertaken using whole plants (or other organisms). The account that follows, therefore, also includes work undertaken with whole plants, which may be particularly significant in providing a basis for future work to increase or modulate secondary product formation in organ (or cell) cultures.

Expression of homologous or heterologous genes encoding key metabolic enzymes. Attempts to increase end-product formation by increasing the activity of an individual enzyme of a pathway (for example, by introducing additional gene copies) is unlikely to succeed. As indicated above, Metabolic Control Theory recognises that overall limitation of a metabolic pathway does not rest with a single rate-limiting enzyme; though Stitt (1995) has, nevertheless, made a distinction between 'regulatability' (the potential for the activity of an enzyme to respond to regulatory mechanisms *in vivo*) and 'regulatory capacity' (the potential for an alteration in the activity of an enzyme to cause a change in metabolic flux). Increasing the activity of an individual enzyme, for example by genetic manipulation, may transfer its component of metabolic flux limitation to other enzymes that normally have little influence on flux, leaving overall metabolic flux to the end-products of interest largely, and disappointingly, unaffected. Furthermore, if additional copies of an existing gene from the same or a closely-related plant species are introduced, the phenomenon of co-suppression may lead to a downregulation in expression rather than the increased activity that is desired (Flavell, 1994; Meyer and Saedler, 1996).

In studies of tryptophan biosynthesis in yeast, it was found to be necessary to increase the activities of all five enzymes of the pathway by genetic manipulation before an increase in tryptophan synthesis occurred (Niederberger *et al.*, 1992). Examples of such 'buffering' have been reported in plants. Burtin and Michael (1997) studied the effects of overexpression of the *adc* gene, encoding arginine decarboxylase, on the formation of polyamines and pyrrolidine alkaloids in *Nicotiana* plants. The sole apparent effect was an increase in the pool size of agmatine, the amine decarboxylation product of arginine; there was no effect on the levels of polyamines or of alkaloids. It was deduced that other activities were (or had become) rate-limiting, buffering completely the effects of overexpression of *adc*.

In contrast, however, in hairy root cultures of *Nicotiana rustica* transformed with a yeast ornithine decarboxylase gene (*odc*), specific lines displayed an enhanced nicotine content, though only within a particular period towards the end of the culture cycle (Hamill *et al.*, 1990). It is not certain whether this was a direct result of the increase in ornithine decarboxylase activity; it is possible that it reflected some uncharacterised pleiotropic effect of enhanced activity. In many organisms, ornithine decarboxylase is subject to complex and sensitive regulation, associated with its role in polyamine metabolism (Michael *et al.*, 1996; Hayashi and Murakami, 1995).

Phenylalanine ammonia-lyase (PAL), the initial enzyme of the phenylpropanoid pathway, may provide apparent exceptions to the generalisation that increasing the activity of a specific pathway enzyme may have little overall effect on total end-product formation. Here, there is correlative evidence for a relationship between enzyme activity levels and phenylpropanoid end-products (particularly chlorogenic acid) in transgenic plants of *Nicotiana tabacum* transformed with exogenous PAL genes and expressing – as a result of co-suppression – a range of PAL enzyme activities both below and up to *ca.* 2.5-fold higher than normal levels (Bate *et al.*, 1994; Howells *et al.*, 1996). The results suggest that, within this range, increased PAL activities may increase end-product levels, though fine control of metabolic flux is, nevertheless, exerted by enzymes of the individual phenylpropanoid branch pathways. Increasing PAL activities by transformation with exogenous PAL genes may, therefore, be worthwhile (at least as an enabling or supporting strategy) in cultures established for the purpose of producing phenylpropanoid-derived compounds. There are, however, alternative approaches based on the expression of pathway transcription factors (see below).

In contrast to the upregulation of existing enzyme activities, the introduction by heterologous gene expression of new enzymes may offer opportunities to make new compounds. An example of this is the introduction of a bacterial lysine decarboxylase (*ldc*) gene from *Hafnia alvei* into hairy root cultures of *Nicotiana* (Fecker *et al.*, 1992), following the observation that feeding cadaverine, the decarboxylation product of lysine, stimulated the formation of the piperidine alkaloid, anabasine, in *Nicotiana* hairy root cultures (Walton *et al.*, 1988). In a subsequent study, the effectiveness of *ldc* transgene expression was increased by using an *rbcS* leader sequence to target expression to the plastid, where lysine is biosynthesised (Herminghaus *et al.*, 1996).

Other examples of the use of heterologous gene expression to engineer 'new' metabolic pathways are: the expression of a bacterial gene encoding isochorismate synthase in *Rubia peregrina* hairy root cultures, which increased the biosynthesis of anthraquinones (Lodhi *et al.*, 1996); and the expression of *ubiC*, the bacterial gene encoding chorismate-pyruvate lyase, which caused the formation of 4-hydroxybenzoate (a precursor, *inter alia* of naphthoquinones, which is normally formed via phenylalanine, cinnamic acid and 4-coumaric acid) in plants and cell cultures of *Nicotiana* (Siebert *et al.*, 1996; Li *et al.*, 1997). In this case, 4-hydroxybenzoate did not accumulate but was converted to its glucoside and glucose ester. This illustrates a potential limitation of the approach, since glucose-conjugation is a common response of plants to the presence of potentially-toxic metabolites (see also Section 7.2.3). More generally, however, there is a bright future, as increasing numbers of regio- and stereo-specific enzymes (for example terpene cyclases, cytochrome P_{450} monooxygenases) acting on key biosynthetic intermediates are isolated from bacterial, fungal and plant sources and their genes are cloned and become available for expression in plants and plant organ cultures.

Expression of regulatory genes. Increasing the activities of individual existing secondary-metabolic enzymes offers a stepwise and piecemeal approach to the enhancement of pathway end-products, with no certainty of achieving a worthwhile increment. An alternative strategy seeks to utilise regulatory genes that control the expression of pathway genes *en bloc*, thus increasing the activities of many or all of the enzymes of a metabolic pathway simultaneously. Regulatory genes encode transcription factors, which interact with the basal transcriptional machinery to activate gene expression. The essential elements of a transcription factor are a specific-DNA-binding domain and a transcriptional-activating domain, both of which are often highly-conserved. Thus, transcription factors in plants show substantial homology to transcription factors in other organisms. Well-characterised families of transcription factors include the so-called basic leucine zipper proteins (bZIP), the Myb- and Myc-related transcription factors, the homeodomain proteins and the MADS-box proteins (see Martin, 1996; Katagiri and Chua, 1991).

Up-regulation of a complete secondary-metabolic pathway by expression of a pathway transcription factor was clearly demonstrated, for example, by the effects of expressing a maize anthocyanin regulatory gene, *Lc*, or an *Antirrhinum* anthocyanin regulatory gene, *Delila*, in tobacco or tomato; a *ca.* 20-fold increase in anthocyanin levels in tomato leaves was achieved by expressing *Delila* under the control of a cauliflower mosaic virus (CaMV) 35S promoter, which gives high-level, constitutive gene expression (Mooney *et al.*, 1995). In a very recent study, Grotewold and co-workers (1998) demonstrated that secondary product formation in cultured maize cells can be manipulated by ectopic expression of appropriate regulatory genes; cell lines engineered to express the transcriptional activators, C1 and R, produced cyanidin derivatives similar to the anthocyanin characteristic of differentiated tissues, whereas cell lines expressing the Myb-related regulator, P, accumulated 3-deoxyflavonoids and other phenylpropanoid metabolites, some in a subcellularly-compartmented manner.

Heterologous expression of regulatory genes can also act to reduce the expression of pathway genes, as shown by the inhibition of phenolic-acid and lignin biosynthesis caused by the overexpression in tobacco of two Myb transcription factors (Myb308 and Myb305) from *Antirrhinum* (Tamagnone *et al.*, 1998). It is considered that this is probably the result of competition for DNA-binding between transcription factors, the heterologously-expressed factor being, in such cases, a weaker transcriptional activator than the usual, endogenous transcription factor. There is also some evidence for direct protein-protein interaction between transcription factors (see Jackson *et al.*, 1992). Given the modular structure of transcription factors, with discrete DNA-binding and transcriptional-activating domains, it is feasible, in principle, to tailor the degree of transcriptional activation by engineering the regulatory gene, incorporating a transcriptional-activating domain with the properties required.

A different approach, intervening at a prior stage in the regulatory hierarchy, seeks to use components of signal transduction mechanisms. Biosynthesis of many secondary products shows a clear response to external stress stimuli, such as UV-irradiation, wounding or pathogen attack; this is especially obvious in the case of phytoalexins, for example the isoflavonoid pterocarpans of legumes, but is also true of some constitutively-produced substances, including a number of alkaloids. An account of signal transduction mechanisms associated with plant defence responses is beyond the scope of the present chapter and the reader is referred elsewhere for details (Somssich and Hahlbrock, 1998; Zhu *et al.*, 1996; Dixon *et al.*, 1994).

Jasmonate is of particular interest in the context of cell and organ culture. This compound can stimulate the biosynthesis of several classes of alkaloids (including indole alkaloids, nicotine and benzophenanthridines) responsive to wounding or, in cell cultures, to fungal cell-wall elicitors (Blechert *et al.*, 1995; Mueller *et al.*, 1993; Kutchan, 1995). Jasmonate is a degradation product of linolenic acid, occurring by the successive action of 13-lipoxygenase, allene oxide synthase, allene oxide cyclase and β-oxidation; various derivatives, including conjugates with amino acids and volatile esters, are also physiologically active. At the time of writing, relatively little detailed biochemical characterisation has been achieved of the control of jasmonate synthesis in response to external factors, such as wounding or elicitor substances, or of the mechanisms by which jasmonate and its derivatives exert their effects (see Chapter 4). There is some evidence that a wounding-associated mitogen-activated protein (MAP) kinase may be involved in regulating the biosynthesis of jasmonate (Wasternak and Parthier, 1997; Seo *et al.*, 1995); 'downstream', it may be presumed that jasmonate itself interacts with a binding protein (or proteins) and that ultimately, by a series of further interactions, a transcription factor (or factors) is caused to activate transcription of target genes.

The situation is complicated by probable interaction and 'cross-talk' between signal transduction pathways; for example, ethylene appears to be involved in some jasmonate responses (see Wasternak and Parthier, 1997). Nevertheless, it is anticipated that by a combination of mutant analysis and biochemical and molecular genetic approaches, the components of the jasmonate signal-transduction pathway will be rapidly elucidated and characterised. This will offer interesting new possibilities to control jasmonate-responsive biosynthetic pathways, or perhaps to confer jasmonate-responsiveness upon normally-unresponsive pathways. Essentially similar considerations apply in respect of other signal molecules, such as ethylene (Fluhr and Mattoo, 1996) and salicylate (Durner *et al.*, 1997) and point the way towards rational chemical control of gene expression (Ward *et al.*, 1993), which is potentially applicable to both whole plants and to organ or cell cultures.

Finally, there is the possibility of controlling secondary-product biosynthesis in culture by intervention at the organ-differentiation level. Transcriptional regulators are fundamental to plant morphogenesis, acting in determining, for example, the timing and induction of floral meristems and the differentiation of flower organs (Coen, 1991; Weigel and Nilsson, 1995; Mandel and Yanofsky, 1995; Martin, 1996). Larkin and co-workers (1994) demonstrated that expression of the maize *Lc* gene stimulated the ectopic production of trichomes in *Arabidopsis* flowers. This suggests the possibility that such genes might obviate the need for manipulation at a biosynthetic-pathway level. For example, increasing the number of trichomes on the leaf, or causing their ectopic production on other organs, might provide a relatively simple means to increase the overall production of essential oils in shoot cultures of members of the Lamiaceae, such as *Mentha*. A more ambitious objective is to increase the range of organs that can be generated in culture; in particular, perhaps, to develop methodologies for the production of floral organs and fruits. Achieving these goals requires a detailed understanding of organogenesis and of the hierarchy of regulatory genes and developmental events involved.

7.4 Outlook

There is no doubt that new scientific opportunities will develop for the application of cell and organ cultures in relation to secondary products. These will arise from the increasing number of species that can be grown in culture, from the increasing possibilities for the use of cultures in the heterologous expression of genes encoding enzymes with new catalytic activities and from fundamental molecular-genetic advances in our understanding of organ development, transcriptional control and signal-transduction mechanisms. As a scientific tool for the study of biosynthetic pathways and enzymes, organ cultures offer well-proven and distinct advantages over intact plants.

From a technological viewpoint, however, there remain difficulties. Growth rates are relatively slow (in comparison with microorganisms and particularly in woody species) and sterility problems may still be encountered in large-scale culture. Taking account of capital costs and the cost of growth media, cell and organ culture is not attractive for the majority of commercially-extracted compounds. (Normal plants growing under nonsterile, conditions in the field offer many advantages of robustness and resilience!) This suggests that, in the medium term, the future of cell and organ culture may lie in biotransformations and in exploratory applications to produce new compounds not readily available from normal plant sources. In the longer term, however, there is the possibility that new types of plant cell and organ cultures, expressing pathway genes obtained from a variety of chosen sources and under the control of selected regulatory genes, will be developed with the physiological characteristics needed for sustained commercial production of a wide range of secondary metabolites.

References

Aird, E.L.H., Hamill, J.D., Robins, R.J. and Rhodes, M.J.C. (1988) Chromosome stability in transformed hairy root cultures and the properties of variant lines of *Nicotiana rustica*, in *Manipulating Secondary Metabolism in Culture* (eds. R.J. Robins and M.J.C. Rhodes), Cambridge University Press, Cambridge, pp. 137-44.

Akiu, S., Suzuki, Y., Fujinuma, Y., Asahara, T. and Fukuda, M. (1988) Inhibitory effect of arbutin on melanogenesis: biochemical study in cultured B16 melanoma cells and effect on the UV-induced pigmentation in human skin. *Proc. Japan. Soc. Invest. Dermatol.*, **12** 138-39.

Alfermann, A.W., Bergmann, W., Figur, C., Helmbold, U., Schwantag, D., Schuller, I. and Reinhard, E. (1983) Biotransformation of β-methyldigitoxin to β-methyldigoxin by cell cultures of *Digitalis lanata*, in *Plant Biotechnology* (eds. S.H. Mantell and H. Smith), Cambridge University Press, Cambridge, pp. 67-74.

Allison, A.J., Butcher, D.N., Connolly, J.D. and Overton, K.H. (1968) Paniculides A, B and C, bisabonolenoid lactones from tissue cultures of *Andrographis paniculata*. *Chem. Comm.*, p. 1493.

Amselem, J. and Tepfer, M. (1992) Molecular basis of novel phenotypes induced by *Agrobacterium rhizogenes* A4 on cucumber. *Plant Mol. Biol.*, **19** 421-32.

Ando, M., Shimomura, K., Yamakawa, T. and Ishimaru, K. (1997) Polyacetylene production in hairy root cultures of *Wahlenbergia marginata*. *J. Plant Physiol.*, **151** 759-62.

ap Rees, T. and Hill, S.A. (1994) Metabolic control analysis of plant metabolism. *Plant Cell Environ.*, **17** 587-99.

Arens, H., Borbe, H.O., Ulbrich, B. and Stöckigt, J. (1982) Detection of pericine, a new CNS-active indole alkaloid from *Picralima nitida* cell suspension culture by opiate receptor binding studies. *Planta Med.*, **40** 218-23.

Arens, H., Ulbrich, B., Fischer, H., Parnham, M.J. and Römer, A. (1986) Novel inflammatory flavonoids from *Podophyllum versipelle* cell culture. *Planta Med.*, **52** 468-73.

Asada, Y., Li, W. and Yoshikawa, T. (1998) Isoprenylated flavonoids from hairy root cultures of *Glycyrrhiza glabra*. *Phytochemistry*, **47** 389-92.

Bakkali, A.T., Jaziri, M., Ishimaru, K., Tanaka, N., Shimomura, K., Yoshimatsu, K., Homes, J. and Vanhaelen, M. (1997a) Tannin production in hairy root cultures of *Lawsonia inermis*. *J. Plant Physiol.*, **151** 505-508.

Bakkali, A.T., Jaziri, M., Foriers, A., VanderHeyden, Y., Vanhaelen, N. and Homes, J. (1997b) Lawsone accumulation in normal and transformed cultures of henna, *Lawsonia inermis*. *Plant Cell Tiss. Org. Cult.*, **51** 83-87.

Ballica, R., Ryu, D.D.Y. and Kado, C.I. (1993) Tropane alkaloid production in *Datura stramonium* suspension cultures: elicitor and precursor effects. *Biotechnol. Bioeng.*, **41** 1075-81.

Banerjee, S., Rahman, L., Uniyal, G.C. and Ahuja, P.S. (1998) *Agrobacterium rhizogenes* mediated transformation of *Artemisia annua*: production of transgenic plants. *Plant Sci.*, **31** 203-208.

Bate, N.J., Orr, J., Ni, W., Meromi, A., Nadler-Hassar, T., Doerner, P.W., Dixon, R.A., Lamb, C.J. and Elkind, Y. (1994) Quantitative relationship between phenylalanine ammonia-lyase levels and phenylpropanoid accumulation in transgenic tobacco identifies a rate-determining step in natural product synthesis. *Proc. Natl. Acad. Sci. USA*, **91** 7608-12.

Bavage, A.D., Davies, I.G., Robbins, M.P. and Morris, P. (1997) Expression of an *Antirrhinum* dihydroflavonol reductase gene results in changes in condensed tannin structure and accumulation in root cultures of *Lotus corniculatus* (bird's foot trefoil). *Plant Mol. Biol.*, **35** 443-58.

Berlin, J. (1997) Secondary products from plant cell cultures, in *Biotechnology*, 2nd edn. (series eds. H. Kleinkauf and H. van Dühren), Vol. 7. Products of Secondary Metabolism (eds. H.J. Rehm and G. Reed), VCH Publishers, Weinheim, Germany, pp. 593-640.

Berlin, J., Knobloch, K.-H., Höfle, G. and Witte, L. (1982a) Biochemical characterisation of two tobacco lines with different levels of cinnamoyl putrescines. *J. Nat. Prod.*, **45** 83-87.

Berlin, J., Witte, L., Hammer, J., Kukeschke, K.G., Zimmer, A. and Pape, D. (1982b) Metabolism of *p*-fluorophenylalanine sensitive and resistant tobacco cell cultures. *Planta*, **155** 244-50.

Berlin, J., Wray, V., Mollenschott, C. and Sasse, F. (1986) Formation of α-peltatin-A-methylether and coniferin by root cultures of *Linum flavum*. *J. Nat. Prod.*, **49** 435-39.

Berlin, J., Mollenschott, C., Herminghaus, S. and Fecker, L.F. (1998) Lysine decarboxylase transgenic tobacco root cultures biosynthesize novel hydroxycinnamoylcadaverines. *Phytochemistry*, **48** 79-84.

Bhojwani, S. and Razdan, M.K. (1983) *Plant Tissue Culture: Theory and Practice*, Elsevier, Amsterdam.

Biondi, S., Lenzi, C., Baraldi, R. and Bagni, N. (1997) Hormonal effects on growth and morphology of normal and hairy roots of *Hyoscyamus muticus*. *J Plant Growth Reg.*, **16** 159-67.

Blechert, S., Brodschelm, W., Hölder, S., Kammerer, L., Kutchan, T.M., Mueller, M.J., Xia, X.Q. and Zenk, M.H. (1995) The octadecanoic pathway: signal molecules for the regulation of secondary pathways. *Proc. Natl. Acad. Sci. USA*, **92** 4099-105.

Breuling, M., Alfermann, A.W. and Reinhard, E. (1985) Cultivation of cell cultures of *Berberis wilsonae* in 20-l airlift bioreactors. *Plant Cell Rep.*, **4** 220-23.

Bringi, V., Kadkade, P.K., Prince, C.L., Schubmehl, B.J., Kane, E.J. and Roach, B. (1995) Enhanced production of taxol and taxanes by cell cultures of *Taxus* species. US Patent 5,407,816.

Broomhead, A.J. and Dewick, P.M. (1990) Aryltetralin lignans from *Linum flavum* and *Linum capitatum*. *Phytochemistry*, **29** 3839-44.

Brown, J.T. and Charlwood, B.V. (1986) Differentiation and monoterpene biosynthesis in plant cell cultures, in *Secondary Metabolism in Plant Cell Cultures* (eds. P. Morris, A.H. Scragg, A. Stafford and M.W. Fowler), Cambridge University Press, Cambridge, pp. 68-74.

Burtin, D. and Michael, A.J. (1997) Overexpression of arginine decarboxylase in transgenic plants. *Biochem. J.*, **325** 331-37.

Charlwood, B.V., Charlwood, K.A. and Molina-Torres, J. (1990) Accumulation of secondary compounds by organised plant cultures, in *Secondary Products from Plant Tissue Culture* (eds. B.V. Charlwood and M.J.C. Rhodes), Clarendon Press, Oxford, pp. 201-26.

Coen, E.S. (1991) The role of homeotic genes in flower development and evolution. *Annu. Rev. Plant Physiol. Plant Mol. Biol.*, **42** 241-79.

Dixon, R.A. and Gonzales, R.A. (eds.) (1994) *Plant Cell Culture: A Practical Approach*. IRL Press, Oxford.

Dixon, R.A., Harrison, M.J. and Lamb, C.J. (1994) Early events in the activation of plant defense responses. *Annu. Rev. Phytopathol.*, **32** 479-501.

Doran, P.M. (ed.) (1997) *Hairy Roots, Culture and Applications*, Harwood Academic Publishers, Amsterdam.

Drossard, J., Liao, Y.C. and Fischer, R. (1998) Production of recombinant antibodies in plant suspension cultures. *Recombinant Proteins from Plants*, **3** 143-54.

Durner, J., Shah, J. and Klessig, D.F. (1997) Salicylic acid and disease resistance in plants. *Trends Plant Sci.*, **2** 266-74.

Fahn, A. (1979) *Secretory Tissues in Plants*, Academic Press, London.

Farnsworth, N.R. (1985) The role of medicinal plants in drug development, in *Natural Products and Drug Development* (eds. P. Kroogsgard-Larsen, S. Brogger Christensen and H. Koford), Munksgaard, Copenhagen, pp. 17-30.

Fecker, L.F., Hillebrandt, S., Rügenhagen, C., Herminghaus, S., Landsmann, J. and Berlin, J. (1992) Metabolic effects of a bacterial lysine decarboxylase gene expressed in hairy root culture of *Nicotiana glauca*. *Biotechnol. Lett.*, **14** 1035-40.

Fell, D.A. (1992) Metabolic control analysis: a survey of its theoretical and experimental development. *Biochem. J.*, **286** 313-30.

Fischer, R., Drossard, J., Liao, Y.C. and Schillberg, S. (1998) Characterization and applications of plant-derived recombinant antibodies. *Recombinant Proteins from Plants*, **3** 129-42.

Flavell, R.B. (1994) Inactivation of gene expression in plants as a consequence of specific sequence duplication. *Proc. Natl. Acad. Sci. USA*, **91** 3490-96.

Fluhr, R. and Mattoo, A.K. (1996) Ethylene biosynthesis and perception. *Crit. Rev. Plant Sci.*, **15** 479-523.

Ford, Y.Y., Ratcliffe, R.G. and Robins, R.J. (1998) *In vivo* nuclear magnetic resonance analysis of polyamine and alkaloid metabolism in transformed root cultures of *Datura stramonium*: evidence for the involvement of putrescine in phytohormone-induced dedifferentiation. *Planta*, **205** 205-13.

Franssen, M.C.R. and Walton, N.J. (1999) Biotransformations, in *Chemicals from Plants, Perspectives on Plant Secondary Products* (ed. N.J. Walton and D.E. Brown), Imperial College Press, World Scientific Publishers, Singapore, pp. 277-325.

Fujita, Y. (1988) Industrial production of shikonin and berberine, in *CIBA Foundation Symposium 137: Application of Plant Cell and Tissue Culture*, Wiley, New York, pp. 228-38.

Fujita, Y. and Tabata, M. (1987) Secondary metabolites from plant cells: pharmaceutical application and progress in commercial production, in *Secondary Metabolites from Plant Cells: Pharmaceutical Application and Progress in Commercial Production.* (eds C.E. Green, D.A. Somers, W.P. Hackett and D.D. Biesboer), Alan R. Liss, New York, pp. 169-85.

Fujita, Y., Tabata, M., Nishi, A. and Yamada, Y. (1982) New medium and production of secondary compounds with the two-staged culture method, in *Plant Tissue Culture* (ed. A. Fujiwara), Maruzen, Tokyo, pp. 399-400.

Fukui, H., Hasan, A.F.M.F., Ueoka, T. and Kyo, M. (1998) Formation and secretion of a new brown benzoquinone by hairy root cultures of *Lithospermum erythrorhizon. Phytochemistry*, **47** 1037-39.

Furze, J.M., Rhodes, M.J.C., Parr, A.J., Robins, R.J., Whitehead, I.M. and Threlfall, D.R. (1991) Abiotic factors elicit sesquiterpenoid phytoalexin production but not alkaloid production in transformed roots of *Datura stramonium. Plant Cell Rep.*, **10** 111-14.

Gasse, F., Buchholz, M. and Berlin, J. (1983) Selection of cell lines of *Catharanthus roseus* with increased tryptophan decarboxylase activity. *Z. Naturforsch.*, **38c** 916-22.

Ghosh, B., Mukherjee, S. and Jha, S. (1997) Genetic transformation of *Artemisia annua* by *Agrobacterium tumefaciens* and artemisinin synthesis in transformed cultures. *Plant Sci.*, **122** 193-99.

Graser, G., Witte, L., Robins, D.J. and Hartmann, T. (1998) Incorporation of chirally deuterated putrescines into pyrrolizidine alkaloids: a reinvestigation. *Phytochemistry*, **47** 1017-24.

Grotewold, E., Chamberlin, M., Snook, M., Siame, B., Butler, L., Swenson, L., Maddock, S., St. Clair, G. and Bowen, B. (1998) Engineering secondary metabolism in maize cells by ectotopic expression of transcription factors. *Plant Cell*, **10** 721-40.

Gupta, R. (1991) Agrotechnology of medicinal plants, in *The Medicinal Plant Industry* (ed. R.O.B. Wijseker), CRC Press, Boca Raton, pp. 43-57.

Haberlandt, G. (1902) Culturversuche mit isolierten Pflanzenzellen Sitzungsberichte. *Math. Naturw. Kl. Kais. Akad. Wiss. Wien*, **111** 69-92.

Hagimori, M., Matsumoto, T. and Obi, Y. (1983) Effect of mineral salts, initial pH and precursors on digitoxin formation by shoot-forming cultures of *Digitalis purpurea* (L.) grown in liquid medium. *Agric. Biol. Chem.*, **47** 565-71.

Hallard, D., Geerlings, A., van der Heijden, R., Cardoso, M.I.L., Hoge, J.H.C. and Verpoorte, R. (1997) Metabolic engineering of terpenoid indole and quinoline alkaloid biosynthesis in hairy root cultures, in *Hairy Roots, Culture and Applications* (ed. P.M. Doran), Harwood Academic Publishers, Amsterdam, pp. 43-49.

Hamill, J.D. and Lidgett, A.J. (1997) Hairy root cultures: opportunities and key protocols for studies in metabolic engineering, in *Hairy Roots, Culture and Applications* (ed. P.M. Doran), Harwood Academic Publishers, Amsterdam, pp. 1-30.

Hamill, J.D., Parr, A.J., Robins, R.J. and Rhodes, M.J.C. (1986) Secondary product formation by cultures of *Beta vulgaris* and *Nicotiana rustica* transformed with *Agrobacterium rhizogenes*. *Plant Cell Rep.*, **5** 111-14.

Hamill, J.D., Robins, R.J. and Rhodes, M.J.C. (1989) Alkaloid production by transformed roots of *Cinchona ledgeriana*. *Planta Med.*, **55** 354-57.

Hamill, J.D., Robins, R.J., Parr, A.J., Evans, D.M., Furze, J.M. and Rhodes, M.J.C. (1990) Overexpressing a yeast ornithine decarboxylase gene in transgenic roots of *Nicotiana rustica* can lead to enhanced nicotine accumulation. *Plant Mol. Biol.*, **15** 27-38.

Hartmann, T. (1985) Principles of plant secondary metabolism. *Plant Syst. Evol.*, **150** 13-34.

Hartmann, T., Ehmke, A., Eilert, U., von Borstel, K. and Theuring, C. (1989) Sites of synthesis, translocation and accumulation of pyrrolizidine alkaloid *N*-oxides in *Senecio vulgaris* (L.). *Planta*, **177** 98-107.

Hashimoto, T. and Yamada, Y. (1992) Biosynthesis of scopolamine and its application for genetic engineering of medicinal plants, in *Plant Tissue Culture and Gene Manipulation for Breeding and Formation of Phytochemicals* (eds. R. Oono, T. Hirabayashi, S. Kiruchi, H. Handa and S. Kahwara), National Institute of Agrobiological Resources, Tsukuba, Japan, pp. 255-59.

Hashimoto, T., Yukimune, Y. and Yamada, Y. (1986) Tropane alkaloid production in *Hyoscyamus* root cultures. *J. Plant Physiol.*, **124** 61-75.

Hashimoto, T., Matsuda, J., Okabe, S., Amano, Y., Yun, D.J., Hayashi, A. and Yamada, Y. (1990) Molecular cloning, tissue- and cell-specific expression of hyoscyamine 6β-hydroxylase, in *Progress in Plant Cellular and Molecular Biology* (eds. H.J.J. Nijkamp, L.H.W. van der Plas and J. van Aartrijk), Kluwer, Dordrecht, pp. 775-80.

Hashimoto, T., Hayashi, A., Amano, Y., Kohno, I., Iwanari, H., Usuda, S. and Yamada, Y. (1991) Hyoscyamine 6β-hydroxylase, an enzyme involved in tropane alkaloid biosynthesis, is localized at the pericycle of the roots. *J. Biol. Chem.*, **266** 4648-53.

Hayashi, S. and Murakami, Y. (1995) Rapid and regulated degradation of ornithine decarboxylase. *Biochem. J.*, **306** 1-10.

Herminghaus, S., Tholl, D., Rügenhagen, C., Fecker, L.F., Leuschner, C. and Berlin, J. (1996) Improved metabolic action of a bacterial lysine decarboxylase gene in tobacco hairy root cultures by its fusion to a *rbcS* transit peptide coding sequence. *Transgen. Res.*, **5** 193-201.

Hilton, M.G., Jay, A., Rhodes, M.J.C. and Wilson, P.D.G. (1995) Growth and monoterpene production by transformed shoot cultures of *Mentha citrata* and *Mentha piperita* in flasks and fermenters. *Appl. Microbiol. Biotechnol.*, **43** 452-59.

Hippolyte, I., Marin, B., Baccou, J.C. and Jonard, R. (1992) Growth and rosmarinic acid production in cell suspension cultures of *Salvia officinalis*. *Plant Cell Rep.*, **11** 109-12.

Howells, P.A., Sewalt, V.J.H., Paiva, N.L., Elkind, Y., Bate, N.J., Lamb, C. and Dixon, R.A. (1996) Overexpression of L-phenylalanine ammonia-lyase in transgenic tobacco plants reveals control points for flux into phenylpropanoid biosynthesis. *Plant Physiol.*, **112** 1617-24.

Inoue, K., Nayeshiro, H., Inouye, H. and Zenk, M.H. (1981) Anthraquinones in cell suspension cultures of *Morinda citrifolia*. *Phytochemistry*, **20** 1693-700.

Ionkova, I., Kartnig, T. and Alfermann, W. (1997) Cycloartane saponin production in hairy root cultures of *Astragalus mongholicus*. *Phytochemistry*, **45** 1597-600.

Jackson, D., Roberts, K. and Martin, C. (1992) Temporal and spatial control of expression of anthocyanin biosynthetic genes in developing flowers of *Antirrhinum majus*. *Plant J.*, **2** 425-34.

John, P. (1992) *Biosynthesis of the Major Crop Products*, John Wiley and Sons, Chichester, UK.

Katagiri, F., Chua, N.-H. (1991) Plant transcription factors: present knowledge and future challenges. *Trends Genet.*, **8** 22-27.

Kesselring, W. (1985) Pflanzenzellkulturen (PZK) zur Auffindung neuer, therapeutisch relevanter Naturstoffe und deren Gewinnung durch Fermentationsprozesse, in *Pflanzliche*

Zellkulturen (ed. Bundesminister für Forschung und Technologie) Projekttröger Biotechnologie, KFA, Jülich, pp. 111-29.

Kittipongpatana, N., Hock, R.S. and Porter, J.R. (1998) Production of solasodine by hairy root, callus and cell suspension cultures of *Solanum aviculare* Forst. *Plant Cell Tiss. Org. Cult.*, **52** 133-43.

Komali, A.S. and Shetty, K. (1998) Comparison of the growth pattern and rosmarinic acid production in rosemary (*Rosmarinus officinalis*) shoots and genetically transformed callus cultures. *Food Biotech.*, **12** 27-41.

Kutchan, T.M. (1995) Alkaloid biosynthesis: the basis for metabolic engineering of medicinal plants. *Plant Cell*, **7** 1059-70.

Kutchan, T.M. and Zenk, M.H. (1993) Enzymology and molecular biology of benzophenanthridin alkaloid biosynthesis. *J. Plant Res.*, **3** 165-73.

Kutchan, T., Dittrich, H., Bracher, D. and Zenk, M.H. (1991) Enzymology and molecular biology of alkaloid biosynthesis. *Tetrahedron*, **47** 5945-54.

Larkin, H.C., Oppenheimer, D.G., Lloyd, A.M., Paparozzi, E.T. and Marks, M.D. (1994) Role of the *GlabrousI* and transparent testa *Glabra* genes in *Arabidopsis* trichome development. *Plant Cell*, **6** 1065-76.

Li, S.-M., Wang, Z.-X., Wemakor, E. and Heide, L. (1997) Metabolization of the artificial secondary metabolite 4-hydroxybenzoate in *ubiC*-transformed tobacco. *Plant Cell Physiol.*, **38** 844-50.

Liu, C.Z., Wang, Y.C., Ouyang, F., Ye, H.C. and Li, G.F. (1997) Production of artemisinin by hairy root cultures of *Artemisia annua* (L.). *Biotechnol. Lett.*, **19** 927-29.

Liu, C.Z., Wang, Y.C., Ouyang, F., Ye, H.C. and Li, G.F. (1998) Production of artemisinin by hairy root cultures of *Artemisia annua* (L.) in bioreactor. *Biotechnol. Lett.*, **20** 265-68.

Lodhi, A.H., Bongaerts, R.J.M., Verpoorte, R., Coomber, S.A. and Charlwood, B.V. (1996) Expression of bacterial isochorismate synthase (EC 5.4.99.6) in transgenic root cultures of *Rubia peregrina*. *Plant Cell Rep.*, **16** 54-57.

Lutterbach, R. and Stöckigt, J. (1992) High-yield formation of arbutin from hydroquinone by cell-suspension cultures of *Rauwolfia serpentina*. *Helv. Chim. Acta*, **75** 2009-11.

Lutterbach, R., Ruyter, C.M. and Stöckigt, J. (1994) Isolation and characterization of a UDPG-dependent glucosyltransferase activity from *Rauwolfia serpentina* Benth. cell suspension cultures. *Can. J. Chem.*, **72** 51-55.

Mandel, M.A. and Yanofsky, M.F. (1995) A gene-triggering flower formation in *Arabidopsis*. *Nature*, **377** 522-24.

Martin, C. (1996) Transcription factors and the manipulation of plant traits. *Curr. Opin. Biotech.*, **7** 130-38.

Matsubara, K. and Fujita, Y. (1991) Production of berberine, in *Plant Cell Culture in Japan* (eds. A. Komamine, A. Misawa and F. DiCosmo), CMC Co., Tokyo, pp. 39-44.

Matsuda, J., Okabe, S., Hashimoto, T. and Yamada, Y. (1991) Molecular cloning of hyoscyamine 6β-hydroxylase, a 2-oxoglutarate-dependent dioxygenase from cultured roots of *Hyoscyamus niger*. *J. Biol. Chem.*, **266** 9460-64.

McGarvey, D.J. and Croteau, R. (1995) Terpenoid metabolism. *Plant Cell*, **7** 1015-26.

Mehmetoglu, U. and Curtis, W.R. (1997) Effects of abiotic inducers on sesquiterpene synthesis in hairy root and cell-suspension cultures of *Hyoscyamus muticus*. *Appl. Biochem. Biotechnol.*, **67** 71-77.

Merkli, A., Christen, P. and Kapetanidis, I. (1997) Production of diosgenin by hairy root cultures of *Trigonella foenum-graecum* (L.). *Plant Cell Rep.*, **16** 632-36.

Meyer, P. and Saedler, H. (1996) Homology-dependent gene silencing in plants. *Annu. Rev. Plant Physiol. Plant Mol. Biol.*, **47** 23-48.

Michael, A. and Spena, A. (1995) The plant oncogenes, *Rol* A, B and C, from *Agrobacterium rhizogenes*, in *Methods in Molecular Biology*, Vol. 44 (ed. K.M.A. Gartland and M.R. Davey) Humana Press Inc., Totowa, N.J., pp. 207-22.

Michael, A., Furze, J.M., Rhodes, M.J.C. and Burtin, D. (1996) Molecular cloning and functional identification of a plant ornithine decarboxylase cDNA. *Biochem. J.*, **314** 241-48.

Miele, L. (1997) Plants as bioreactors for biopharmaceuticals: regulatory considerations. *Trends Biotechnol.*, **15** 45-50.

Mizukami, H., Terato, T., Miura, H. and Ohashi, H. (1983) Glucosylation of salicyl alcohol in cultured plant cells. *Phytochemistry*, **22** 679-80.

Mizukami, H., Terato, T. and Ohashi, H. (1985) Partial purification and characterization of UDP-glucose: salicyl alcohol glucosyltransferase from *Gardenia jasminoides* cell cultures. *Planta Med.*, **46** 104-107.

Mizukami, H., Terato, T., Amano, A. and Ohashi, H. (1986) Glucosylation of salicyl alcohol by *Gardenia jasminoides* cell cultures. *Plant Cell Physiol.*, **27** 645-50.

Mizukami, H., Terato, T. and Ohashi, H. (1987) Effect of substituent groups on the glucosyl formation of xenobiotic phenols by cultured cells of *Gardenia jasminoides. Plant Sci.*, **48** 11-15.

Mooney, M., Desnos, T., Harrison, K., Jones, J., Carpenter, R. and Coen, E. (1995) Altered regulation of tomato and tobacco pigmentation genes caused by the *Delila* gene of *Antirrhinum. Plant J.*, **7** 333-39.

Mueller, M.J., Brodschelm, W., Spannagl, E. and Zenk, M.H. (1993) Signalling in the elicitation process is mediated through the octadecanoid pathway leading to jasmonic acid. *Proc. Natl. Acad. Sci. USA*, **90** 7490-94.

Mühlbach, H.P. (1998) Use of plant cell cultures in biotechnology, in *Biotechnology Annual Rev. 4* (ed. M.R. El-Gewely), Elsevier, Amsterdam, pp. 113-76.

Mukundan, U. and Hjortso, M. (1990) Effect of fungal elicitor on thiophene production in hairy root cultures of *Tagetes patula. Appl. Microbiol. Biotechnol.*, **33** 145-47.

Murakami, Y., Shimomura, K., Yoshihira, K. and Ishimaru, K. (1998a) Polyacetylenes in hairy root cultures of *Trachelium caeruleum* (L.). *J. Plant Physiol.*, **152** 574-76.

Murakami, Y., Omoto, T., Asai, I., Shimomura, K., Yoshihira, K. and Ishimaru, K. (1998b) Rosmarinic acid and related phenolics in transformed root cultures of *Hyssopus officinalis. Plant Cell Tiss. Org. Cult.*, **53** 75-78.

Niederberger, P., Prasad, R., Miozzari, G. and Kacser, H. (1992) A strategy for increasing an *in vivo* flux by genetic manipulation: the tryptophan system of yeast. *Biochem. J.*, **287** 473-79.

Nishikawa, K. and Ishimaru, K. (1997) Flavonoids in root cultures of *Scutellaria baicalensis. J. Plant Physiol.*, **151** 633-36.

Noguchi, M., Matsumoto, T., Hirata, Y., Yamamoto, K., Katsuyama, A., Kato, A., Azechi, S. and Kato, K. (1977) Improvement of growth rates of plant cell cultures, in *Plant Tissue Culture and its Bio-Technological Application* (eds. W. Barz, E. Reinhard and M.H. Zenk), Springer Verlag, Berlin, pp. 85-94.

Nussbaumer, P., Kapetanidis, I. and Christen, P. (1998) Hairy roots of *Datura candida x D. aurea*: effect of culture medium composition on growth and alkaloid biosynthesis. *Plant Cell Rep.*, **17** 405-409.

Paniego, N.B. and Giulietti, A.M. (1996) Artemisinin production by *Artemisia annua* (L.)-transformed organ cultures. *Enzyme Microbial Technol.*, **18** 526-30.

Parr, A.J., Payne, J., Eagles, J., Chapman, B.J., Robins, R.J. and Rhodes, M.J.C. (1990) Variation in tropane alkaloid accumulation within the Solanaceae and strategies for its exploitation. *Phytochemistry*, **29** 2545-50.

Petersen, M. and Alfermann, A.W. (1993) Plant cell cultures, in *Biotechnology*, 2nd edn. (series eds. H.J. Rehm and G. Reed), Vol. 1. Biological Fundamentals (ed. H. Sahm) VCH Publishers, Weinheim, pp. 577-614.

Petersen, M. and Seitz, H.U. (1985) Cytochrome P_{450}-dependent digitoxin 12β-hydroxylase from cell cultures of *Digitalis lanata. FEBS Lett.*, **188** 11-14.

Petersen, M. and Seitz, H.U. (1988) Reconstitution of cytochrome P_{450}-dependent digitoxin 12β-hydroxylase from cell cultures of *Digitalis lanata* Ehrh. *Biochem. J.*, **252** 537-43.

Petersen, M., Alfermann, A.W., Reinhard, E. and Seitz, H.U. (1987) Immobilization of a 12β-hydroxylase from cell suspension cultures of *Digitalis lanata* Ehrh. *Plant Cell Rep.*, **6** 200-203.

Petersen, M., Seitz, H.U. and Reinhard, E. (1988) Characterisation and localisation of digitoxin 12β-hydroxylase from cell cultures of *Digitalis lanata* Ehrh. *Z. Naturforsch.*, **43c** 199-206.

Petersen, M., Dombrowski, K., Gertlowski, C., Häusler, E., Karwatzki, B., Meinhard, J. and Alfermann, A.W. (1992) The use of plant cell cultures to study natural product biosynthesis, in *Plant Tissue Culture and Gene Manipulation for Breeding and Formation of Phytochemicals* (eds. R. Oono, T. Hirabayashi, S. Kiruchi, H. Handa and S. Kahwara), National Institute of Agrobiological Resources, Tsukuba, Japan, pp. 297-310.

Petersen, M., Husler, E., Meinhard, J., Karwazki, B. and Gertlowski, C. (1994) The biosynthesis of rosmarinic acid in suspension cultures of *Coleus blumei. Plant Cell Tiss. Org. Cult.*, **38** 171-79.

PerazaSanchez, S.R., GamboaAngulo, M.M., ErosaLopez, C., RamirezErosa, I., EscalanteErosa, F., PenaRodriguez, L.M. and LoyolaVargas, V.M. (1998) Production of 19(*S*)-epimisiline by hairy root cultures of *Catharanthus roseus. Nat. Prod. Lett.*, **11** 217-24.

Pilgrim, H. (1970) Untersuchungen zur Glycosidbildung in pflanzlichen Gewebekulturen. *Pharmazie*, **25** 568.

Pradel, H., DumkeLehmann, U., Diettrich, B. and Luckner, M. (1997) Hairy root cultures of *Digitalis lanata*: secondary metabolism and plant regeneration. *J. Plant Physiol.*, **151** 209-15.

Pras, N. (1992) Bioconversion of naturally occurring precursors and related synthetic compounds using plant cell cultures. *J. Biotechnol.*, **26** 29-62.

Rhodes, M.J.C. (1990) Properties of transformed root cultures, in *Proceedings of the Phytochemical Society of Europe*, Vol. 30, Secondary Products from Plant Tissue Culture (eds. B.V. Charlwood and M.J.C. Rhodes), Clarendon Press, Oxford, pp. 201-27.

Rhodes, M.J.C., Parr, A.J. and Walton, N.J. (1997) Studies of secondary pathways in transformed roots, in *Hairy Roots, Culture and Applications* (ed. P.M. Doran), Harwood Academic Publishers, Amsterdam, pp. 31-41.

Rijhwani, S.K. and Shanks, J.V. (1998a) Effect of elicitor dosage and exposure time on biosynthesis of indole alkaloids by *Catharanthus roseus* hairy root cultures. *Biotechnol. Prog.*, **14** 442-49.

Rijhwani, S.K. and Shanks, J.V. (1998b) Effect of subculture cycle on growth and indole alkaloid production by *Catharanthus roseus* hairy root cultures. *Enzyme Microbiol. Technol.*, **22** 606-11.

Robbins, M.P., Hartnoll, J. and Morris, P. (1991) Phenylpropanoid defence responses in transgenic *Lotus corniculatus.* I. Glutathione elicitation of isoflavan phytoalexins in transformed root cultures. *Plant Cell Rep.*, **10** 59-62.

Robbins, M.P., Thomas, B. and Morris, P. (1995) Phenylpropanoid defence responses in transgenic *Lotus corniculatus.* II. Modelling plant defence responses in transgenic root cultures using thiol and carbohydrate elicitors. *J. Exp. Bot.*, **46** 513-24.

Roberts, E.H. and Street, H.E. (1955) The continuous culture of excised roots. *Physiol. Plantarum*, **8** 238-62.

Roberts, S.G. and Shuler, M.L. (1997) Large-scale plant cell culture. *Curr. Opin. Biotechnol.*, **8** 154-59.

Robins, R.J. and Rhodes, M.J.C. (1986) The stimulation of anthraquinone production by *Cinchona ledgeriana* cultures with polymeric adsorbents. *Appl. Microbiol. Biotechnol.*, **24** 35-41.

Robins, R.J., Rhodes, M.J.C., Parr, A.J. and Walton, N.J. (1990) The biosynthesis of bitter compounds, in *Bitterness in Foods and Beverages* (ed. R.L. Rousseff), Elsevier, Amsterdam, pp. 49-79.

Routien, J.B. and Nickel, L.G. (1952) Cultivation of Plant Tissue. US Patent 2,747,334.

Ruyter, C.M. and Stöckigt, J. (1989) Novel natural products from plant cell and tissue cultures: an update. *GIT Fachz. Lab.*, **4** 283-93.

Salem, K.M.S.A. and Charlwood, B.V. (1995) Accumulation of essential oils by *Agrobacterium tumefaciens*-transformed shoot cultures of *Pimpinella anisum. Plant Cell Tiss. Org. Cult.*, **40** 209-15.

Santos, P.M., Figueiredo, A.C., Oliveira, M.M., Barroso, J.G., Pedro, L.G., Younis, S.G., Deanus, A.K.M. and Scheffer, J.C. (1998) Essential oils from hairy root cultures and from fruits and roots of *Pimpinella anisum. Phytochemistry*, **48** 455-60.

Sasaki, K., Udagawa, A., Ishimaru, H., Hayashi, T., Alfermann, A.W., Nakanishi, F. and Shimomura, K. (1998) High forskolin production in hairy roots of *Coleus forskohlii. Plant Cell Rep.*, **17** 457-59.

Schiel, O., Jarchow-Redecker, K., Piel, G.W., Lehmann, J. and Berlin, J. (1984) Increased formation of cinnamoyl putrescines by fed-batch fermentation of cell suspension cultures of *Nicotiana tabacum. Plant Cell Rep.*, **3** 18-20.

Schlatmann, J.E., ten Hoopen, H.J.G. and Heijnen, J.J. (1996) Large-scale production of secondary metabolites by plant cell cultures, in *Plant Cell Cultures Secondary Metabolism* Towards Industrial Application (eds. F. Dicosma and M. Misawa), CRC Press, Boca Raton, pp. 11-52.

Schübel, H., Ruyter, C.M. and Stöckigt, J. (1989) Improved production of raucaffricine by cultivated *Rauwolfia* cells. *Phytochemistry*, **28** 491-94.

Scott, A.I. (1994) Genetically-engineered synthesis of natural products. *J. Nat. Prod.*, **57** 557-73.

Scragg, A.H. (1992) Bioreactors for mass cultivation of plant cells, in *Plant Biotechnology, Comprehensive Biotechnology* (ed. M. Moo-Young), Second supplement (eds. M.W. Fowler and G.S. Warren), Pergamon Press, Oxford, pp. 45-62.

Seitz, H.U., Seitz, U. and Alfermann, A.W. (1985) *Pflanzliche Gewebekultur: ein Praktikum*, Fischer, Stuttgart.

Seo, S., Okamoto, M., Seto, H., Ishizuka, K., Sano, H. and Ohashi, Y. (1995) Tobacco MAP kinase: a possible mediator in wound signal transduction pathways. *Science*, **270** 1988-92.

Sevon, N., Hiltunen, R. and Oksman-Caldentey, K.-M. (1998) Somaclonal variation in transformed roots and protoplast-derived hairy root clones of *Hyoscyamus muticus. Planta Med.*, **64** 37-41.

Siebert, S., Sommer, S., Li, S., Wang, Z., Severin, K. and Heide, L. (1996) Genetic engineering of plant secondary metabolism: accumulation of 4-hydroxybenzoate glucoside as a result of the expression of the bacterial *ubiC* gene in tobacco. *Plant Physiol.*, **112** 811-19.

Signs, M.W. and Flores, H.E. (1990) The biosynthetic potential of plant roots. *Bioessays*, **12** 7-13.

Somssich, I.E. and Hahlbrock, K. (1998) Pathogen defence in plants: a paradigm of biological complexity. *Trends Plant Sci.*, **3** 86-90.

Spencer, A., Hamill, J.D. and Rhodes, M.J.C. (1993) *In vitro* biosynthesis of monoterpenes by *Agrobacterium*-transformed shoot cultures of two *Mentha* species. *Phytochemistry*, **32** 911-19.

Spieler, H., Alfermann, A.W. and Reinhard, E. (1985) Biotransformation of β-methyldigitoxin by cell cultures of *Digitalis lanata* in airlift and stirred tank reactors. *Appl. Microbiol. Biotechnol.*, **23** 1-4.

Stafford, A. (1998) Lecture presented at the symposium, 'Future Trends in Phytochemistry', of the Phytochemical Society of Europe, at Rolduc (Kerkrade, NL), May 12–15, 1998.

Stafford, A.M. and Pazoles, C.J. (1997) Harnessing phytochemical diversity for drug discovery: the Phytera approach, in *Phytochemical Diversity, a Source of New Industrial Products* (eds. S. Wrigley, M. Hayes, R. Thomas and E. Chrystal), The Royal Society of Chemistry, London, pp. 179-89.

Stitt, M. (1995) The use of transgenic plants to study the regulation of plant carbohydrate metabolism. *Aust. J. Plant Physiol.*, **22** 635-46.

Stojakowska, A. and Kisiel, W. (1996) Production of parthenolide in organ cultures of feverfew. *Plant Cell Tiss. Org. Cult.*, **47** 159-62.

Stöckigt, J. (1993) Biotransformations with plant cells. *Agro. Food Industry Hi-Tech.*, **4** (6) 25-28.

Street, H.E. (1977) Cell (suspension) cultures: techniques, in *Plant Tissue and Cell Culture* (ed. H.E. Street), Blackwell Scientific Publishers, Oxford, pp. 61-102.

Subroto, M.A., Kwok, K.H., Hamill, J.D. and Doran, P.M. (1996) Co-culture of genetically transformed roots and shoots for synthesis, translocation and biotransformation of secondary metabolites. *Biotechnol. Bioeng.*, **49** 481-94.

Subroto, M.A., Mahagamasekera, M.G.P., Kwok, K.H., Hamill, J.D. and Doran, P.M. (1997) Co-culture of hairy roots and shooty teratomas, in *Hairy Roots, Culture and Applications* (ed. P.M. Doran), Harwood Academic Publishers, Amsterdam, pp. 81-88.

Suga, T. and Hirata, T. (1990) Biotransformation of exogenous substrates by plant cell cultures. *Phytochemistry*, **29** 2393-406.

Suzuki, T., Yoshioka, T., Tabata, M. and Fujita, Y. (1987) Potential of *Datura innoxia* cell suspension cultures for glycosylating hydroquinone. *Plant Cell Rep.*, **6** 275-78.

Tabata, M., Ikeda, F., Hiraoka, N. and Konoshima, M. (1976) Glucosylation of phenolic compounds by *Datura innoxia* suspension cultures. *Phytochemistry*, **15** 1225-29.

Tabata, M., Umetani, Y., Ooya, M. and Tanaka, S. (1988) Glucosylation of phenolic compounds by plant cell cultures. *Phytochemistry*, **27** 809-13.

Tamagnone, L., Merida, A., Parr, A., Mackay, S., Cullianez-Macia, F.A., Roberts, K. and Martin, C. (1998) The AmMYB308 and AmMYB330 transcription factors from antirrhinum regulate phenylpropanoid and lignin biosynthesis in transgenic tobacco. *Plant Cell*, **10** 135-54.

Tanaka, S., Hayakawa, K., Umetani, Y. and Tabata, M. (1990) Glucosylation of isomeric hydroxybenzoic acids by cell suspension cultures of *Mallotus japonicus*. *Phytochemistry*, **29** 1555-58.

Toppel, G., Witte, L., Riebesehl, B., Borstel, K.V. and Hartmann, T. (1987) Alkaloid patterns and biosynthetic capacity of root cultures from some pyrrolizidine alkaloid-producing *Senecio* species. *Plant Cell Rep.*, **6** 466-69.

Ulbrich, B., Wiesner, W. and Arens, H. (1985) Large-scale production of rosmarinic acid from plant cell cultures of *Coleus blumei* Benth., in *Primary and Secondary Metabolism of Plant Cell Cultures* (eds. K.H. Neumann, W. Barz and E. Reinhard), Springer, Berlin, pp. 293-303.

Ushiyama, K. (1991) Large scale cultivation of ginseng, in *Plant Cell Cultures in Japan* (eds. A. Komamine, A. Misawa and F. DiCosmo), CMC, Tokyo, pp. 92-98.

Van der Esch, S.A. (1996) Lecture presented at the symposium, 'Future Trends in Phytochemistry', of the Phytochemical Society of Europe, at Rolduc (Kerkrade, NL), May 12–15, 1998.

Venkat, K. (1996) Opportunities and challenges to large-scale plant tissue cultures: the paclitaxel story. Paper presented at Meeting of the German Botanical Society, Düsseldorf (Germany), August 25–31, 1996.

Wagner, F. and Vogelmann, H. (1977) Cultivation of plant tissue cultures in bioreactors and formation of secondary products, in *Plant Tissue Culture and its Bio-Technological Application* (eds. W. Barz, E. Reinhard and M.H. Zenk), Springer Verlag, Berlin, pp. 244-52.

Wahl, J. (1985) Adaption konventioneller Fermenter zur Züchtung von Pflanzenzellen zum Zwecke der Gewinnung von Naturstoffen, in *Pflanzliche Zellkulturen* (ed. Bundesminister für Forschung und Technologie), Projekttröger Biotechnologie, KFA, Jülich, pp. 35-43.

Walton, N.J., Robins, R.J. and Rhodes, M.J.C. (1988) Perturbation of alkaloid production by cadaverine in hairy root cultures of *Nicotiana rustica*. *Plant Sci.*, **54** 125-31.

Ward, E.R., Ryals, J.A. and Miflin, B.J. (1993) Chemical regulation of transgene expression in plants. *Plant Mol. Biol.*, **22** 361-66.

Wasternak, C. and Parthier, B. (1997) Jasmonate-signalled gene expression. *Trends Plant Sci.*, **2** 302-307.

Weigel, A. and Nilsson, O. (1995) A developmental switch sufficient for flower induction in diverse plants. *Nature*, **377** 495-500.

Westcott, R.J., Cheetham, P.S.J. and Barraclough, A.J. (1994) Use of organized viable vanilla plant aerial roots for the production of natural vanillin. *Phytochemistry*, **35** 135-38.

Westphal, K. (1990) Large scale production of new biologically-active compounds in plant cell cultures, in *Progress in Plant Cellular and Molecular Biology* (eds. H.J.J. Nijkamp, L.H.W. van der Plas and J. van Aartrijk), Kluwer, Dordrecht, pp. 601-608.

Wewetzer, A. (1998) Callus cultures of *Azadirachta indica* and their potential for the production of azadirachtin. *Phytoparasitica*, **26** 47-52.

White, P.R. (1934) Potentially unlimited growth of excised tomato roots in culture. *Plant Physiol.*, **9** 586-600.

Whitelam, G.C. and Cockburn, W. (1996) Antibody expression in transgenic plants. *Trends Plant Sci.*, **8** 268-72.

Wilson, P.D.G. (1997) The pilot-scale cultivation of transformed roots, in *Hairy Roots, Culture and Applications* (ed. P.M. Doran), Harwood Academic Publishers, Amsterdam, pp. 169-78.

Wink, M. (1993) Quinolizidine alkaloids, in *Methods in Plant Biochemistry* (ed. P. Waterman), Academic Press, San Diego, Vol. 8, pp. 197-239.

Wysokinska, H. and Rozga, M. (1998) Establishment of transformed root cultures of *Paulownia tomentosa* for verbascoside production. *J. Plant Physiol.*, **152** 78-83.

Yokoyama, M. (1996) Industrial application of biotransformations using plant cell cultures, in *Plant Cell Culture Secondary Metabolism Towards Industrial Application* (eds. F. DiCosmo and M. Misawa), CRC Press, Boca Raton, pp. 79-122.

Yokoyama, M. and Yanagi, M. (1991) High-level production of arbutin by biotransformation, in *Plant Cell Culture in Japan* (eds. A. Komamine and M. Misawa) CMC, Tokyo, pp. 79-91.

Yukimune, Y., Tabata, H., Higashi, Y. and Hara, Y. (1996) Methyljasmonate-induced overproduction of paclitaxel and baccatin III in *Taxus* cell suspension cultures. *Nature Biotechnol.*, **14** 1129-32.

Zenk, M.H. (1991) Chasing the enzymes of secondary metabolism: plant cell cultures as a pot of gold. *Phytochemistry*, **30** 3861-63.

Zenk, M.H., El-Shagi, H. and Schulte, U. (1975) Anthraquinone production by cell suspension cultures of *Morinda citrifolia*. *Planta Med.* (Suppl.) pp. 79-101.

Zenk, M.H., El-Shagi, H., Arens, H., Stöckigt, J., Weiler, E.W. and Deus, B. (1977) Formation of indole alkaloids serpentine and ajmalicine in cell suspension cultures of *Catharanthus roseus*, in *Plant Tissue Culture and its Bio-Technological Application* (eds. W. Barz, E. Reinhard and M.H. Zenk), Springer Verlag, Berlin, pp. 27-43.

Zhong, J.J. and Yoshida, T. (1995) High density cultivation of *Perilla frutescens* cell suspensions for anthocyanin production: effects of sucrose concentration and inoculum size. *Biotechnol. Bioeng.*, **38** 653-58.

Zhu, Q., DrogeLaser, W., Dixon, R.A. and Lamb, C.J. (1996) Transcriptional activation of plant defense genes. *Curr. Opin. Genet. Devel.*, **6** 624-30.

Index